Enhancing Art, Culture, and Design With Technological Integration

Mehdi Khosrow-Pour
Information Resources Management Association, USA

A volume in the Advances in Media, Entertainment, and the Arts (AMEA) Book Series

IGI Global
DISSEMINATOR OF KNOWLEDGE

Published in the United States of America by
IGI Global
Information Science Reference (an imprint of IGI Global)
701 E. Chocolate Avenue
Hershey PA, USA 17033
Tel: 717-533-8845
Fax: 717-533-8661
E-mail: cust@igi-global.com
Web site: http://www.igi-global.com

Library of Congress Cataloging-in-Publication Data

Names: Khosrow-Pour, Mehdi, 1951- editor.
Title: Enhancing art, culture, and design with technological integration / Mehdi Khosrow-Pour, editor.
Description: Hershey : Information Science Reference, [2017]
Identifiers: LCCN 2017034508| ISBN 9781522550235 (hardcover) | ISBN 9781522550242 (ebook)
Subjects: LCSH: Social sciences--Research. | Humanities--Research. | Industrial design.
Classification: LCC H62 .E5925 2018 | DDC 300--dc23 LC record available at https://lccn.loc.gov/2017034508

This book is published in the IGI Global book series Advances in Media, Entertainment, and the Arts (AMEA) (ISSN: 2475-6814; eISSN: 2475-6830)

British Cataloguing in Publication Data
A Cataloguing in Publication record for this book is available from the British Library.

All work contributed to this book is new, previously-unpublished material.
The views expressed in this book are those of the authors, but not necessarily of the publisher.

For electronic access to this publication, please contact: eresources@igi-global.com.

Advances in Media, Entertainment, and the Arts (AMEA) Book Series

ISSN:2475-6814
EISSN:2475-6830

Editor-in-Chief: Giuseppe Amoruso, Politecnico di Milano, Italy

MISSION

Throughout time, technical and artistic cultures have integrated creative expression and innovation into industrial and craft processes. Art, entertainment and the media have provided means for societal self-expression and for economic and technical growth through creative processes.

The **Advances in Media, Entertainment, and the Arts (AMEA)** book series aims to explore current academic research in the field of artistic and design methodologies, applied arts, music, film, television, and news industries, as well as popular culture. Encompassing titles which focus on the latest research surrounding different design areas, services and strategies for communication and social innovation, cultural heritage, digital and print media, journalism, data visualization, gaming, design representation, television and film, as well as both the fine applied and performing arts, the AMEA book series is ideally suited for researchers, students, cultural theorists, and media professionals.

COVERAGE

- Cultural Heritage
- Music & Performing Arts
- Humanities Design
- Design of Interiors
- Cross-Media Studies
- Digital Heritage
- Design Tools
- Drawing
- Popular Culture
- New Media Art

IGI Global is currently accepting manuscripts for publication within this series. To submit a proposal for a volume in this series, please contact our Acquisition Editors at Acquisitions@igi-global.com or visit: http://www.igi-global.com/publish/.

The Advances in Media, Entertainment, and the Arts (AMEA) Book Series (ISSN 2475-6814) is published by IGI Global, 701 E. Chocolate Avenue, Hershey, PA 17033-1240, USA, www.igi-global.com. This series is composed of titles available for purchase individually; each title is edited to be contextually exclusive from any other title within the series. For pricing and ordering information please visit http://www.igi-global.com/book-series/advances-media-entertainment-arts/102257. Postmaster: Send all address changes to above address.

Titles in this Series

For a list of additional titles in this series, please visit:
https://www.igi-global.com/book-series/advances-media-entertainment-arts/102257

Promoting Global Competencies Through Media Literacy
Melda N. Yildiz (New York Institute of Technology, USA) Steven S. Funk (Montana State University – Billings, USA) and Belinha S. De Abreu (National Telemedia Counil, USA)
Information Science Reference • ©2018 • 300pp • H/C (ISBN: 9781522530824) • US $195.00

Selfies as a Mode of Social Media and Work Space Research
Shalin Hai-Jew (Kansas State University, USA)
Information Science Reference • ©2018 • 327pp • H/C (ISBN: 9781522533733) • US $185.00

Exploring Journalism Practice and Perception in Developing Countries
Abiodun Salawu (North-West University, South Africa) and Toyosi Olugbenga Samson Owolabi (Lagos State University, Nigeria)
Information Science Reference • ©2018 • 339pp • H/C (ISBN: 9781522533764) • US $185.00

Digital Innovations in Architectural Heritage Conservation Emerging Research ...
Stefano Brusaporci (University of L'Aquila, Italy)
Engineering Science Reference • ©2017 • 152pp • H/C (ISBN: 9781522524342) • US $115.00

Music as a Platform for Political Communication
Uche Onyebadi (Texas Christian University, USA)
Information Science Reference • ©2017 • 309pp • H/C (ISBN: 9781522519867) • US $195.00

Media Law, Ethics, and Policy in the Digital Age
Nhamo A. Mhiripiri (Midlands State University, Zimbabwe & St. Augustine University, Tanzania) and Tendai Chari (University of Venda, South Africa)
Information Science Reference • ©2017 • 330pp • H/C (ISBN: 9781522520955) • US $195.00

For an entire list of titles in this series, please visit:
https://www.igi-global.com/book-series/advances-media-entertainment-arts/102257

IGI Global
DISSEMINATOR OF KNOWLEDGE

701 East Chocolate Avenue, Hershey, PA 17033, USA
Tel: 717-533-8845 x100 • Fax: 717-533-8661
E-Mail: cust@igi-global.com • www.igi-global.com

Table of Contents

Detailed Table of Contents

This chapter examines some of the ideas that Theodor Adorno elucidated around the term culture industry, compiling mainly the ideas published in the text Aesthetic Theory of 1970. The term culture industry is also contextualized in the chapter with the reflections that Adorno previously exposed in 1947. A dialog is created with the proposal of the North American theoretician and artist Martha Rosler to understand the chronological development of art before, during, and after Adorno. Regarding the relation between art and autonomy, the ideas of Adorno offer elements to understand contemporary art production. This way, the author also discusses contemporary new media art manifestations, which are analyzed in key terms such as autonomy/ culture industry in relation to the proposals of the Brazilian theoretician Arlindo Machado. Lastly, the chapter offers an approach to the artistic institution analyzing the museum in relation with Adorno's ideas.

During the twentieth century, the modern media was born and viewed as an industrial factory-model machine. These powerful media such as film, radio, and television transmitted culture to the passive masses. These art forms were divorced of ritual and authenticity and were reproduced to reinforce their prowess. In the twenty-first century post-media condition, a process of convergence and evolution toward a social consciousness, facilitated by a many-to-many social network strategy, is underway. Web 2.0 technologies are a catalyst toward an emergence of a collectivist aesthetic consciousness. As the prophecy of a post-industrial society becomes fulfilled, a post-media society emerges whose quest is for knowledge dependent upon economy that barters information. This chapter identifies a conceptual model of this recent paradigmatic shift and to identify some of the possibilities that are emerging.

As social entities, people could come together and create regular relations and institutions. Concepts such as group, community, society, social movement, etc. are about the social dimension of man. However, according to the conditions of the social, political, religious, and physical environment in which the person lives, their needs and problems can change. As a natural consequence of this, social characteristics of social movements can change. It is inevitable to value new social movements in this respect. Because new social movements in the West are born from a critical intellectual atmosphere against modernity. This situation has developed in an economic system based on the service sector rather than economic-order-based on heavy industry. On the other hand, the Eastern world, especially the Islamic world, has a repertoire of social movements that brings different problems to the agenda because it has different conditions. In this respect, new observations and analyses of Turkey and its surrounding will provide important contributions to the theory of social movements.

The concept of interaction is foundational in technology interface design with its presuppositions being taken for granted. But the interaction metaphor has become ambiguous to the extent that its application to interface design contributes to misalignments between peoples' expected and actual experience with computer-enhanced actions. This chapter re-examines the presuppositions governing human-computer interaction with the motivation of strengthening weaknesses in their foundational concepts, and contributing a theoretical framework to designing for artistic as well as mundane experience. It argues for abandoning the interaction metaphor to refocus design discourse toward the intermediation and mediation roles of technology interfaces. Remediation (i.e., representation of one medium in another) is proposed as a conceptual model that more precisely describes the human-to-computer actions.

In this chapter, the author examines the levels of a cognitive process (visceral, behavioural, and reflection design) in experience-centred design and their applicability in developing effective products. The study further explores the role of computer-aided tools (CAD) in engaging users as the fundamental participants in the creative process. This was achieved by developing two products digitally (CAD models) based on Norman's three design levels, and evaluated using a semantic differential scale to test the emotive response towards the designs. The findings show that CAD is a viable tool for gaining insight into users' perceptions towards a design idea. Also, the results revealed that the process of supporting or sustaining cultural values through design also enhances the affective quality (reflective level) of the design.

This chapter delves into the creation of text and its visualization through typographic design inside three-dimensional virtual worlds known as the metaverse, with a particular focus upon the way in which these environments may place the usage of text within a context that stands in contradiction to its traditional one – namely legibility. Readability can be displaced through the usage of text and typography as playful devices that may manifest as visual structures, the contents of which are meant to be understood through means other than reading – as it is understood in the traditional sense of the word. Instead, such spaces are aimed to bring about states of heightened engagement, wonder, and "play" by being immersed within the typographic conglomerations that constitute their essences. This subject is explored through the discussion of three such typographic playgrounds that the authors created in the metaverse.

The experience of creating in technology-mediated spaces through an avatar form brings the phenomenological experience of the body into the act of creating itself. Ways of explaining this creative process and its relationship to imaginative experience highlights a phenomenology of virtual practice of artists working in this realm. This chapter explores the phenomenology of the imagination in relation to embodied experience and considers the implications of phenomenology as a methodological approach and presents the case for investigating artistic and imaginative experience through adapting phenomenological research methods.

Although online shopping has become a popular and convenient instrument for companies to buy and sell products, the design of these web-shops does not always offer the rich multisensory experiences that physical retailing offers. In the chapter, the authors argue that introducing audio-visual contents in the design could provide dynamic multisensory information to offer more engaging experiences to the consumer. Despite existing controversies regarding universalism of the emotional experiences induced by perceptual processes, the authors present evidence that suggests cultural modulations of videos experiences. Spanish and Finnish participants interacted with audiovisual products depicting videos of a culturally loaded brand design. Content analysis of participants' verbalizations helped to identify categories and subcategories that defined the representation of the video elements and their relative weight depending on the cultural background of the viewer. Although results indicate common elements affecting viewers of the two countries, they differ in the relative weight to global aesthetics features.

This chapter practically investigates how rapid prototyping technology can be effectively adopted to enhance ceramics production. The researcher used an experimental approach that analysed the conventional methods used in ceramics production and then introduced computer-aided design tools (CAD) and rapid prototyping technology into the development process. After that, four ceramic products were developed to test the viability of the rapid production process. Finally, the process was evaluated and compared to the conventional methods. The findings revealed that the introduction of CAD and rapid prototyping technology into the development process reduces the development time and also helps to ensure the accuracy of the prototypes produced. Besides, visualising and validating the design digitally (CAD models) also helps to identify possible faults at the early stage of the development process. Therefore, the ideas explored in this chapter will give insight to ceramics designers and artisans on how they can effectively eliminate bottlenecks in the production process.

Preface

The constant changing landscape of technology as both a creative and unifying tool generates the need for knowledge resources that will empower academics, researchers, and students all over the world.

Enhancing Art, Culture, and Design With Technological Integration is a vital reference source that will meet these needs by exploring some of the latest coverage on technology utilization in society, industry, and design, with discussions centered on topics such as: social movements, virtual communities, product development, and mobile learning.

This reference source is organized into 12 chapters contributed by global experts, drawing on their experiences, observations, and research findings. A brief description of each of the chapters can be found in the following paragraphs.

In Chapter 1, the author examines some of the ideas that Theodor Adorno elucidated around the term culture industry, compiling mainly the ideas published in the text Aesthetic Theory of 1970. This way, the author also discusses contemporary new media art manifestations, which are analyzed in key of terms such as autonomy in relation to the proposals of the Brazilian theoretician Arlindo Machado. Lastly, the chapter offers an approach to an artistic institution by analyzing a museum in relation with Adorno's ideas.

In Chapter 2, the author discusses the 20th century's media as an industrial factory-model machine before its transformation into the 21st century post-media condition. Web 2.0 technologies are presented as a catalyst toward an emergence of a collective aesthetic consciousness, where a post-media society emerges with a quest for knowledge dependent upon an economy that barters information. This chapter identifies a conceptual model of this recent paradigmatic shift and to study some of the possibilities that are emerging.

In Chapter 3, the authors evaluate the transformation and change of characteristics in social movements within a person's social, political, religious, and physical environments. As new social movements in the west are born from a critical intellectual atmosphere against modernity, they are developed in an economic system based on the service sector rather than an economic order based on heavy industry. In this respect, the authors analyze Turkey and its surroundings to provide important contributions to the theory of social movements.

In Chapter 4, the author re-examines the presuppositions governing human-computer interaction with the motivation of strengthening weaknesses in their foundational concepts. The chapter proposes a theoretical framework for designing for artistic as well as mundane experiences. It argues for abandoning the interaction metaphor to refocus design discourse toward the intermediation and mediation roles of technology interfaces.

In Chapter 5, the author evaluates the levels of a cognitive process in experience-centered design and their applicability in developing affective products. The chapter further explores the role of computer-aided tools in engaging users as the fundamental participants in the creative process by developing two products digitally based on Norman's three design levels. The results revealed that the process of supporting or sustaining cultural values through design also enhances the affective quality of the design.

In Chapter 6, the author attempts to delve into the creation of text and its visualization through typographic design inside three-dimensional virtual worlds, focusing on the way in which these environments may place the usage of text within a context that stands in contradiction to its traditional one. Such spaces are aimed to bring about states of heightened engagement by being immersed within the typographic conglomerations that constitute their essences. This subject will be explored through the discussion of three example typographic playgrounds created by the author.

In Chapter 7, the author explores the role of virtual worlds as spaces for creative practice challenges and enhancing understanding of the phenomena of imagining. The chapter also explains the creative process and its relationship to imaginative experience, and highlights a phenomenology of virtual practice of artists working in this realm. The author explores the phenomenology of the imagination in relation to embodied experience and considers the implications of phenomenology as a methodological approach and presents the case for investigating artistic and imaginative experience through adapting phenomenological research methods.

In Chapter 8, the authors consider the architecture of virtual spaces and buildings within multiple social activities ranging from entertainment to education, economics, and medicine, as well as the opportunities to develop novel and imaginative forms of buildings. The chapter identifies the methods to better meet the requirements and needs of the users by utilizing formulae adapted to the potentialities of the world of information. The authors propose a possible starting point with the real-world architecture complemented by the virtual and exclusive domain of virtual architecture.

In Chapter 9, the author presents how different processes were developed for collectively producing a series of explorative soundscapes through interface creation and mechanical artifacts using specific constraints influenced by theories of art, design, and architecture. The chapter will also present some examples of different prototypes and iterations that brought together an editor and the final expression of the artwork into one surface of interaction and execution using a virtual cityscape as an iterative ground for sound and music explorations. Additionally, the chapter will also discuss the parameters of the framework, the sounds, and the final visual expression to match the artistic intention.

In Chapter 10, the authors argue that introducing culturally-aware audio-visual contents in web-shop design could provide dynamic multisensory information to offer more engaging experiences to the consumer. Content analysis of participants' verbalizations with audiovisual products helped to identify categories and subcategories that defined the representation of the video elements and the categories' relative weight depending on the cultural background of the viewer.

In Chapter 11, the author investigates how rapid prototyping technology can be effectively adopted to enhance ceramics production. The findings revealed that the introduction of computer-aided design tools and rapid prototyping technology into the development process reduces the development time and helps to ensure the accuracy of the prototypes produced. The ideas explored in this chapter give insight to ceramics designers and artisans on how to effectively eliminate bottlenecks in the production process.

In Chapter 12, the author studies the role of mobile learning as a spontaneous learning model, providing a more flexible and collaborative technology-based learning and creating new opportunities for enhancing student learning experiences. The chapter presents the development of a game to assist in teaching and learning, aiming to help students acquire knowledge in the field of geometry. Findings show that by using the game, students have improved around 35% the hits of correct responses to the classification and differentiation between edge, vertex and face in 3D solids.

The comprehensive coverage this publication offers is sure to enhance its reader's understanding of the latest topics and discoveries pertaining to technological integration in art, culture, and design. Furthermore, the contributions included within this publication will be instrumental to the expansion of knowledge offerings in diverse fields and will inspire future research and innovation.

Chapter 1
Culture Industry:
A Contemporary Revision of the Term of Theodor Adorno

Claudia Sandoval Romero
Academy of Fine Arts, Austria

ABSTRACT

This chapter examines some of the ideas that Theodor Adorno elucidated around the term culture industry, compiling mainly the ideas published in the text Aesthetic Theory of 1970. The term culture industry is also contextualized in the chapter with the reflections that Adorno previously exposed in 1947. A dialog is created with the proposal of the North American theoretician and artist Martha Rosler to understand the chronological development of art before, during, and after Adorno. Regarding the relation between art and autonomy, the ideas of Adorno offer elements to understand contemporary art production. This way, the author also discusses contemporary new media art manifestations, which are analyzed in key terms such as autonomy/ culture industry in relation to the proposals of the Brazilian theoretician Arlindo Machado. Lastly, the chapter offers an approach to the artistic institution analyzing the museum in relation with Adorno's ideas.

INTRODUCTION

Since its mere emergency art has been asked about its autonomy in relation with its social role and with the industry. At the same time art has always been the deposit of romantic ideals and the dreamed place from where society is criticized. Theodor Adorno was one of the pioneer authors who reflected about the relation of art industry

DOI: 10.4018/978-1-5225-5023-5.ch001

with communication media such as the TV, which in 1947, time of Adorno's first reflections, was getting consolidated1. The time of Adorno appears similar to ours of the consolidation of the internet, making his reflexions pertinent for a critique of the media art of our time.

This way, strongly contemporary Adorno's ideas about the *Hitlerian strategies* (Adorno, 1947, p. 16-17) of the mass media, help us to understand the idealism of Adorno's art conceptions, through the approach to his historical moment and his own personal life.

The present article briefly describes what Theodor Adorno reflected about the relation between art and the culture industry in his posthumous and probably most influential publication *Aesthetic Theory* of 1970. Subsequently the term is contextualized with previous reflections of the author since 1947.

Together with the research of the North American theoretician and artist Martha Rosler, the ideas of Adorno are shown in a parallel to the moments of art before, during and after Adorno's time. At the same time that contemporary new media art manifestations are analyzed in the article in key of the couple autonomy/culture industry, this time in relation to the proposals of the Brazilian theoretician Arlindo Machado.

Lastly, the article offers an approach to the artistic institution analyzing the museum in relation with Adorno's ideas.

CULTURE INDUSTRY INSIDE THE AESTHETIC THEORY OF 1970

In the first section some of the historical conceptions of the culture industry will be brought together in order to create a compilation of the ideas that illustrate the complexity of the term proposed by Theodor Adorno.

This way, the idea of culture industry will be described as a factor that made possible for art to become a consumer good, understanding this process also in key of a commercialization of the catharsis in the reception of an art piece.

Culture Industry and Consumer Goods

Theodor Adorno explains in *A Critic to the Culture industry*, section of the book *Aesthetic Theory* published in 1970, that as far as art corresponds to a social manifested necessity, it transforms itself mostly in a business governed by the profit, which persists as long as it is profitable. By doing so, art makes itself aside, confirming being nothing but something already dead (Adorno, 1970, p. 34).

Adorno exposes the means used by the culture industry in order to transform art pieces into commodities when he makes clear that the 'naïve' people of the culture industry, avid of commodities, locate themselves closer to art, perceiving how art is inadequate to accompany the process of social life. Adorno argues that the creation of this proximity to art only intensifies the culture industry, as well as, at the same time, here the idea of the immediatism of art to the society is planned to deceive (Adorno, 1970, p. 376). On the other hand, the culture industry defends that art suffered a process in which it ceased being what it is and lost its specificity, becoming consumer goods in the shape of catharsis and art pieces themselves (Adorno, 1970, p. 34).

In the following paragraphs we will then explain how for Adorno the art and its catharsis became consumer goods.

Art Pieces as Commodities

For Adorno indubitable symptoms of the tendency of the culture industry are the passion for what is palpable, as well as for not letting any work be what it is, accommodating them, while diminishing the distance in relation to the spectator (Adorno, 1970, p. 32).

For the author considering art *vested interests* means to classify subjectively art inside the commodities. Adorno specifies the complexity in the relation with art as goods when he hopes that *at least* art was simply consumable, then, this way, the relation with art would be based on the mere relation with commodities but for Adorno in a time of super production the value of use of art is also problematic since it is submitted to the secondary delight of prestige, of fashion and then, finally, submitted to the own character of commodity. This way, of the autonomy of art just remains the fetishistic character of commodity, being this a regression to the archaic fetishism in the origin of art (Adorno, 1970, p. 32).

On the other hand, the culture industry's praxis advocates for a servile respect for empirical details, this is, for the shape, allying this way the praxis of the culture industry with the ideological manipulation (Adorno, 1970, p. 336). A fact that makes believe in the importance of art based on its technique, and on its final shape, which is equal to its fetishistic character. And for Adorno, the sublimation of the form is a vehicle of ideology.

For the author a type of art that prevails in the culture industry establishes its significance on its *value of exposition*, instead of its *auratical*, *cultuai*, value (Adorno, 1970, p. 73).

At the same time that the culture industry transforms art into commodity, it also feigns the catharsis in the shape of a consumable aesthetical stimulus, as it will be explained as follows.

Catharsis as a Commodity

For Adorno the catharsis experienced through the appreciation of art pieces is taken and administrated by the culture industry and, by doing so, the aesthetical experience is no more sublime but one that makes possible its commercialization.

According to the theoretician, the real catharsis of art consists in the loss of oneself in the perception of the own limits and finitude not in a particular satisfaction of the 'I' or in the pleasure proposed by the culture industry.

For Adorno the idea of a profound catharsis is nonsense, is madness, for the culture industry (Adorno, 1970, pp. 364-365).

In the same line of Hegel's conception about the freedom to the object, Adorno argues that the aesthetical sublimation, guarantied in the past the dignity of the spectator who, in a spiritual experience, becomes a subject through his/her alienation to the object of art. And this conception opposes the *philistines* aspirations that demand art to give something (Adorno 1970, p. 32).

This *something to give* is understood like the reductionist process in which art becomes a thing and a medium for the use of the psychological processes of the spectator (Adorno, 1970, p. 33).

Secondly, to create a market from art some of the strategies used by the culture industry are the progressive subjective differentiation of the creators of art, as well as the intensification and diffusion of the domain of the aesthetical stimulus (Adorno, 1970, p. 354). Therefore, art becomes inferior art and entertainment, as well as in this context art is a manifestation of the obvious. This process legitimates the ideology of the culture industry, an ideology that made art popular and closer to the public. But here for Adorno, the popular art is just a strategy that the culture industry uses to profit. Conversely to this idea of the art made popular and closer to the public, the notion of noble art has been, since Baudelaire, an accomplice of the privilege (Adorno, 1970, p. 354-355).

Lastly, concerning the transformation of the catharsis into a commodity, for Adorno, art transforms itself into a consumable product when, in an effort for not to be a commodity, it becomes a spiritual matter available at all the levels of life. Being this, as mentioned before, another kind of commodity that is taken and administrated by the culture industry (Adorno, 1970, p. 354).

This way, art pieces are turned into consumer goods and developed certain characteristics within the spirit of the culture industry that Adorno explains as follows.

Spirit of the Commodities

One of the characteristics of art as a commodity is its ridiculous seriousness. Adorno points to the fact that the culture industry is nowadays the only space where the spirit of seriousness of art appears, being this a distorted seriousness, clarifying that when art is serious is posing of not being modern art, turning art itself into something ridiculous (Adorno, 1970, p. 65). Simultaneously, for Adorno, if art pieces were absolute responsible, it would make art sterile.

The complexity of this hypothesis appears when Adorno exposes that at the same time art cannot be responsible but if it is exclusively irresponsible, art is nothing but a game. This way, for the author, it is only in front of the culture industry that art keeps its seriousness (Adorno, 1970, p. 65), but a fake new one that makes possible the commercialization of art pieces.

Regarding the irresponsibility of art, Adorno explains that to be delirious is for art a critic to the social praxis, a mimetic residue of art, and the price of art's impermeability. But in this barbarian moment, the irresponsibility of art transforms it in mediocrity and is only used by the culture industry as profit in the shape of *the fun* of art; this is, in the shape of a cultivated puerility of art (Adorno, 1970, p. 181). In a cycle, the puerility, which shares the objectives of the anti-intellectualism of art, degenerates directly in culture industry and is then fed by this, being, in this movement, preserved (Adorno, 1970, p. 309).

The question that therefore would follow is: what do we need art for? and if so, then, how to make it possible outside of the culture industry? this is, autonomous?

The Necessity of Art

According to Adorno, the necessity of art within the context of the culture industry is a creation of the culture industry.

For the author, the first time that the masses were confronted with art was through the mechanical medium of reproducibility offered by the culture industry.

The increase of the necessity of art for the masses becomes thus suspicious, since the culture industry hides behind it. For Adorno, the fact that the culture industry validates art, not only raises doubts, but would also not be sufficient to assure the justification of art, since it emerges from an area external to art (Adorno, 1970, p. 34).

In this context, clarifies Adorno, the one who defends art equalizes it to the ideologies of the culture industry (Adorno, 1970, p. 34).

Regarding the necessity of art, Adorno explains that the aesthetical necessity is a concept slightly vague and unarticulated, while at the same time the practices of the culture industry didn't produce the many changes they want us to believe, and that we are taking for granted.

For Adorno the necessity of art is in great size ideology. For him, it would be possible to live without art, not only objectively, but also in the consumer's psyche, who is suggested –constantly and without cost- to change their taste, since he/she follows the way of the smallest resistant: The shape of culture that matters to the culture industry is one without complexion and simple enough to be able to profit from it. The cultural goods are here a naïve and affirmative image of the culture (Adorno, 1970, p. 361).

Autonomy of Art

As a solution for the autonomy of art inside the context where the culture industry reigns, Adorno describes the place of artworks as its own. Not the place of art in relation to any social function.

For Adorno when art crystallizes as a specific thing in itself -instead of opposing the existent social rules or instead of qualifying itself as *socially useful*-, criticizes the society because of its simple existence.

This, according to the author, has been reproved for puritans of all confessions because what exists purely, completely structured according to its immanent law, exercises a critic without words, and denounces the degradation of society. This is, the society of exchange where everything exists only for something else.

This way, the asocial aspect of art is a certain denial of a certain society. But furthermore, for the autonomous art applies the idea that from the distance, art leaves intact the society from which it feels horror (Adorno, 1970, p. 253).

At the same time for Adorno, art keeps itself alive only because of its strength of resistance: Its contribution to society is not the communication with society, but a resistance to it (Adorno, 1970, p. 254).

Thanks to the complexity of this scenario, the autonomy of art is conceived by Adorno as a utopia, considering the fact that art emerges in relation to the manipulation offered by the culture industry (Adorno, 1970, p. 33).

Even though for the author the autonomy of art is an utopia, the author envisions a solution, a space where the culture industry cannot reign, when he claims that all these reflections can begin with the fact that, in reality, something aspires objectively to art, beyond the veil woven by institutions and fake necessities. For the author this would be then an art that evidences what the veil hides (Adorno, 1970, p. 34).

Regarding the value of art, although Adorno recognizes that provocative effects of art are quickly dissipated, he moreover identifies that artworks offer a practical effect in the transformation of counsciousness (Adorno, 1970, p.272)

These possible solutions to the autonomy of art were only conceived by the Adorno of the *Aesthetic Theory* in 1970.

Previous to this text, Adorno pronounced overwhelmingly the relation between culture industry and art, with less than few possible solutions.

Following, a brief tracking of the term culture industry will be exposed, as well as some more reflections about the term and its implications that were exposed prior to the significant publication of 1970.

THE TERM CULTURE INDUSTRY

Brief Trajectory of the Term

The term Culture industry didn't appear the first time in Adorno's posthumous publication

Aesthetic Theory, in 1970. As Adorno himself mentions in *Résumé über Kulturindustrie* -a text that was published in 1968 based on the radio conferences broadcasted in 1962 in Germany-, the first time that the term 'culture industry' was used, was in the book *Dialektik der Aufklärung* published by Adorno and Horkheimer in Amsterdam in 1947(2).

According to Adorno (1968, p. 61), the authors initially used the term *mass culture* that was subsequently substituted by for 'culture industry'. And they did so intending to exclude the defenders of the culture industry who aim it to be a sort of culture emerging spontaneously from the masses.

For Adorno the culture industry differs from this radically: the culture industry is the deliberated integration from above, of its consumers. It forces the union of domains, separated since thousand years, of superior art and inferior art, prejudicing both: superior art experience and the frustration of its seriousness on benefit of the speculation about their effect.

Through civilized domestication inferior art loses the element of resistant and rude nature, inherent to it when the social control is not absolute (Adorno, 1968, p. 61).

The origin of the culture industry can be located, according to Adorno, from the economical point of view when merchants where looking for new possibilities of application of the capital in more developed countries (Adorno, 1968, p. 63).

In addition to this, Adorno recognizes that authors like Brecht und Suhrkamp reflected in the decade of 1930s about the fact that cultural commodities of the industry are oriented towards its commercialization and not according to art's own content or adequate configuration (Adorno, 1968, p. 62).

Subject Means Object

Another aspect reflected for Adorno since the text published in 1968 is the concern about the state of consciousness and unconsciousness of millions of people, which is defined by the culture industry.

In terms of the author, for the culture industry the masses are nothing but a secondary factor, a calculation element, and an accessory to the machine. The consumer is in this system not a king, as the culture industry would like to make believe. He/she is not the subject but the object of that industry. And at the same time the culture industry reinforces the mentality of the masses, by taking this mentality as something given a priori and immutable (Adorno, 1968, p. 62).

For Adorno the interest of the culture industry in people only takes place when they are considered clients or employees (Adorno, 1947, p. 15).

At the same time, in the publication of 1968, Adorno exposes a diminishment in the spiritual character of art pieces inside the culture industry, as explained in the next section.

Individualization and the Spirit of Creations

For the theoretician, every practice of the culture industry transfers the motivation of profit to spiritual creations, and from the moment these commodities assure the position of its producers in the market, this motivation contaminates the spiritual creations. (Adorno, 1968, p. 62)

The products of the spirit, according to the author, once inscribed inside the style of the culture industry, are not only commodities as they are integrally commodities (Adorno, 1968, p. 64). For Adorno the culture industry forces the spiritualization of distraction (Adorno, 1968, p. 13)

At the same time that in this process, every product is presented like individual, considering this a kind of individuality that contributes to strengthen the ideology (Adorno, 1968, p.62): An ideology that defends the system of the 'vedettes' taken from the individualist art and its commercial exploitation.

In opposition, when it comes to defend themselves from the critics, the promoters of the culture industry have the pleasure to claim that what they provide is not art but an industrial product (Adorno, 1968, p. 66).

The culture industry transforms the fabrication of a simple good will in public relations, without the relation with the producers or with any particular object of sales. It looks for a client to sell him/her a total and not critic agreement (Adorno, 1968, p. 64)

Autonomy of Art Seen by the Adorno of 1947 and 1968

Regarding the autonomy of art, Adorno claims since his previous publications to the *Aesthetic Theory* in 1970, that pure art pieces that deny society's mercantile character, because of following its own rules, become also a commodity:

Until XVIII c the protection of the maecenas defended artists from the market, but these artists were instead subjected to the maecenas and his wishes. For the author, the freedom of the great modern artwork lives from the anonymity of the market (Adorno, 1947, p. 21).

The autonomy of art almost never existed in a pure form and was always marked by connections of effect. Nevertheless, with the culture industry, this autonomy sees itself in the limit of being abolished. And this is a process that occurs with or without the conscious will of its promoters (Adorno, 1968, p. 63)(3).

Critique

The main points of criticism distinguished in Adorno's early reflections are, first of all, the exhortation to not to take culture industry seriously. It is something suspicious, since in the name of the social role of the culture industry judgments are repressed or eliminated about its quality, about its authenticity or falsity, as well as questions about the aesthetical level of its message.

For the author, to take seriously the proportion of its unquestioned role means to take it critically serious and not to curve oneself in front of its monopoly (Adorno, 1968, p. 68).

In our current time authors like Laymert Garcia dos Santos propose the same debate in publications like *To Politicize the New Technologies*, 2002, this time in relation with the internet, the digital era, the genetic manipulation and nanotechnology. For the author it is necessary to politicize the debate about technology and its relation with the capital, instead of leaving this debate to the transnational industries, as the establishment prescribes.

For Adorno the justification of the culture industry is contradictory once the *lawyers* of the culture industry defend it with the excuse of the *not intention to be art*.

Once again, the ideology excuses the responsibility in relation to what maintains the business alive. But at the same time this type of no-art is also referred to the fact that even the worst movie is presented objectively as an art piece inside the culture industry.

The criticism of Adorno focuses in media like radio, cinema and the TV, pointing how the products they offer, like the sentimental mail and horoscope, are intended to be inoffensive and democratic because they obey to a demand. But these demands were previously stipulated.

The defenders also claim that these manifestations bring all kinds of benefits like the diffusion of information and advices, while, at the same time, it acts as a reliever of tension.

But this information is poor or meaningless, like proves a sociological study about the level of political information found in these media (Adorno, 1968, p. 67). At the same time that in case these media disseminate informations, these are simply futilities, or even worse, they are shameless conformist standards of behavior. This all, only concludes for Adorno in the idea that *it is now more real than it never intended to be, that the world wants to be deceived* (Adorno, 1968, p. 68).

As a consequence of these processes, what is firstly affected is the formation of individuals as explained in the next paragraphs.

A Critique to the Lack of Formation of Individuals

In the previous reflections to the text *Aesthetic Theory* Adorno presents a more strengthened connection within the political implications of the culture industry.

For him, the critique to the culture industry in relation to the impossibility of creation of individuality is based on the imperative of the culture industry according to the idea of *you must subjugate*, without indicating to what.

This way the culture industry impels to subjugate to whatever is already given and to what -as a reflection of its power and omnipresence-, everybody, -this is, the other-, thinks.

As a conclusion of this process, through the ideology of the culture industry, conformism substitutes consciousness, and this conformism is created in key of the culture industry's ruminations. This means that these are not rules for a happy life, neither a new kind of moral responsibility, but the exhortations to conform to what is behind the interests of the power (Adorno, 1968, p. 69).

As a critic to the conformism that the culture industry creates, Adorno asseverates that without confessing it, the spectators of these media sense that their lives become intolerable if they don't hold to satisfactions which, truly, are not any satisfaction (Adorno, 1968, p. 68).

In this context, the one who doubts of the potency of monotony is a maniac. While at the same time, as the author provocatively argues, *nobody will be cold or have hunger: whoever feels hunger or cold, will end in a concentration camp*. This sentence, emerged from the hitlerian Germany, could *shine as a slogan* of the culture industry, where to be the outsider is the most serious guilt (Adorno, 1947, pp. 16-17).

The compensatory satisfaction that the culture industry offers to people when it awakes in them the sensation of comfort, of a world in order, frustrates them in their own happiness, which it deceptively supplies (Adorno, 1968, p. 70).

Adorno concludes about the lack of individualization that it impedes the formation of autonomous and independent individuals, able to judge and decide consciously (Adorno, 1968, p. 70). As they constitute the previous condition to democracy, is here easy to imagine what for this is convenient.

In general terms, it is noticeable how the tone and ideas of Adorno in *Dialektik der Aufklärung*, 1947, is softened in the text of 1968 *Résumé über Kulturindustrie*, and become more abstract in the late *Aesthetic Theory* of 1970. At the same time that the harsh critiques, even more the ones that made a connection between the ideals of the culture industry and the ones of the Hitlerian Germany, tend to dissapear in Adorno's discourse of the last times.

Also, Adorno is important to us because, as it was already mentioned, he is one of the pioneers in interpreting the art industry nowadays as we will as follows explain.

AN APPROACH TO ADORNO

Adorno contributed with tools of analysis to understand the relation between art and industry in the current time. Diverse contemporary authors like Martha Rosler and Arlindo Machado have also reflected about this topic, and this way it is possible to bring them together and complete the trace of history allowing them to dialog and describe the context since 1947 until the present days.

The Autonomy of Art: Adorno and Martha Rosler

As a way to complete Adorno's ideas about the autonomy of art, here is exposed the reflexion of the artist and theoretician Martha Rosler, building a past-present-and-future chart in relation to Adorno's time.

Before Adorno's Time

(...) social negativity was a central element of the Frankfurt School, as exemplified by Adorno's insistence, against Brecht and Walter Benjamin, pointing to the fact that art in order to be appropriately negative must remain autonomous, above partisan political struggles (Rosler, 2010).

But different to Adorno, for Martha Rosler the alliance between artists and social phenomenon during the Industrial Revolution pointed to the liberation of the media of the time benefiting artistic freedom.

Artists working in a variety of media and cultural registers, from high to low, expressed positions on the political ferment of the early Industrial Revolution. One might find European artists exhibiting robust support for revolutionary ideals or displaying identification with provincial localism, with the peasantry or with the urban working classes, especially using fairly ephemeral forms (such as the low-cost prints available in great numbers) (...) New forms of subjectivity and sensibility were defined and addressed in different modalities (the nineteenth century saw the development of popular novels, mass-market newspapers, popular prints, theater, and art), even as censorship, sometimes with severe penalties for transgression, was sporadically imposed from above (Rosler, 2010).

Even though it is a complex labor to define clear limits between tendencies, to this respect, Rosler argues that precisely only a couple of hundred years ago, -*as the old political order crumbled under the changes wrought by the Industrial Revolution, and direct patronage and commissions from the Church and aristocrats declined*-became possible for painters and sculptors to be independent enough to be critic.

But this criticism would not exclude the fact that a type of *art that exhibits an imperfect allegiance to the ideological structures of social elites has often been poorly received.* And this would mean that to step outside the ambit of patronage or received opinion, meant *losing one's livelihood or, in extreme situations, one's life* (Rosler, 2010).

For Adorno the autonomy of art was seen as almost inexistence as far as pure art pieces that deny society's mercantile character, because of following its own rules, become also commodities (Adorno, 1947, p. 21).

Contrary to what Martha Rosler points, the patronage for Adorno signified a solution for the autonomy of art from the market, which at the same time meant an opposite way of losing its autonomy in order to satisfy the patrons' wills.

For Adorno, until XVIII c the protection of the maecenas defended artists form the market, but these instead where subjected to the maecenas and their wishes. The freedom of the great modern artwork lives of the anonymity of the market (Adorno, 1947, p. 21).

During Adorno's Time

For Rosler the time of Adorno, first half of the twentieth century,

Consisted on a prodigious industrialization and capital formation, witnessed population flows from the impoverished European countryside to sites of production and inspired millenarian conceits that impelled artists and social critics of every stripe to imagine the future. We may as well call this modernism. And we might observe, briefly, that modernism (inextricably linked, needless to say, to modernity) incorporates technological optimism and its belief in progress, while anti-modernism sees the narrative of technological change as a tale of broad civilizational decline, and thus tends toward a romantic view of nature (Rosler, 2010).

Adorno interprets in this context a new death of art, as he perceived how through the process in which art persists as long as it is profitable, art makes itself aside, confirming being nothing but something that is already dead (Adorno, 1970, p. 34).

To complete the trace of this aspect for the current time, and in words of Rosler, in 2008 the real estate market brought down much of the world economy, taking the art market with it (Rosler, 2010). What in relation to the capitalized idea of art for Adorno this would mean that what was as well brought down during the economical fall, was not just a market created around art pieces, but the very same art which grew until the current time in relation to its market. It is peculiar to perceive how since Hegel's death of art in the beginning of the XIX century, authors like Adorno and Rosler keep reflecting about new deaths of art.

Martha Rosler's perspective adds clarity to the last years of life of Adorno and the ones that followed his death in 1968. For the theoretician the decade of 1960s consisted in

A robust moment, if not of outspoken criticality in art, then of artists unrest, while the culture at large, especially the civil rights/youth culture/counterculture/antiwar movement, was more than restive, attempting to re-envision and remake the cultural and political landscape. Whether they abjured or expressed the critical attitudes that were still powerfully dominant in intellectual culture, artists were chafing against what they perceived as a lack of autonomy, made plain by the grip of the market, the tightening noose of success (though still nothing in comparison to the powerful market forces and institutional professionalization at work in the current art world). In the face of institutional and market ebullience, the 1960s saw several forms of revolt by artists against commodification, including deflationary tactics against glorification. One may argue about each of these efforts, but they nevertheless asserted artistic autonomy from dealers, museums, and markets, rather than, say, producing fungible items in a signature brand of object production (Rosler, 2010).

As Rosler manifests, the market forces, as well as the institutional professionalization of the art world, were stronger than the fight against the lack of autonomy. But it was only possible in this context that media art manifestations appeared to question the lack of autonomy. Like in the cycle of *work's production, absorption and neutralization*, the search for autonomy is followed by the scope of the culture industry. Like this, it declares art what was supposed to be anti-art, looking for new means of commerce.

This way, Adorno and Rosler agree with the fact that what was left of the robust political, cultural and artistic manifestations of 1960 was nothing but the development of new strategies of the culture industry in order to maintain the commercialization of art.

The fact that the culture industry changes its strategies with time in order to maintain art as a consumer good, which was observed by Adorno, didn't alter after his lifetime. On the contrary it became more complex as explained as follows.

After Adorno's Life

To illustrate the complexity of the autonomy of art after the time of Adorno, it is pertinent to consider what Rosler specifies about the late 1960s when in the United States

President Johnson's Great Society included an expansive vision of public support for the arts. In addition to direct grants to institutions, to critics, and to artists, nonprofit, artist-initiated galleries and related venues received Federal money. This led to a great expansion of the seemingly uncapitalizable arts like performance, and video, whose main audience was other artists. Throughout the 1970s, the ideological apparatuses of media, museum, and commercial gallery were deployed in attempts to limit artist's autonomy, bring them back inside the institutions, and recapitalize art. (…) Art educators began slowly adopting the idea that they could sell their departments and schools as effective in helping their students find gallery representation by producing a fresh new line of work (Rosler, 2010)

With the apparition of new patronages like the own government art becomes more restricted of its autonomy, being this the new faces of the culture industry. This way its autonomy becomes every time a more intricate dance of interests, at the point that nowadays it is pertinent to ask -without this being a rhetorical question-, like the seminar of Univ.-Prof. Dr. Violetta Waibel at the University of Vienna: *Do artworks exist? How are they possible?*

For Rosler the only possible solution is to consider that *these, then, are not abandonments of art world participation but acceptance that these institutions are the proper – perhaps the only - platform for artists* (Rosler, 2010).

The theoretician finds then a positive general solution on the interpretation of the changes between art and industry that affects not just art, but also the way of actively live in society, based on the fact that the new movements occupy *a middle region between 'individual and collective'* and so have the possibility of engineering a different relationship to society, state, and capital. For the author the new forms of communication can produce *a radically new form of democracy* (Rosler, 2010).

This conclusion of the present time would be a solution that contradicts what Adorno reflected about the lack of autonomy in the creation of individuals at the end of the 1960s, also in terms of individuals who keep alive the apparatus called democracy without any critical formation, being this maintained by the information spread through the culture industry (Adorno, 1968, p. 70).

On the other hand Rosler provokes with the idea about *whether choosing to be an artist means aspiring to serve the rich.* With the new global economy relations the idea here exposed by Rosler as typical of the United States can be generalized. But also Adorno explores this axis, noticing that this relation is not new when he claims that the notion of noble art has been, since Baudelaire, an accomplice of the privilege (Adorno, 1970, pp. 354-355). And at the same time, in Adorno's words no infamy is softened when it is declared as such (Adorno, 1968, p. 68).

Regarding the autonomy of art, Rosler points to a context where as she explains, *how long can the academy of art success without the galleries?* At the same time, for the theoretician, artists are stubborn and there are always artworks, or art *actions*, that are situated outside the art world or that *cross-list* themselves in and outside the golden ghettos. For the author, the art market doesn't mean it all: *It is not the market alone, after all, with its hordes of hucksters and advisers, and bitter critics, that determines meaning and resonance: there is also the community of artists and the potential counterpublics they implicate* (Rosler, 2010).

It is important to follow the trace in the search of both authors regarding the autonomy of art, as something superior to a commercial intention; this is, headed towards an intention inside its own manifestation. And it is important because it could give us the clues to understand the complexity of the time we live, ultimate objective of art's expression.

Finally, since the decade of 1960s and with the apparition of a strong tool like the mass media to control and distribute art, new ways to make art also emerged as a consequence and as a contrary position to this tendency. As Martha Rosler mentions even inside the strengthened culture industry through the use of the media, also appeared a

So-called 'dematerialization': the production of low-priced, often self-distributed multiples;(...); the development of multimedia or intermedia and other ephemeral forms such as smoke art or performances that defied documentation; dance based on ordinary movements; the intrusion or foregrounding of language, violating a foundational modernist taboo, and even the displacement of the image by words in Wittgensteinian language games and conceptual art; the use of mass-market photography; sculpture made of industrial elements; earth art; architectural deconstructions and fascinations; the adoption of cheap video formats; ecological explorations; and, quite prominently, feminists overarching critique... (Rosler, 2010).

This strengthen of control of the culture industry would shape the time that followed Adorno, this is, our current time.

It is then interesting to analyze the relations of the subsequent new media art and culture industry, as the final stage where this relation is manifested.

Culture Industry and New Media Art

Continuing with the exposition of manifestations after the time of Adorno, the Brazilian theoretician Arlindo Machado reflected about the new media art and its relation with the culture industry.

Here the analysis is brought to the new media art, considering it a contemporary type of art that deals more accurately with the technological evolution stage of our time, helping us to describe the relation between last manifestations within the art field with the autonomy of art, as well as with the culture industry.

What was already distinguished by Adorno in relation to the art of his time and the culture industry does not relieve at the present time. In words of Arlindo Machado, in the case of the art made with new media technology, what promised to be a period of experimentation and discoveries was soon revealed as the trivialization of routines already crystallized in the art history: the return to conformism and integration to dominating values (Machado, 2007, p. 55).

For the new art manifestations the rule seems to be today marked by a strong standardization, by a general uniformity, as if what is at stake is a sort of aesthetic of merchandising, in which every piece must show a merely demonstration of the qualities of the hardware or the potential of the software (Machado, 2007, p. 55).

Adorno already intuited this perspective in the fetishist character of art, which the practice of culture industry, in a servile way, respects. To be more precise, the process suffered of what was seen by Adorno as the culture industry admiring the empirical details -or technique-, which allies in a big success with the ideological manipulation through the utilization of those elements (Adorno, 1970, p. 336).

For Machado, in the art of the new media what seems to happen is, in most of the cases, the subtle but undeniable lose of the more radical perspective of art nowadays. And he adds: today, when one visits any event of electronic art, digital music or interactive writing, as well as when one browses any magazine dedicated to those specialties, it is not necessary a big effort to confirm that the aesthetical discussion was entirely substituted for a technical discussion, and that matters like algorithms, hardware and software, took majorly the place of creative ideas, of subversion of rules and re-invention of life (Machado, 2007, p. 56).

To conclude about the idea of great value given to the empirical details in the new media art, Machado manifests that with the boom of electronic technologies, art seems reduced -except from, naturally, some few powerful and disturbing experiences- to certain professional skills, while at the same time the technical ability took the place of the most radical attitudes. And he completes that in the environment of the relation between art and technology; only few events could overtake the inevitable industrial framing, looking to confront the most profound inquiries of our time (Machado, 2007, p. 56).

For Machado, the description of an artist who, in Adorno's terms, was taken as a *vedette* (Adorno, 1968, p. 66), would nowadays correspond to a generation of uninformed yuppies that today produces multimedia pieces, uses devices of no-linear edition, diagrams its homepages in internet but never saw a movie of Vertov, never read Artaud, never heard about Beckett, or touched a *bicho* of Lygia Clark (Machado, 2007, p. 56).

As well as Adorno saw a conflict when he perceived that the technique made possible that even the worst movie were presented objectively as an art piece (Adorno, 1968, p. 68), for the present time the same consideration is applicable:

According to Machado we must define with more severe and rigorous criteria to separate the weed from the wheat in the slippery territory of the technological aesthetics, in order to differentiate and privilege works that mark our time, works that bring an effective contribution and gilder, works that, finally, point to perspectives of invention, freedom and knowledge. (Machado, 2007, p. 56)

And, as well as Adorno pointed to the danger of declaring art every manifestation done inside the culture industry, for Machado, the challenge of the new media art doesn't lay in a mere naïve apology of the current possibilities of creation. New media art, on the contrary, must trace a clear difference between what is an industrial production of pleasant stimulus for the mass media, and on the other side, the search for an ethic and an aesthetic for the electronic age. (Machado, 2007, p. 17)

More precisely, regarding the difficult production of autonomous art inside the new media, it is important to point to the fact that nowadays manifestations like the net.art or web art are cheaply produced and freely exposed to the public of the wide spectrum of the virtuality. Art is being produced like never before, and without the intervention not only of interests of the culture industry and also of the artistic institution.

This way, the ultimate solution for the most radical shapes of art can exist only inside its non-commercial character, as it was exposed in Martha Rosler's, as well as in Adorno's ideas.

And by doing so it questions more effectively not just the culture industry but in a broader extension, the whole society, being this the role that art is called to play.

On the other hand, the question *How can there be poetry after Auschwitz, or, indeed, pace Adorno, after television?* (Rosler, 2010) is explained by the similarity that Adorno found between the hitlerian ideals and the television (Adorno, 1947, pp. 16-17), as well as with the art emerged within the culture industry.

It will not be enough to keep mentioning the importance of the context in which Adorno produced his critiques to the art within the culture industry: His context is the one of a person who lived and thought Auschwitz and who even lost his friend and intellectual partner Walter Benjamin in 1940, who committed suicide in the border between Spain and France, trying to escape from the Nazis. (Gagnebin, p. 1982).

In this personal context of Adorno, it is intuitively understandable how his search for a place for art turned to be a capital search for answers to oppose the horror of human cruelty.

By doing so, Adorno insisted in the autonomy of art, maintaining it as a manifestation that emerges pure from any human dimension, apart in its own character and justified by its mere existence.

Perhaps trying to find a place where to rest from the horror, Adorno fought for a utopian untouched place called art.

CULTURE INDUSTRY AND THE MUSEUM

Adorno´s ideas on culture industry comprised the mass media of the time –radio, film, magazines and television-, as well as the artistic institution as such, like the museum, as explained in relation to the catharsis phenomenon, experienced in the reception of an art work and the lack of autonomy of art works. For Adorno art suffered a process in which it ceased being what it is and lost its specificity, becoming consumer goods in the shape of catharsis and art pieces themselves (Adorno, 1970, p. 34).

It is curious however to notice how carefully Adorno did not comprise his criticism within particular artistic structures and did not connect openly culture industry with the museum. Nevertheless his approach referred to this artistic institution, as well as galleries, art fairs, biennals, the explosion of cultural and artistic turism, and other artistic venues as they too, in the search for continuity face the necessity of dealing with art pieces as commodity, as we will explain in the following chapter.

But how has the culture industry developed within its highest aspect, this is, the museum? In this order of ideas, does the artistic institution turn the spectator into a client? And until what extent is the museum still the ultimate secure place where to preserve the legacy of mankind and support its production insofar as it is defined as the place that brings together art and the public?

The history of the museums is in itself not the most transparent one, and sciences like museography and museology aim to deal with such historical and irresolvable questions.

In the following section we will point to certain characteristics of the museum in order to deepen in the continuation of Adorno's ideas regarding the culture industry and how it evolved until our current time.

A Brief History of the Museum

The wide span from the middle of the 16th century to the beginning of the 18th century is regarded as the *century of the collection* and at the same time the birth of the modern museum (Flügel apud Herzig, 2010, p.52). Public circulation as in today's museum took place in early private collections and art chambers, but the admission was only allowed to a limited number of visitors and selected persons. The entrance fee was usually paid in the form of a gift and thus represented a contribution to the completion of the collection.

In a first wave in the 15th and 16th century, art and wonder chambers were founded as very early forms of the museum in the centers of early-modern Europe. These centers were Florence, Paris, Madrid, Prague and London. Objects of different origin and purpose - works of art, antiques, books, natural materials, curiosities, exotica or technical devices were presented here (Baur apud Breuer, 2012 p.60).

This way, a general opening of aristocratic art collections, libraries and gardens did not begin until the second half of the fifteenth century. The universities also began with their collections of research and science: In Pisa, in 1543, a botanical garden with free admission for visitors was built next to the university collection. The university collection was also known as *hortus medicus*, *hortus simplicum*, and *theatrum anatomicum*. (Roth apud Herzig, 2010, p.52). A century later, around 1661, the University of Basel opened the first public museum at the north of the Alps. By acquiring the Amerbach Collection with a large number of paintings, drawings, engravings, coins and antique rarities, the viewer in Basel could for the first time have a comprehensive insight into art. (Grote apud Herzig, 2010, p.53).

A second museum boom begins in Europe in the middle of the eighteenth century, which has its origins in the gradual opening of private, mostly princely collections.

The roots of the collection activity of the modern museum date back to the 17th and 18th century. The methods and insights of the natural sciences and the philosophy of the Early Enlightenment paved the way for a new quality of museum collecting.(Breuer, 2012 p.69)

The opening up of the private art collections, which had been held up to this time, aroused an immense interest in art and newly created values among the population. The demand for art and education on part of the bourgeoisie inspired Napoleon's government to use art as a political instrument. For example, the histories of Jacques Louis David and others were propaganda instruments for glorifying the emperor as well as a medium to represent the nation, and as learning material and source of knowledge for the formation of the population. (Wolff apud Herzig, 2010, p.54).

On the other hand the cultural policy after the example of Louis XIV became a guide for many sovereign personalities, such as Napoleon's collectivist activity or Hitler's dream of a *Reichsmuseum*. (Wolff apud Herzig, 2010, p.95).

The creation of a separate area for public art appreciation strengthened the belief in the autonomy of art. (Grote apud Herzig, 2010, p.56) But this autonomy had nothing to do with the one that Adorno thought as for Adorno the place of art works as its own and not the place of art in relation to any social function, that is, not inside a Museum as a commodity. It is in light of this that Adorno conceived the autonomy of art as a utopia. Considering the fact that the appreciation of artworks inside a museum is the result of the dynamics of the culture industry (Adorno, 1970, p. 33).

The history of the museum is marked by the gradual opening of the museum walls being this the transformation of princely collections to the modern and public museum. (Raffler and Fluegel apud Breuer, 2012 p.59).

This way, to deal with museology means to deal with an almost elitist form of history of the ideas, regardless of the ever-renewed discussion about the term museum. This is so as the principles of order, genres and species, as well as names and concepts change with time. They have been defined and have changed under specific historical conditions.

In this respect it is pertinent to ask for the difficulties of the museums trying to approach the spectators in every historical period since the 15th century to art, when art pieces do not play any role in their life. Even more, when the social development of the present moves away at an ever-faster pace from the historical cultures, their cultures of origin, and the distance between them become then ever greater and ever alien. (Fluegel Apud Breuer, 2012 p.60)

Adorno addressed the fact that through this process the culture industry transforms art into commodity, not letting any art work be what it is, accommodating it, while diminishing the distance in relation to the spectator (Adorno, 1970, p. 32). Or even more, for Adorno in a time of super production, the value of use of art is also problematic since it is submitted to the secondary delight of prestige, of fashion and then, finally, submitted to the own character of commodity. This way, of the autonomy of art just remains the fetishistic character of commodity, being this a regression to the archaic fetishism in the origin of art (Adorno, 1970, p. 32).

In key of Adorno's ideas on culture industry, which constitute an early approach to the institutional critique of our time, Kenneth Hudson in the article *The Museum Refuses to Stand Still* (2004) suggests that *A very important feature of the majority of the museums today, in contrast to what characterized them in the mid-1940s, is the extent to which they have become visitor-centered. This amounts to saying that, as good shopkeepers, museum directors are slowly coming to think of the customers first and the goods on sale second* (Apud Gorlenko, 2010 p.24)

In the same direction Brian O'Doherty in *Museums in Crisis* (1972) described the fact that

The more the objects of private collecting reverted to public hands, the more inevitable it became that the economic interests behind that collecting would move to control the public houses of art. Museums are, or will soon be treated like entities in a conglomerate. Moreover, with a different sponsor for each ambitious exhibition, the museum becomes subject to a carrousel of sponsors, each feeling entitled to have its interests acknowledged. (Apud Gorlenko, 2010 p.28)

On the other hand according to Isabella Herzig, one must not forget that all the collections contribute to secure the cultural heritage. The museums are no longer able to cope with this task by themselves, and they are certainly going to be dependent on private collectors. Only through the financial aid and through the loan of financially potent collectors, as well as exhibiting contemporary works of non-established artists, can the growing demand for new exhibitions be satisfied. (Herzig 2010, p.140).

For our current time also Lisa Dennison from Sotheby's in New York agrees that the future of art museums is endangered. Her main argument lays in the shift of power and influence from the art museums towards the auction houses and private galleries:

[…] museums are in difficult position today because they don't seem to have the power to speak with the same authority that they had in the past. More and more people are learning about art from galleries and auction houses. Museums need to build their permanent collections, but what we are seeing is a phenomenon where collectors don't want to gift their art in perpetuity to a museum that may put it in the basement storerooms. Collectors want their collections to be living, breathing entities – and this makes museums less relevant to them at this time. (Apud Gorlenko, 2010 p.27)

This situation is explained by Gludowacz, van Bennigsen and van Hagen (2010) in their text *Global Art* as:

The public museums are under increasing financial pressure and becoming more and more dependent on collectors and gallery owners, all of whom are pursuing their own – primarily economic – interests. For this reason many museums are putting on shows aimed at drawing in large numbers of people. Success, so the reasoning goes, consists in having as many visitors to your exhibition as spectators at an international football game, and letting them set the cash registers ringing. This is a pretty dangerous trend. According to the authors, it must be possible for public museums to show important artists who don't have great popular appeal, and who are perhaps reserved and difficult to understand. That really would be exciting. And it's really the only way for public museums to fulfill their mission of educating the public. (Apud Gorlenko, 2010 p.27)

What is therefore the future of the museum and what is the museum to convey to future generations? Peter Weibel sees the future art museum as an organic place, which is subject to a constant transformation and becomes a place of pilgrimage more than an eventual site of the general display of art. Using the media, the virtual way to store information, a *universe of the visual* is to be created. The dream of a *Cyberspacemuseum*, a kind of *universal metamuseum*, is intended to open up unprecedented access and possibilities, to allow virtual visits to museums everywhere and at any time, thus the circle of history of collection closes: the wonder chambers with their all-encompassing view of the world were abandoned, only to return to a technical and practical possibility of connection (Apud Herzig, 2010, p.72) with their public, that the museum is invited to experience.

In light of this, the mere idea of the digital artwork could seem to turn in the direction of Adorno's ideas criticizing the fetishism of the art piece used by the culture industry in order to profit. Finally would art pieces lose their materiality and become available to everyone. Nevertheless, such process of bringing art pieces closer to the spectator would be criticized from Adorno's perspective, as again, this would negate a utopian impulse, turning it everyday life and thus low art. As well as it would indicate the same old use of art for personal pleasure, a particular satisfaction of the *I* directing the reception of artworks more to its consume than to the loss of oneself in the perception of the own limits and finitude, expected by Adorno. (Adorno, 1970, pp. 364-365)

REFERENCES

Adorno, T. W. (1968). Résumé über kulturindustrie. In T. W. Adorno (Ed.), *Ohne leitbild- parva aesthetira* (pp. 60–70). Frankfurt am Main, Germany: Suhrkamp Verlag.

Adorno, T. W. (1970). *Äathetische theorie*. Frankfurt am Main, Germany: Suhrkamp Verlag.

Adorno, T. W., & Horkheimer, M. (1947). *La industria cultural. Iluminismo como mistificación de las masas*. Retrieved June 26, 2011, from http://www.cholonautas.edu.pe/modulo/upload/Adorno%20Horkheimer.pdf

Breuer, N. (2012). *Soziale Netzwerke als Brücke zur Kunst*. Vienna: Thesis.

Gagnebin, J.M. (1982). *Walter Benjamin os Cacos da História*. Sao Paulo, Brazil: Editora Brasiliense s.a.

Gorlenko, K. (2010). *Art to Collect Art: Acquisition Policies of Museums of Modern Art from 1980s to the Present*. Vienna: Thesis.

Herzig, I. (2010). *Sammlungspraxis im Wandel der Zeit*. Vienna: Thesis.

Machado, A. (2007). *Arte e mídia*. Rio de Janeiro, Brazil: Jorge Zahar Editor.

Rosler, M. (2010). Take the money and run? Can political and socio-critical art 'survive'? *e-Flux Journal, 12*. Retrieved June 26, 2011, from http://www.e-flux.com/journal/view/107

KEY TERMS AND DEFINITIONS

Autonomy of Art: Is defined by Adorno as the own place of artworks. Not the place of art in relation to any social function. For Adorno when art crystallizes as a specific thing in itself—instead of opposing the existent social rules nor qualifying itself as *socially useful*—it criticizes the society because of its simple existence. This is the autonomy of art, an art that leaves intact the society from which it feels horror.

Catharsis: Loss of oneself in the perception of one's own limits and finitude. It is not a particular satisfaction of the 'I', neither pleasure nor consumerism.

Commodities: Commodities are merchandise. The term refers to commercialization, being in this context the will to make art profitable.

Culture Industry: Adorno's term culture industry comprises the mass media of the time—radio, film, magazines, and television—as well as the artistic institution as such, like the museum, as well as galleries, art fairs, biennals, the explosion of cultural and artistic tourism, and other artistic venues.

Museum: The history of the museums is in itself not the most transparent one; it is a non-profit institution that needs the public to remain working, is permanent but in risk of closure and in the service of society and its development, open to the public, which acquires, conserves, researches, communicates, and exhibits the tangible and intangible heritage of humanity and its environment for the purposes of education, study, and enjoyment.

New Media Art: Type of art that deals more accurately with the technological evolution stage of our time.

Utopian Art: Artworks whose place consists in their own, and not in relation to any social function, that is, not inside a museum. In light of this, the autonomy of art is conceived by Adorno, as a utopia.

ENDNOTES

1 Electromechanical broadcasts began in Germany in 1929, but were without sound until 1934, according to Wikipedia. http://en.wikipedia.org/wiki/History_of_television

2 The text will deepen in the last ideas reflected by Adorno since 1962 that were published in the text *Résumé über Kulturindustrie*, 1968, and the first manuscript will disregarded as, assumedly, it was enough revisited until the last ideas of the author.

3 As an isolated aspect, Adorno mentions in 1947 the political implication of an art being produced inside a system that supports the investments in technical effects, but which, at the same, doesn't find the resources for the abolishment of hunger (Adorno, 1947, p. 11).

Chapter 2
A Collective Consciousness Model in a Post-Media Society

Sherry Mayo
Westchester Community College, USA

ABSTRACT

During the twentieth century, the modern media was born and viewed as an industrial factory-model machine. These powerful media such as film, radio, and television transmitted culture to the passive masses. These art forms were divorced of ritual and authenticity and were reproduced to reinforce their prowess. In the twenty-first century post-media condition, a process of convergence and evolution toward a social consciousness, facilitated by a many-to-many social network strategy, is underway. Web 2.0 technologies are a catalyst toward an emergence of a collectivist aesthetic consciousness. As the prophecy of a post-industrial society becomes fulfilled, a post-media society emerges whose quest is for knowledge dependent upon economy that barters information. This chapter identifies a conceptual model of this recent paradigmatic shift and to identify some of the possibilities that are emerging.

INTRODUCTION

This is an investigation into Enzensberger's (1974) essay on the *Consciousness Industry* to provide a historical theoretical backdrop on today's social network post-media society. In the age of media convergence it is useful to look to social forecasters, sociologists, and media theorists of the past to determine how these frameworks were successful and how the present differentiates from them. The premise of this work is that today's reality is multifaceted, interdependent, global, technology-driven, and exchanged through our collective aesthetic consciousness.

DOI: 10.4018/978-1-5225-5023-5.ch002

The various developments of Web 2.0 technologies and the emergence of a global social network is a paradigm shift in our consciousness, economies of varying scales, and social dynamics. All of these massive changes are happening at the human-computer-interface (HCI) and the resultant end-user created content that exists in the social network "cloud" is in need of harnessing. This flurry of creativity and connectivity shows the possibility of an emerging collective consciousness, which could evolve into a creative knowledge-based economy for many willing countries and their citizens. This discussion presents the following objectives: a reframing of Enzensberger's (1974) *Consciousness Industry,* theory which is a description of current media conditions and trends, draw-backs to the social network and post-media in the present, and an aspirational conceptual model arguing for an emergent collective consciousness.

THE CONSCIOUSNESS INDUSTRY REFRAMED FOR A POST-INDUSTRIAL SOCIETY

All these forms of media are constantly forming new connections both with each other and with older media like printing, radio, film, televisions, telephone, teletype, radar, and so on. They are clearly coming together to form a universal system (Enzensberger, 1974, p. 68).

It could be just Freud's (1917) melancholia at play but Americans today still structure their institutions and corporations in keeping with the factory-model aesthetic. The habitual seating of students in gridded rows in schools, all facing one direction under the flag, and learning through passive transmission of information from teacher to student still perpetuates much of our educational practices. Most workers still mentally construct an image of themselves like Lucille Ball in the chocolate factory (*I Love Lucy,* Second Season, 1952-53), only willing to complete predefined tasks under micro-management. Our most elite intellectual centers remain conservatively tied to notions of corporate practices and adherence with market-forces that wag the dog. And some, only some, of our nations multi-national corporations actually understand the global kaleidoscopic, and create the age that we're living in.

Just like our addiction to post-modernist aesthetics that popular piety understood as quotation, pastiche, revisionist, feminist, multicultural, and layered; our relentless adherence to an industrial society is more about comfort than reality. Enzensberger (1974) long ago complained about the division of labor between the producer and the consumer and was very concerned about the lack of interactivity of the receiver in the power paradigm constructed by the media. Today, much of the labor has been offloaded to the consumer in the new participatory interactive age of new media.

However, he also looked both ahead towards new media and behind to Walter Benjamin. Benjamin (1936) understood mechanical reproduction and foreshadowed an age of many-to-many: a participatory democratic mediated society. He was concerned about the loss of aura, loss of tradition and heritage, through the endless reproduction of images from photography through film. This divorce of ritual practice around the art experience was surpassed by the functionality of the image/art object as a political transferrable sign system that could propagate agendas and build consumerism. In an industrial society aesthetics were defined by the producers of the media and sold to consumers by virtue of a capitalist model. In today's modern blended model where the media has gone convergent while thinking has gone divergent the strategy has not much changed except for the fact that users generate content.

But, what of the consciousness of the media? Durkheim was concerned about the division of labor in society and the formation of a collective consciousness (Harms, J., 1981).

"For Durkheim's perspective, changes in the social structures through which individuals associate affect the faculty of reason and the formation of "collective consciousness," which is the base for normative and moral action" (Harms, J. 1981, p. 395).

Division of labor is concerned with human relations and social consciousness. According to Durkheim who was writing at the turn of the 20th Century, a mechanical society was one in which everyone had the same beliefs and a strong collective consciousness. There was little diversity as opposed to a modern society where there was great diversity and a weak collective consciousness. Durkheim also referred to these modern societies as organic and dynamic and having high moral density.

Collective consciousness serves two functions: It provides a framework that meaningfully orders the individual's sensations and experiences, and it supports a common view or "collective representation" of the external world or "milieu" which makes collective action possible. (Harms, J. 1981, p.398).

Technology can increase dynamic density by facilitating communications and increasing its rapidity and distance between nodes on a network. This brings consciousnesses together increasing the social density, the moral interactivity, and the opportunity for division of labor. Durkheim makes a salient point that merely increasing population density and size is not enough to increase the division of labor. Russia and China had failed to do so previously. They must intimately interact with one another to create a high dynamic of moral, social, density. This is a prelude to social media, which I believe Durkheim would have been a big fan of.

New media has arrived inventing the last avant-garde of artists/designers who code and subvert the interactive experience of the Internet. These "Leonardos" or artist-scientists explore the human-computer interface (HCI) as a cultural zone for social transformation (Manovich, 2001). These early adopters of PCs, digital imaging, web coding, multimedia design, and the Internet led the way to distributed knowledge, collaboration, and social organization facilitated by technology (Rhizome.org). These new media technologies have developed the Internet into a convergence media machine (Jenkins, 2008). In today's post-industrial post-media society, humans are actively participating in the creation of the media/the Internet through social networking. These Web 2.0 technologies, built upon a peer-to-peer, many-to-many strategy, facilitate end-users to help build web content that is then mined for a variety of purposes. This meta-data includes the end-users' biographical, geographical, financial, and even pictures of Mom and dog, Froto. The emergence of meta-data provides great possibilities for information mining and distribution that equips new cultural producers with power unforeseen. In addition to social networking sites such as Facebook, Twitter, YouTube, Instagram, Snap Chat, etc. that came onto the scene since 2006, the Internet has evolved into a cloud of images, video, radio, live and on-demand broadcasting, shopping malls, the blog-o-sphere and now AI. And while this new social cloud model is seemingly a cornucopia, it has caused global crisis on a multitude of levels.

EVERYONE IS A MANIPULATOR

The collapse of the media through technology convergence causes Virilio's (2005) shipwreck or information accident. Whereby, journalists get laid-off, content becomes free, everyone is a broadcasting amateur, ISIS launches a region-wide overthrow of dictatorships with a tweet, and the Russians hack the U.S. election. 21st-century democracy is a buzzword for Enzensberger's (1974) dream of a participatory democratic media. Despite his ridicule, it just might be that the chaos of the media and cultural convergence, a "mash-up," is lending some credence to McLuhan's (1964) return of the "tribal village." Thinking about Islam for a moment and its' aesthetic heritage, which is devoid of representational imagery, consisting of beautiful decorative abstractions is an example of the mash-up havoc. To depict figurative imagery is offensive in this culture, yet the onset of digital technologies has delivered a miasma of hyper-realities that move across screens and project Westernization upon their children. Certainly, the subversive abilities enhanced by mobile technology have played an important role in facilitating communication among dissidents, insurgents, terrorists, and revolutionaries alike. In the era of TMI (too much information) how does one navigate the miasma of the cloud, maintain ones' privacy, dignity, reputation, and achieve sustainability?

The Snowden affair (2013) exposed the NSA and the five-eyes network, and that the American government was actively spying on private citizens without consent. This creates a rupture in our "safe havens" or spherical spaces within which we live. A disillusionment with separation between the public spectacle and personal privacy has become pervasive.

Transitional objects, new themes, secondary themes, multiplicities and new media step between the two partners; the symbiotic space, once intimate and filled with a single motif, opens up into a multiple neutrality, where freedom is only granted along with foreignness, indifference and plurality (Sloterdijk, 2011).

The above installation consists of five tondo paintings done over seven months under a surveillance mirror. These self-portraits imagine a life under personal, spherical surveillance within the banality of the everyday. Interspersed amongst the paintings are federal convex mirrors that place the viewer looking at the paintings under spherical surveillance so that they can see themselves seeing the art works. In addition, there are five security cameras whereby the surveillance footage is kept from the viewer. This installation was an attempt to conceptualize an artist undergoing self-surveillance in order to explore the boundaries of her "safe haven" and it's possible rupture by external forces. Essentially, in this spherical model everyone is

Figure 1. Safe Haven, STEAM exhibition, Arts Exchange of Arts Westchester. (© 2014, Sherry Mayo. Used with permission.)

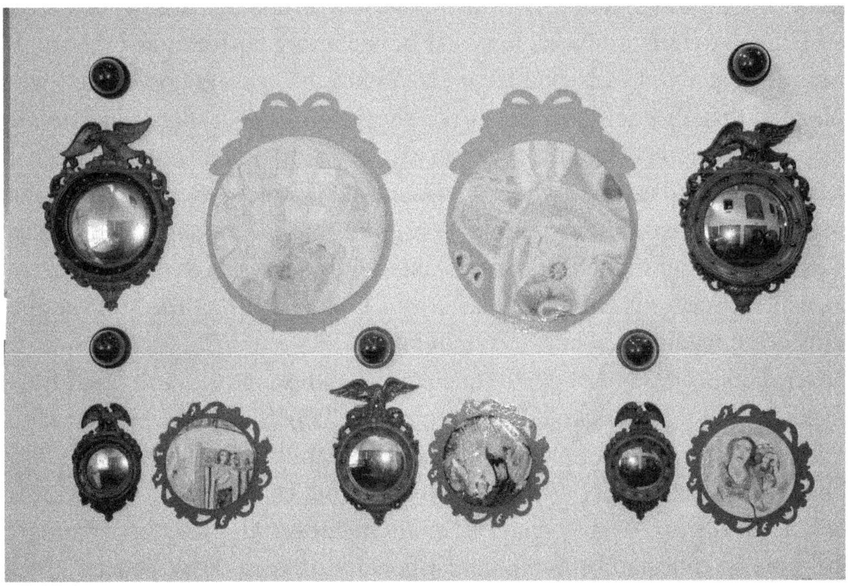

looking at everyone else, whilst also being looked upon. This piece explored what the sphere could symbolize beyond tradition and the Hedeggerian notion of Daesin, as being-in is being-with others (Sloterdijk, 2011, p. 626).

Upon a return to the tribal village is a return to a seamless reality of all things simultaneous without restrictions of time and space, a holistic and embodied experience of a non-linear cybernetic reality. Is humanity less isolated because of reality TV and access to *Orange is the New Black*? Do we gain the same sense of cultural exposure and sharing by watching the *Food Network?* Is post-media connecting us through HCI or simply exposing/shocking us into an awareness of diversity amongst humans? Will this bring peace, tolerance, and understanding or outrage, violence, and censorship? These are turbulent times. We see the crisis in Syria which has killed over 300,000 people and have millions fleeing their country and acts of outrageous violence domestically such as the 2016 Orlando shooting at *Pulse* (49 dead), a gay night club. We look abroad and see attacks frequently such as the coordinated multi-location attack in Paris in 2015 (128 dead per CNN, http://www.cnn.com/2015/11/13/world/paris-shooting/index.html) or the recent shooting in Notre Dame or in England at the London Bridge this past year (8 dead).

Apart of global controversy is one party's claim to democratic freedom of information, as construed as colonialist imperalism by another; and it is really becoming blurry as to who is acting as the imperalist and who is the agent of democracy. This suggests that everything is not "twitterific" in the opaqueness of the cloud, especially since Donald Trump has become twitterpated.

It's true that access to digital technologies has widened extensively since the 1990s and that the general public (middle-class and above in the West) is enjoying the mass consumption of electronica from iphone to ipad and earbuds, while attempting to fulfill their advertising-based fantasies of becoming post-human (Hayles, 1999). While, yes, just about everyone has moved onto the opaque post-digital whereby the cloud is this abstract icon that is the keeper of all data that is invisible to the end-users. And while there still exists an extreme digital divide between developed and underdeveloped nations; whole new micro-economies are emerging around cell phones and solar panels in even remote regions that have transformed village life and created some new strategies for sustainability (Schwartz & Ishofsky, 2011). So again, while everyone is on Facebook, Twitter, LinkedIn, YouTube, etc., the question remains is the media democratized (Figure 2)?

Enzensberger (1974) thought the question was who controls the media? But today almost everyone who is counted in a democracy is participating and qualifies as a manipulator of the media. Lately, it has been is that fake news? Does that have any truthiness? Enzensberger (1974) also thought it would be a good thing for consumers to become producers, so they could examine their everyday and learn from it. This has become a reality and the millennial generation trusts user generated content

Figure 2. I want you in my room, December 16, 2009. (© 2009, Stephanie Hough, Ireland. Used with permission.)

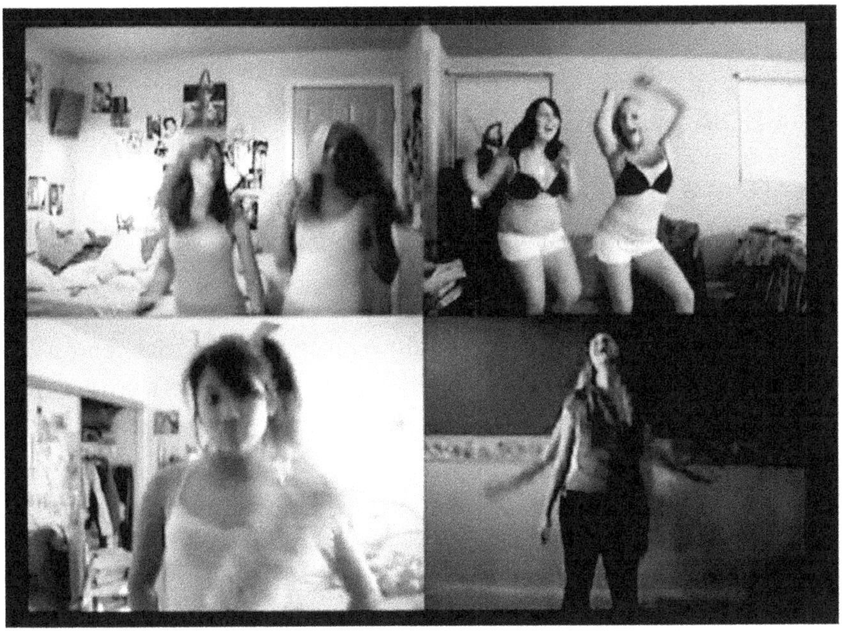

more than any other kind (see https://blog.bufferapp.com/user-generated-content). It holds more credibility than branded media.

In events such as the death of Conrad Roy (2017), who committed suicide endorsed and encouraged by Michelle Carter, the cell phone was the weapon. Miss Carter ends up texting him to get back into his car and kill himself after he has second thoughts. We've gone from the first webcam suicides to now killing people by texting them. If cyber-bullying wasn't enough, the Comedy Central show, *Tosh.O,* makes fairly evident that society is on a wicked learning curve on how to utilize, manipulate, and maintain boundaries with social media tools.

In *Tosh.O* all the silly manipulators of the media provide free content for a comedian to harvest, exploit, and capitalize upon. So big media is still producing or capitalizing on participatory media. Moviemaking wasn't truly democratized by Apple and YouTube – content just got cheaper. Now with Perispcope or Instagram or Faceback, even, you can send live feed of yourself online in real time. Your chances of making it onto *Tosh.O* are minute, never mind getting a movie contract with a studio whom spies your clip out of the zillion online. The importance of a show such as *Tosh.O* is in that it showcases an aesthetic and cultural shift facilitated by the use of Web 2.0 technologies. The shock and awe of what people will put up online voluntarily for free is truly unsettling, morally offensive at times. The show itself presses the boundaries of taste, appropriateness, and morality towards the limit of getting away with it. It reflects who we are and how we are living in this new brand of tribalism and desperation for human connectivity at the HCI.

POST-MEDIA AND ITS DISCONTENTS

Since Bell (1973), it was foretold that the professional class with technical skills would increase in power and that access to power would be through education. Ultimately, knowledge becomes king in Bell's (1973) post-industrial society. Did this pan out? Enzensberger (1974), Bell (1973), and more recently Gardner (2009), point towards a critical need for moral self-regulation and social control through self-organization and collective action. Seems like a really cost effective way to have individuals utilize available free technologies in order to harmonize, aggregate, and act collectively and sustainably.

However, in reflecting on search engine products such as *Google Optimization* and *Chrome* it is apparent that aggregation technology or the ability to "bubble-up" online is less than democratic. The last (2014) controversy involved large corporations such as, the cable companies, being able to buy extra speed for their materials to be streamed onto consumers. Fortunately, the *Open Internet Order* was passed by the FCC maintaining net neutrality. However, the threat to implement differential treatment is again questioning the democratic nature of the Internet. Various Republican advocates are still making arguments for creating fast lanes for paying clients. There is no economy capable of a big brother monitoring system that would truly regulate large global cybernetic systems. But then again, governments are more active in managing economy and information than ever before and placing "us" under surveillance. This Bell (1973) also predicted. This questions the viability of free market trade and freedom of information exchange.

What about the role of education? How are the digital natives doing in the cloud? If education would truly serve as the access point to power in society, it would be tasked with enabling students to synthesize and vett information from a diverse array of data sets to make-meaning, solve problems, and create knowledge. In the situations facilitated by teachers, students would garner necessary technical and research skills to support the aforementioned goal.

Today there are amazing amounts of online educational resources and gaming sequences that "drill and practice" kids. However, these young digital natives aren't learning to manipulate the media through living on screens – they are surface players. They are conditioned to follow pictorial semiotic sign systems that tell them how to behave. That's game mechanics. Hayles' (1999) vision of a flickering of identity in the post-human age might be coming to fruition. These young children are texting incessantly and playing gaming apps neither wholly present here nor there.

But, the good news are that these "kids" are also apart of "Gen C: The YouTube Generation" a new cohort defined across the age spectrum by Google that are "deeply concerned about creation, curation, connection, and community." (see https://www.thinkwithgoogle.com/consumer-insights/introducing-gen-c-the-youtube-generation/). The blend is made up of 80% millennial but not restricted by them. I guess that's why 26% of the adult population is on Pinterest. These are people who live on the Internet, make decisions based on what their peers say, watch YouTube, and define themselves within social media.

If education is to have a role in brokering access to knowledge and power, then American society is currently leaving most children behind. The middle-class's consumer power isn't accomplishing participation in the media; but perpetuating an older one-to-many/passive transmitter-receiver power paradigm of yesteryear. What's worse than television for a young flickering mind (Oppenheimer, 2004)? It's a portable individual TV (Postman, 1985). Spending too much time on screens is destroying social skills, capability for information absorption, and human connectivity. Children are increasingly exiting K-12 education without literacy in the U.S. and few have real technical skills required to survive in the digital age (Partnership for 21st Century Skills, 2011).

However, there is much being done about this and much that we can do about it. In taking a Science, Technology, Engineering, Arts, and Mathematics (STEAM) approach to engaging children with utilizing these new interactive devices constructively and creatively you can turn a student into a maker. Arts technology integrated arts education is one strategy to use to facilitate such a curriculum. In an article entitled, "Learning 21st-Century Skills Through STEAM Digital Artmaking," Liao, C., Motter, J.L., and Patton, R. M. (2016) discuss the benefits of a STEAM curriculum especially for girls and minority students. They developed a course in animated digital stories and machinima movies and game design. One of the outcomes was that these students felt more confident and likely to take STEM classes in the future.

In our Center for the Digital Arts, we offer a wide array of technology that facilitates art and design education and have a perfect environment to support a STEAM curriculum. We offer courses to youth from ages 7-17 and then college credit courses and other non-credit courses to adults. The technology supports courses in 2D and 3D animation, Graphic Design, Digital Imaging, Filmmaking, Stop Action Animation, 3D printing, Robotics, Game Design, Digital Illustration, UX/UI, and more. It has been our experience that children want to learn more and more about this once they get started. They want to design their own environment, understand the interface and how it works, and learn how to code.

EMERGENCE OF AN AESTHETIC CONSCIOUSNESS

Where the media (in the guise of cultural industries') were regarded as the social technology of ideological control in the modern industrial era, the creative industries may be regarded as the social technology of distributed innovation in the era of knowledge-based complex systems (Hartley, 2008, p.8).

Gladwell established the "tipping point" in 2002, which was a new spin on Dawkins' (1976) "meme" theory. It depicts ideas as if they were viruses that can cause epidemic social transformation, e.g. Facebook with 1.94 billion active users (First Quarter, 2017). According to Pink (2005), "the wealth of nations and the well-being of individuals now depend on having artists in the room," (p. 69). The real opportunity is in treating the creative industries as a knowledge industry (Hartley, 2008). This relates to the U.K.s strategy toward a knowledge economy or information economy (Aldrick, 2011), which is supported by Bell's (1973) predictions regarding the emergence of a professional class of workers who would be computational-based cyberfactory workers that would demand services from the service class. Gardner (2006) advocates for integrated thinking across disciplines and "five minds" for the future; that correlates with Pink's (2005) view that a creative knowledge economy and critical consciousness are necessary for economic sustainability. This is further supported by a white paper on how art supports sustainable integrative thinking in the brain and is necessary for long-term health (Bolwerk, Mack-Andrick, Frieder, Arnd Dörler, & Malhöfner, 2013). And these are just a few reasons why STEAM strategized integrated curriculum design is much needed, see www.steam.com. Why would artists/designers/makers be of great value in a knowledge economy? How could an art object or asset be turned into a knowledge product stimulating growth as suggested by Hartley (2008)?

What if the art object or asset was actually a knowledge capsule, as depicted in Figure 3? Inside it would contain a nucleus of artist/designer/maker backstory and other cultural DNA such as: art historical reference points, gallery/museum/media distribution systems, hybridization across media and disciplines, social networks, critical theoretical frameworks, mass media, cultural influences, etc. All this data contained within a creative product that is then jettisoned into space both real and virtual.

What if these creative knowledge capsules were distributed and organized into collections online and in physical space? Some would be included and known, others outside established perimeters and interacting with other objects. Some objects would strategically interact or message with one another or try to eclipse, overlap or exclude others. Certain outsider objects would draw relationships and signal to objects across boundaries and vice versa. Meaning-making would remain

Figure 3. Genome model of collection distribution. (© 2011, Sherry Mayo. Used with permission.)

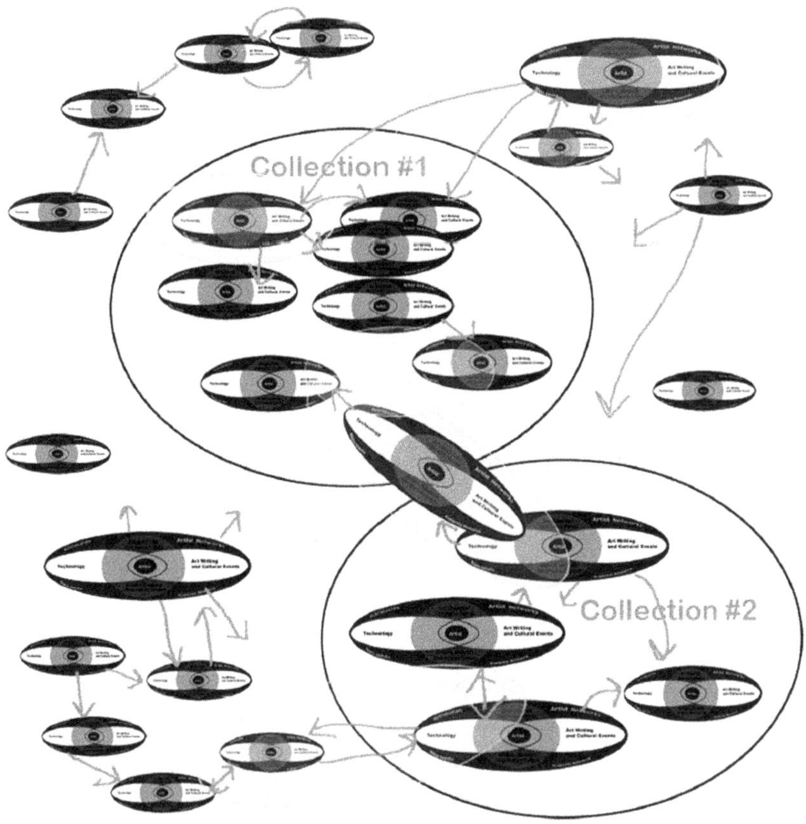

an active process for the viewer, whereby, they are creating relationships amongst and between objects that they then interact with situationally. Akin to intracellular signaling, certain outsider objects would permeate boundaries and enter into dialog with sequestered collections. Some objects would resonate louder or larger than others dependent upon the viewer's response and the frequency of viewer response to the same objects.

The opportunity, as modeled by Hartley (2009, p. 61), is in turning the creative citizen/artist/designer or end-user into an agent of demand for creative products, services, and knowledge. The University has adopted the corporate model to become a creative industry to churn out these workers for the creative economy (Martiniano, C., 2017). In order for this model to work there needs to be an evolution of a social consciousness and one that evokes a collective aesthetic consciousness of a sustainable world. The U.S. is behind the caboose of this global initiative to harvest creativity amongst popular piety and its' anti-intellectualist spirit continues to thwart investment in the arts and a knowledge-based economy. Instead there is a process of scientization of creativity going on whereby, "the current corporatized creativity discourse "scientizes" creativity or rationalizes and institutionalizes its processes and does so in order to monetize its creative productions for the marketplace" (Martiniano, C., 2017, p.162-163). Although new developments in advanced manufacturing technology such as: 3D printing, biometrics, wearables, etc. are beginning to shift things. As technology integration becomes pervasive and all kinds of hybridized smart products emerge; the value of multimodal literacies in arts technology integration becomes apparent. These emerging markets are also creating a demand for workers with 21st Century skills, which include web design, social media marketing and networking, blog writing, Internet PR and more.

This demand and the desire to scientize creativity in order to monetize it has led to advocacy for teaching creativity across disciplines to all students. This will be an imperative in building a creative economy or so we think. According to Martiniano (2017) American children's creativity has been on the decline in recent years (p. 167). Nevertheless, the quest to teach creativity pushes on to both facilitate the student's ability to associate abstractly and create new things. But, here is what's new. We no longer wish to produce the individual creative – that mad genius we all love to study about that made things in some garret. Instead, as well sorted out by the UK Department for Culture, we want to breed creatives that are not self-sufficient but anti-egoist, and desires community (Martiniano, C., 2017, p. 180). The folks that outlined these specifications haven't seen *Project Runway*. The Creative Economy only has use for creative individuals who can produce innovative marketable ideas collaboratively. However, there is also the need for discipline in this new creative individual that may seem at first at odds with freedom but made me think of practices such as Marina Abramovics'. Intense art requires intense discipline and I do not think this is a new attribute, as does the author of these new requirements for creative practitioners.

However, there exists the reality of the social network and the vast numbers of Americans participating and spending increasing hours daily online[1]. This has led to a flattening and homogenization of digital aesthetics existing in the cloud. The visual data sets all look and feel the same. In fact, it's just one big remix or mash-up culture (Lessig, 2007). That said flat media such as: video, web pages, print, film is being out competed by 3D gaming environments that are hyper-real and hyper-kinetic. Digital natives are demanding real-time manipulation of 3D virtual environments with a new hyperbolized 3D aesthetic that almost looks creepily real and now AI is entering the scene. This is combined with wearable and transparent interface designs in biometrics that enable us to operate our cell phones with a thumbprint or a blink.

These new technology developments continue to shift our collective consciousness and the future of what our world(s) will look and feel like.

The art of the art is in the selection, edit, or composite of various samples or digital visual culture distributed across the Internet. Sweeny (2005) refers to an aesthetics of cloning; whereby, digital visual culture is replicated infinitely and reinterpreted, recreated, revised by the end-user. This speaks to a collectivist aesthetic. So how does society leverage end-user created content to spur on a collectivist aesthetic for creative industries, yield a dynamic knowledge economy, and support a creative class?

In Krauss (1979), she utilized a semiotic grid to define what sculpture is by what it is not. In examining aesthetic consciousness in this way it appears that for collective consciousness to occur empathy and aesthetic knowledge to appreciate art and design is needed. The expansion of art to include visual, performance, design, media, communications, sound industries, etc., embraces the largess of creative production and its' post-medium condition (Krauss, 1999). Without aesthetic consciousness, the human has dreaming, automatism, flow, and non-sense but not meaning-making that is obtained through consciousness. In the above discussion, a distinctive aesthetic in digital visual culture emerges, the mash-up being one example that occurs in both audio/visual materials.

The persons contained in one another in the shared space located themselves in such a way that they illuminate and pervade and surround one another, without being harmed by the clarity of their difference. One could say that they are as invisible as air to one another---but an air in which they lie for one another; each one inhales and exhales what the others are—the perfect conspiration.... (Sloterdijk, p. 607).

A BEAD THEORY MODEL OF COLLECTIVE CONSCIOUSNESS

The model, Figure 4, refers to Sloterdijk's (2011) extensive investigation of the sphere and interfacial relations. This bead model for the transmission of thoughts through the inhalation and exhalation of air facilitates collective consciousness. Art celebrates diversity through self-expression and studio practice. This field, facilitated by arts technology, creates our digital visual culture (Sweeney, R, 2010). Our culture is evolving through technology, media convergence, and geospatial collapse. This has connected everyone and strained the crisis of separation and sameness (Shirky, C., 2009) and caused desperation for "heimat" or notion of belonging to a place and "daesin" or being-in by being-with others. It is with urgency that new models for higher education are developed that help students navigate signs, symbols, the human-computer interface and, most importantly critical consciousness.

Figure 4. Bead Theory Model, 3-point perspective. (© 2014, Sherry Mayo. Used with permission.)

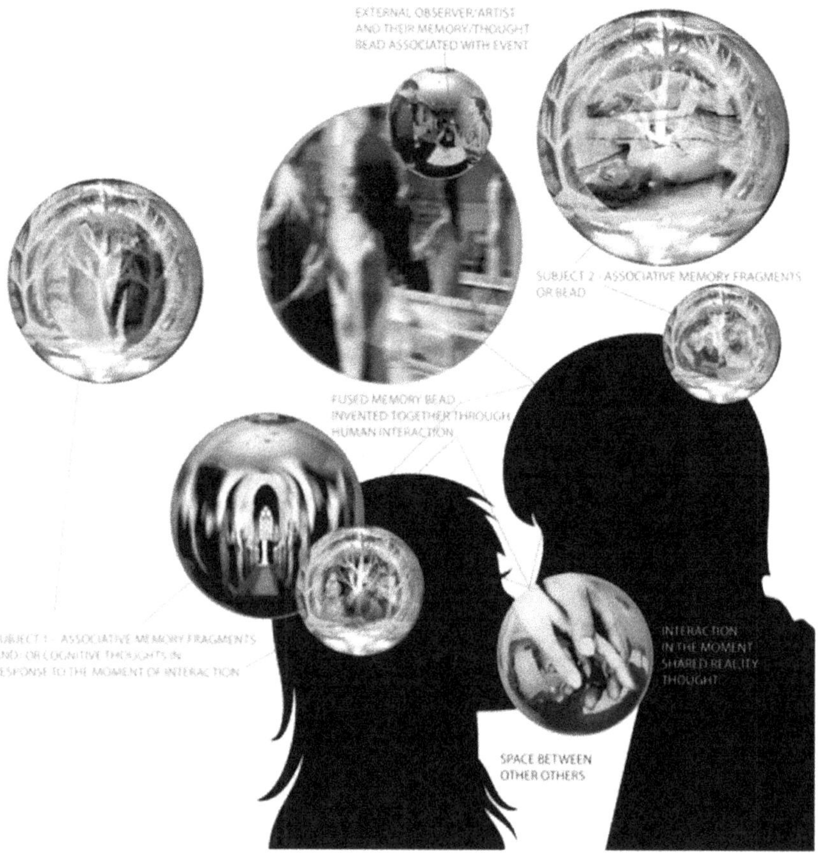

The representation of the sphere has meant many things in art and iconography, for example, such as the world. In a Jesuit manuscript, Hesius' *Emblemata Sacra* (1636) the sphere is a stand-in for faith, or the mind's ability to believe in what it cannot apprehend (in Hesius, 2002, p.88). Consciousness is this way, a phenomenon that cannot be physiologically proven and must be accepted in some ways on faith (Maley, C.J. & Piccinini, G, 2017). This provides the mind with the potential for infinite reflection upon itself within its environment.

This idea of infinite self-reflection within the sphere of the mind connects, for me, with reflexive consciousness. Reflexive consciousness is an automatic process, in which stimuli trigger the formation of thoughts about the self. This type of self-awareness is about the self no matter what it is cogitating on and happens in a loop. It differs from mindfulness, which would not attach to any thought but watch them as if on display. It separates itself from reflexive consciousness or self-reflection and is a key to self-control. It is important in the formation of a collective aesthetic consciousness to understand these features in order to facilitate them.

"In approaching the theme of consciousness, one becomes aware of the intimate interrelatedness of the inner awareness of the human and the outer world of phenomena that formulates the context for all thought and the content for meaning" (Brinkman, 1998, p. 135). The intersections of art, science, technology, and culture exist across disciplines (Alexenberg, 2008). In my mind, the cornerstone of learning is the self-portrait. As individuals are challenged by situational problem-solving that is multimodal and transdisciplinary; it makes relationships with others and the world (Peppler, K., 2009). It is at the point of collaboration, whereby, interrelatedness occurs and the self transforms through esoteric cosmology. Eventually, the evolution of cognitive thinking yields the ability to tell one's story until it's the tale they want to tell (Keagan, R., 1982; Brunner, 1996). In a post-digital society the catalyst for change isn't technical, its conscious humanism (Brinkman, 1998; Hayles, 2010; Turkle, 2009).

"Syncretic thinking is associative and non-linear: associative thought and non-linear inquiry inevitably breaches boundaries and subverts protocols" (Ascott, R. in Alexenberg, M., p. 49, 2008). The value of utilizing an art education model is that the arts have always been constructivist, invested in self-expression, and rooted in critique. In other words, inviting a group of students to tell their story in response to a situated problem is a strategy that can be broadly applied and integrated across disciplines.

If there is a technological revolution in art it lies not simply in the global connectivity of person to person, mind to mind (significant as that is), but in its power to provide for the release of the self, release from the self, the fictive "untitled self" of Western philosophy. It lies in our ability to be many selves, telematically in many places at the same time, our self-creation leading to many personas and serial identities. (Ascott, R., 2008, p.204).

This model will be complemented by examples of art objects aimed at developing multimodal meta-thinking, multimedia literacy, and critical consciousness of our digital visual culture. Essential in the problem-solving process is for students to obtain the ability to retrieve vetted information that they then can synthesize to construct new knowledge. In addition to utilizing contemporary art within a digital visual culture as site for this work; students may build the skills to negotiate collaboration, appreciate other points-of-view, and access other cultures to gain a better understanding of themselves. The ultimate goal of this collective consciousness model is to depict arts education as an approach to fostering critical consciousness amongst students (Elkins, 2013).

In Figure 4 a man and a woman are engaged in an exchange. They are communicating and generating polysensual data sets. The data is then generating thought beads, some of which are shared, some of which are real, some of which are generated, and some kept to the individual for themselves. The couple in Figure 4 hold hands, for example, and each takes a cognitive journey to retrieve previous hand-holding experiences that they've had before. The associations based on the space between the couple and the place and time in which they touch generate many beads, which aggregate for each individual and generate a new memory of that particular hand-holding event. Will the couple generate the same exact memory of how the event felt; how the place smelled and sounded; how the light was in the room; what they spoke, ate, drank? She will appear in his beads and he in hers that will carry forward. Will her account be the same as his? More than likely, not. Each thought or memory in time forms a video bead within which you and the other(s) inhabit or achieve daesin. Imagine this on a massive scale whereby many people appear in one another's thought beads. Beads then would continue to transmit and collide -- sometimes attracting each other, at other moments repelling, and at certain points aggregation of many beads would occur.

"In other words, when individuals associate with each other, the contents of their "particular consciousnesses," derived from their experience with the external world, combine to form a conception which is able to frame and order meaningfully the diversity of individual experience (it is thus the basis for collective action) (Harms, J., 1981, p. 399). When applying human reason the individual can rise above the particular and get to the universal toward collective consciousness. Collective consciousness is the base for normative and moral action (Harms, J., 1981). This is of course is based on Durkheim's ([1912] 1965) work and from his perspective change in social structure that occurs through individuals associating and impacting reason and forming a collective consciousness that then could take social action. So these collective consciousness beads would be very powerful indeed.

The art object that the artist makes, as a result of witnessing this exchange, generates unique beads of thought. The artist places on view a resultant work of art based on these beads of memory and shares this art object with other viewers (Barthes, 1973, 1975). In this way the thought beads generated in the space between this couple multiply exponentially and circulate amongst many people across time and space virtually and without plugging anybody in.

In Figure 5 see the potential for generating collective consciousness when one brings the bead theory to scale. With or without Facebook anyone could, through art objects of any type, distribute tremendous amounts of messages or "memes" that could cause a cultural tipping point (Gladwell, 2000) and/or connect people regardless

Figure 5. Bead Theory – Collective Cosmic Consciousness. (© 2014, Sherry Mayo. Used with permission.)

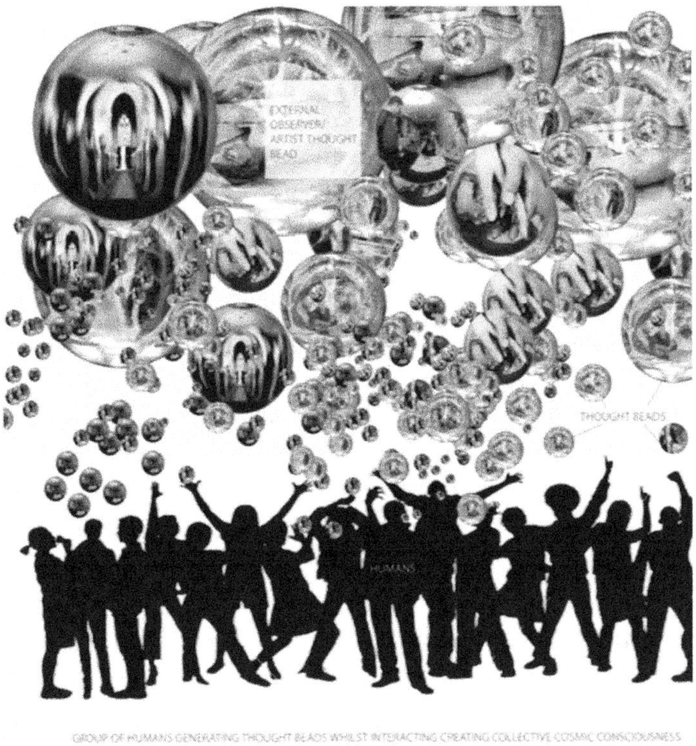

of time, space, and place. In a way the art object brings together a community of people that collaborate on a shared consciousness of where they belong and what they believe in. The art object was once most powerful "insitu" and post-Benjamin (1936) was stripped of this and instead mass reproduced. However, repeating an image increases its power; distributing an image to more people across more borders also increases its power. It's not only the art object that gets collected; it's the object that collects viewers.

I believe that new media, social media, digital visual culture, media convergence, and global diaspora has made feelings of having no-place or a loss of heritage and sense of not belonging, not at all unusual. The opportunity in this is that we are then free to resonate, identify, and connect beyond typical tribal indicators. We begin to find others that we claim as tribe based on the thoughts we exchange, the ideas we generate, the art we love, etc. This provides us with a greater opportunity to find one another because we do not necessarily need to go somewhere to live with other members of our new tribe. We can remain remote with the advent of all our technologies and we can create a heimat anywhere we practice. The bead theory is about a utopian aspiration but a real potential opportunity for people to co-invest in generating power in community through consciousness-raising, through positive thought generation, and through hybridization of ideas.

CONCLUSION

How do we extend this collectivist aesthetic into one that is politically meaningful, knowledge generating, and capital producing? Note in my bead theory model there aren't any wires. The digital natives or GEN-C are communicating in the clouds above through consciousness within a seamless physically embodied digital universe. In conclusion, this brief look at our 21st-Century culture in the Digital Age or Social Network Age or Age of Globalism or Conceptual Age depicts some of the cultural, aesthetic, economic, and educational impact during a paradigm shift.

In my mind the big question is: *What alternate system for the regulation of the struggle for power can be devised that would sustain a moral collective consciousness?* The U.S.

has undergone perhaps its most entertaining election to date. For the first time it has an entrepreneur / CEO in the White House. How would Donald Trump answer this question? From what has gone on so far, I do not see a sustainable moral collective consciousness in modern America. I see extremist chaos and impasse.

Figure 6. Huh? (@ 2017, Tricia McLaughlin, U.S., Used with Permission.)

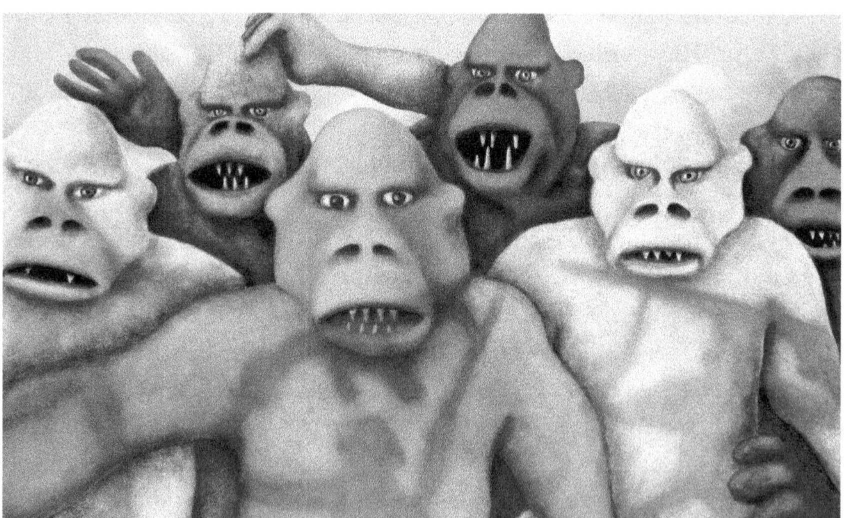

There has been talk about the need for moral self-regulation and control, recently, and forecasted long ago by Daniel Bell (1973). I don't have an answer to this big question but it connects me back to several things in this chapter. The need for discipline in the new creative that is essential in order to generate creative capital or to sustain any art practice. The value of mindfulness in wellness and self-control. Reflexive consciousness which will bring your thoughts back to the self and heighten your self-awareness. The goal of the individual to go from phenomenal consciousness, stemming from particular consciousness, onto universal consciousness. I think there is a lot to be learned from emergent artists right now who are increasingly working collaboratively and conceptually.

In a high tech culture there is a greater demand for moral self-regulation and control. The aforementioned items such as: discipline, mindfulness, and reflexive consciousness all contribute to sustaining self-regulation and control. If you continue on Durkheim's pathway you will go from the personal to the universal and get to collective consciousness and the spiritual. I feel finding a higher power or level of connectivity from collective consciousness a possibility and a goal. This can be very instrumental in moral self-regulation.

And while politicians have lost my like buttons, I do see a possibility for bead theory to play out in communities as a real model. We are used to groups such as Dog Lovers to get together and talk about dogs. But, what if we had them meet-up with a Cat Lovers group and have them shift their discussion to become inclusive? To get people to cross-pollinate socially out of their homogenous zones would facilitate tolerance and increase moral density. We have become so exclusive and divided from one another, aided and abetted by social media that we need to do something about it.

How do we build beads with people who might differ from us on many major issues but we otherwise may greatly enjoy and have many things in common? How do we get to know that? Could we get groups of people across gender, race, creed, age, political affiliation, and interest to show up to a meeting with the objective of cultivating a collective consciousness? This group wouldn't have leaders but facilitators and volunteers that took care of nuts and bolts responsibilities. There would be topics to discuss and questions to respond to and research to be done. Perhaps the group would evolve to a point where local problems and issues would come to them as a reasonable sounding board. A group that would understand how to bring people together with high interactivity and moral reasoning. But, for this to work the group really has to include people who would normally never hang out together.

For instance, I would welcome meeting with folks that have diverse perspectives on education today. A conversation about art and education is always a valuable opportunity.

A classroom has no purpose if it's not for sharing whats that students can wrap their minds around, whats that help them practice cognitive plasticity at a time when their brains are still developing. Whats involve stubborn facts and powerful narratives, valuable models of activist thinking about the most difficult topics facing humanity. And thinking about these whats transforms many students into intelligent members of a diverse democracy. This isn't activism per se, as much as we may want to believe it, or as DeVos wants to claim, but for some students it can be the beginning of an activist life. Such whats can compel one to care...(Cole, A., 2017, p.304)

I think we need to come together now more than ever and start to mingle. The kind of collective consciousness that Durkheim talks about is one of high dynamic density where people are highly and intimately interactive. That sounds wonderful doesn't it – like a French film. The U.S. is approximately 330 million people. Is it possible to get intimately acquainted with us all? Now remember Durkheim cautions that volume doesn't equate with a high degree of division of labor. Meaning just because your population size is large doesn't mean your population is very dynamic or interactive.

The larger issue is a basic pictorial problem of depicting the struggle of the figure in the landscape. The individual navigates the crisis of separation and sameness with hope. From this emerges the need of belonging and connectivity and it is perhaps through this quest one finds the way to a practice that fosters hope and change. And while technology promises to connect everyone everywhere (Shirky, C., 2008), it also provides people with an iLife that is increasingly homogeneous and hermetic. Virtual reality further abstracts life and facilitates a rupture of moral boundaries. This gives me pause to look toward social networking as the way to formulate a collective consciousness.

"Consciousness is a matter of complexity of knowledge, thought and memories (Beichler, J., 2017, p.2). Facebook may be the new "Factory" but it still needs the street to realize community. Is Facebook our collective consciousness then? Are we just pawns in Zuckerberg's plan to steal our memories and create AI with them? Perhaps we won't be needed after they retrieve enough birthday parties, BBQs, and graduations from us. I'm still arguing that we need people to do this job of formulating a collective consciousness. I do think we could utilize social media tools to make it go viral to include absolutely everyone domestically (to have that intimate conversation to scale). But, I also think it needs grassroots footwork to formulate a mass collective consciousness.

A purposeful fluidity between groupings in order to formulate collective consciousness requires the personal to generate the universal and the moral. The higher the participation the greater the interactivity, the morality, and the consciousness. This collective consciousness may then be turned towards collective social action.

Most immanently, there is an epidemic of opioid addicted Americans suffering in this country approximately 59,000 persons of which we lost in 2016 per the *NY Times* see, https://www.nytimes.com/interactive/2017/06/05/upshot/opioid-epidemic-drug-overdose-deaths-are-rising-faster-than-ever.html?_r=0). We're losing each other to the number one killer: drugs. Coping with addiction is a tough job but not only are more people losing the fight; there are more people becoming addicted. Finding heimat and/or daesin and formulating collective consciousness is the way through life's challenges. It's an epidemic worthy of social action. May the beads generate among one another.

Getting back to my big question perhaps we may need more than utopic liberal universalism; I am uncertain if we have embraced integrated interdependence (Bell, D., 2004; Gardner, G., 2006; Pink, D., 2006) which is an awareness we really need. Lastly, I'm totally unclear if I'm living in a democracy in America today.

REFERENCES

Aldrick, P. (2011, June 29). Knowledge economy central to UK growth, says Work Foundation. *The Telegraph.* Retrieved from http://www.telegraph.co.uk/finance/economics/8604057/Knowledge-economycentral- to-UK-growth-says-Work-Foundation.html

Alexenberg, M. (2008). *Educating Artists for the Future: learning at the intersections of art, science, technology, and culture.* Chicago, IL: The University of Chicago Press.

Ascott, R. (2008). Cybernetic, Technoetic, Syncretic: The Prospect for Art. *Leonardo, 41*(3), 204. doi:10.1162/leon.2008.41.3.204

Barthes, R. (1973/1975). *The pleasure of the text* (R. Miller, Trans.). New York: Hill & Wang.

Barthes, R. (1977). Death of the author. In R. Barthes (Ed.), *Image-music-text.* New York, NY: Hill & Wang.

Beichler, J. (2017). *A Physics of Consciousness: The only path to understanding consciousness (since consciousness is its 'self' the path).* Retrieved from https://www.academia.edu/33309104/A_Physics_of_Consciousness_The_only_path_to_understanding_consciousness

Bell, D. (1973). *The coming of post-industrial society: A venture in social forecasting.* New York, NY: Basic Books.

Bolwerk, A., Mack-Andrick, J., & Frieder, R. L., Arnd Dörler, & Malhöfner. (2013). How Art Changes Your Brain: Differential Effects of Visual Art Production and Cognitive Art Evaluation on Functional Brain Connectivity. *PLoS ONE*, *9*(7).

Brinkman, J. (1998). *Buddhist-Christian Studies* (Vol. 18). Honolulu, HI: University of Hawai'i Press.

Brown, K. W., & Ryan, R. M. (2003). The Benefits of Being Present: Mindfulness and Its Role in Psychological Well-Being. *Journal of Personality and Social Psychology*, *84*(4), 822–848. doi:10.1037/0022-3514.84.4.822 PMID:12703651

Brown, K. W., Ryan, R. M., & Creswell, J. D. (2007). Addressing Fundamental Questions about Mindfulness. *Psychological Inquiry*, *18*(4), 272–281. doi:10.1080/10478400701703344

Brunner, J. (1996). The Culture of Education. Cambridge, MA: Harvard College.

Cole, A. (2017, Summer). Class Action: Teaching and Activism. *Artforum*, 304.

Dawkins, R. (1976). *The selfish gene*. New York, NY: Oxford University Press.

Durkheim, E. (1965). *The Elementary Forms of the Religious Life* (J. W. Swain, Trans.). New York: Free Press.

J. Elkins, K. McGuire, M. Burns, & A. Chester (Eds.). (2013). *Theorizing Visual Studies: writing through the discipline*. New York, NY: Routledge.

Enzensberger, H. (1974). *The consciousness industry: On literature, politics and the media*. New York, NY: Seabury Press.

Freud, S. (1964). Mourning and melancholia. In The standard edition of the complete psychological works of Sigmund Freud: Vol. 14. On the history of the psycho-analytic movement, papers on meta-psychology and other works (pp. 237-258). London, UK: Hogarth Press. (Original work published 1917)

Gardner, H. (2006). *Five minds for the future*. Boston, MA: Harvard Business School.

Gladwell, M. (2002). *The tipping point*. London, UK: Little, Brown, and Company.

Harms, J. B. (1981, October). Reason and Social Change in Durkheim's Thought: The Changing Relationship between Individuals and Society. *Pacific Sociological Review*, *24*(4), 393–410. doi:10.2307/1388775

Hartley, J. (2008). From the consciousness industry to creative industries: Consumer-created content, social network markets and the growth of knowledge. In J. Holt & A. Perren (Eds.), *Media industries: History, theory and methods*. Oxford, UK: Blackwell. Retrieved June 29, 2011, from http://cultural-science. org/FeastPapers2008/JohnHartleyBp1.pdf

Hartley, J. (2009). From the consciousness industry to creative industries: Consumer-created content, social network markets and the growth of knowledge. In J. Hartley (Ed.), *The uses of digital literacy* (pp. 39–70). St. Lucia, Australia: University of Queensland.

Hayles, K. (2012). *How We Think: Digital Media and Contemporary Technogenesis*. Academic Press.

Hayles, N. (1999). *How we became posthuman: Virtual bodies in cybernetics, literature, and informatics*. Chicago, IL: University of Chicago. doi:10.7208/chicago/9780226321394.001.0001

Heidegger, M. (1976). Brief über den Humanismus. *Gesamtausgabe*, *9*, 313–364.

Hesius, G. (2002). *Emblemata sacra: De fide, spe, charitate*. Hidesheim, Germany: Georg Olms Verlag.

Hough, S. (2009, December 16). I want you in my room [Web log post]. Retrieved June 29, 2011, from http://stephaniehough.wordpress.com/

Jenkins, H. (2008). *Convergence culture: Where old and new media collide*. New York, NY: NYU Press.

Y. Kafai, K. Peppler, & R. Chapman (Eds.). (2009). *The Computer Clubhouse: constructionism and Creativity in Youth Communities*. New York, NY: Teachers College Press.

Keagan, R. (1982). The Evolving Self: problem and process in human development. Cambridge, MA: Harvard College.

Krauss, R. (1979). Sculpture in the expanded field. *October*, *8*, 30-44.

Krauss, R. (2000). *A voyage on the North Sea: Art in the age of the post-medium condition*. London, UK: Thames and Hudson.

Lessig, L. (2007). *How creativity is being strangled by the law* [Video clip]. Retrieved June 29, 2011, from http://www.youtube.com/watch?v=7Q25-S7jzgs

Lessig, L. (2008). *Remix: Making art and commerce thrive in the hybrid economy*. New York, NY: Penguin Group. doi:10.5040/9781849662505

Liao, C., Motter, J. L., & Patton, R. M. (2016, July). Tech Savy Girls: Learning 21st-Century Skills Through Digital Artmaking. *Art Education.*

Madoff, S. (2009). *Art School (Propositions for the 21st Century).* Cambridge, MA: MIT Press.

Maley, C. J., & Piccinini, G. (2017). The Evolution of Consciousness. In Routledge Handbook of Consciousness. New York: Routledge.

Manovich, L. (2001). *The language of new media.* Cambridge, MA: MIT Press.

Martiniano, C. (2016, Fall). The Scientization of Creativity. *The Journal of the Midwest Modern Language Association, 49*(2), 161–190. doi:10.1353/mml.2016.0033

McLuhan, M. (1964). *Understanding media: The extensions of man.* Cambridge, MA: MIT Press.

Mehta, S., & Nicholas, P. (2011, June 29). Obama highlights manufacturing revival, jobs in Iowa. *Los Angeles Times.* Retrieved from http://www.latimes.com/news/nationworld/nation/la-na-obamaiowa-20110629,0,5563738.story

Miller, D., Costa, E., Haynes, N., McDonald, T., Nicolescu, R., Sinanan, J., ... Wang, X. (2016). *How the World Changed Social Media.* UCL Press.

Miller, P. (2004). *Rhythm science.* Cambridge, MA: MIT Press.

Oppenheimer, T. (2004). *The flickering mind: Saving education from the false promise of technology.* New York, NY: Random House.

Partnership for 21st Century Skills. (2011). *U.S. government framework for 21st-century literacies.* Retrieved June 29, 2011, from http://www.p21.org/index.php?option=com_content&task=view&id=254&Itemid=119

Pink, D. (2005). *A whole new mind: Why rightbrainers will rule the future.* New York, NY: Penguin Books.

Postman, N. (1985). *Amusing ourselves to death: Public discourse in the age of show business.* New York, NY: Penguin Group.

Rheingold, H. (2005). *On collaboration* [Video clip]. Retrieved June 29, 2011, from http://www.ted.com/talks/howard_rheingold_on_collaboration.html

Schwartz, S., & Ishofsky, R. (2011, August 29). How Solar Stores are helping to meet rural Uganda's energy needs. *PV Magazine.*

Shirky, C. (2009). *Here Comes Everybody.* New York, NY: Penguin Press.

Sloterdijk, P. (2011). *Spheres* (Vol. 1). Bubbles Microspherology.

Sullivan, G. (2005). *Art Practice As Research: inquiry in the visual arts*. Thousand Oaks, CA: Sage Publications.

Sweeney, R. (2010). *Inter/Actions/Inter/Sections: art education in a digital visual culture*. Reston, VA: National Art Education Association.

Sweeny, R. (2005). Three funerals and a wedding: Art education, digital images, and an aesthetics of cloning. *Visual Arts Research*, *31*(61), 26–37.

Turkle, S. (1995). *Life on the screen: Identity in the age of the Internet*. New York, NY: Simon & Schuster.

Turkle, S. (2009). *Simulation and Its Discontents*. Cambridge, MA: MIT Press.

Underwood, J. A. (2008). *'The Work of Art in the Age of Mechanical Reproduction' first published in 1936*. England: Penguin Books.

Virilio, P. (2005). *The original accident*. Cambridge, UK: Polity Press.

ENDNOTE

1 2:29 (hrs/min) per day online, 2:21 (hrs/min) per day on mobile per Huffington Post (8/1/2013).

Chapter 3
On the Transformation of Social Movements:
An Analysis From the East-West Axis

Cemile Zehra Zehra Köroğlu
Uşak University, Turkey

Muhammet Ali Köroğlu
Uşak University, Turkey

ABSTRACT

As social entities, people could come together and create regular relations and institutions. Concepts such as group, community, society, social movement, etc. are about the social dimension of man. However, according to the conditions of the social, political, religious, and physical environment in which the person lives, their needs and problems can change. As a natural consequence of this, social characteristics of social movements can change. It is inevitable to value new social movements in this respect. Because new social movements in the West are born from a critical intellectual atmosphere against modernity. This situation has developed in an economic system based on the service sector rather than economic-order-based on heavy industry. On the other hand, the Eastern world, especially the Islamic world, has a repertoire of social movements that brings different problems to the agenda because it has different conditions. In this respect, new observations and analyses of Turkey and its surrounding will provide important contributions to the theory of social movements.

DOI: 10.4018/978-1-5225-5023-5.ch003

INTRODUCTION

When community and social life are taken as a basis, it cannot be said that any society is entirely stable, closed to changes and differences at no time in history. Contrarily, in all societies from the most primitive to the most modern ones, social differences, rebelling to its location, struggling for changing its position and themes like change have been taken into consideration. This emerged as in the form of transformation of discontent, oppression, marginalization into rebellion or a philosophy of one's own life and meaning the world into action socially and massively. From this perspective, it is possible to claim that many social movements have appeared with a variety of purposes and principles in all periods of history. Sometimes economic conditions and sometimes religious, cultural and political conditions have become the driving force of a social movement. Although, sciences, such as politics and sociology allow us to analyze the conceptual meaning of social movements, these sciences often are put forward by the modern era. As a result, most social movements that have been analyzed stem from the modern period. Here, especially, social movements which emerged in Western societies after the 18th century have been taken into account. However, no matter how different the reason, principles, objectives of a social movement; no society is isolated from social movement, demand and change.

Turkey, compared to other Islamic countries around it, has a political system that has given the opportunity to many different social movements to come into the public sphere and struggle there. In contrast, it is not possible to mention a fully political public sphere in other Islamic countries. There is no political area between the state and society. Therefore, emerging social movements have transformed into either rigid political movements targeting to seize the state or movements supporting to the state and having no originality. This structure produces militarism, even terrorism rather than the social movements in the contemporary sense. This is an inevitable end when assessed in terms of Maslow's analysis. Maslow puts the need for security to the second row of the basic hierarchy of needs. Individuals and groups who can not meet basic human needs such as life, property, and self-expression can be easily militarized. It is possible to see many examples of this situation in the Islamic world over the last decade. In countries such as Libya, Syria, Iraq and Lebanon, many groups have emerged against the traditional authoritarian regimes, and all of them have been trying to establish the relationship between state and society by arming. This feeds the chaos and terrorism. This image recalls thirty years wars in the West. As is known, Thomas Hobbes who tries to solve social chaos during these periods, has suggested that social chaos can only be resolved with a stronger and irresponsible state. Today, the same thing is suggested for the Islamic communities. This pushes even the most innocent social movement towards terrorism.

As a typical example of this analysis can be viewed in the Muslim Brotherhood movement in Egypt. Today, Muslim Brotherhood which is the one of the most important social movements of the world is a rooted movement founded in 1928. Since its inception has survived under a severe state repression, and despite this, it was able to create alternative religious, cultural and economic institutions (Davis, Robinson 2015, pp. 80). This movement transformed into a political party, and came to power a short time ago, but it was removed from government through a coup. It has been forced to confront the state violence immediately after this. Very recently, it was declared a terrorist organization by the Egyptian government and some neighboring countries. Naturally, it is not possible the emergence of new social movements in such a society and state equation. Because, in a society where is lack of basic human rights and political rights as the most sacred rights of our civilization; feminists, environmentalists, anti-nuclear and so on. movements are not to be expected.

When assessed in terms social movements, both traditional and modern Turkey provide us with a unique context to study social movements and economic prosperity. As in the past, many secular and religious movements are present in today's Turkey. Ironically, these kinds of movements have not been considered specifically in the general social movement literature. Only by tackling religious social movements or just secular movements in Turkey, we can identify dozens of different social movements. Many social movements can be examined in particular such as Feminist Movement of Turkey, other women's movements, environmental movements (like Bergama and Valley of Fırtına Boycott), movements related to human rights and humanitarian law violations, identity movements and political movements. Likewise, in the category of religious social movements, many individual mystical movements, political and other movements struggle for participation in the public sphere. Although some of the religious social movements express historical continuity, many have emerged in the modern era and have been in constant change ever since. For Turkey, particularly the process of growing political and economic liberalism since 1980s has shaped social movements, especially in terms of religious social movements. From this period, religious communities have received a greater share of economic prosperity and have had the opportunity of better organization to make their voices heard. Departing from this observation, this chapter aims to review the concept of social movement in Turkey within the framework of three main periods: a) religious and social movements during the imperial period; b) religious and social movements between the years of 1923-1950 and; c) recent religious and social movements. With flourishing economic prosperity in the latter period, two social groups, that have turned into social movements and have achieved significant advantages to express themselves in the public sphere, step to the fore, i.e. the Kurdish ethnic movement and Islamic groups. Because classical theories of social movements are insufficient to explain social movements in both categories, a new paradigm is advocated.

SOCIAL AND RELIGIOUS MOVEMENTS

The most rampant discussions about social movements entail the definition of what a social movement actually is, how it should be dealt with and for which social conditions and reasons it has appeared. In general, social movements have been conceptualized as an "organized effort expended by remarkable number of people to change one or more major features of society (or else resisting to be changed). This concept has been coined by Saint-Simon who had to describe the opposition movements in France in the beginning of the 18th century" (Marshall, 1999, p.746). Undoubtedly, the meaning of the concept has developed and has been enriched throughout history. Today, the concept of asocial movement refers to movements outside of the general social, political and cultural body and quite often dissident or divergent identities, cultural features and lifestyle properties come to mind.

Many of today's social movements, like women's movements, gay and lesbian movements, occupy movements and landless movements, thrive on one of the most basic criteria of social movements, i.e. "difference". When difference, such as ethnic, religious, cultural differences or political preferences are taken as a basis, shared lifestyle differences and conspicuous consumption preferences become a way to build a collective platform. Even though it does not automatically turn into mass and political action, any collective body consisting of a group of integrated individuals by any purpose, meets the requirements of a social movement. Although not all movements come out in the public sphere with a cultural and political struggle, all lifestyles and understandings that differ from the main social body indirectly express a social challenge and struggle for participation. Especially, when religious differences are at the core of a social movement, these challenges and struggles become manifest.

In this respect, religious social movements are social movements that struggle for acceptance of an identity, lifestyle, or express religious and humanitarian demands in the public and cultural sphere. From the perspective of classical social movement paradigms, religious social movements have mostly been evaluated as fundamentalist movements and have not been a subject for scientific examination.

When assessed in terms of geography and tradition of Islam, it is possible to see many social movements arise with a wide variety of purposes since the early stages of the history of Islam until today. Their big part has disappeared in the historical process. In the emergence of the first social movements expressed through concepts such as sect and military division, cultural, social and political conditions are stated to be decisive (Kutlu,2010, pp.43-45). In studies conducted in terms of Denominations History, social movements in the Islamic tradition are divided into political, theological and jurisprudential categories (Zehra, 2010, pp.22). It is possible to talk about many social movements in each category. Today, in the general framework, two main Islamic movements that divide Islamic society in two parts in all respects, culture and tradition have remained. These two main movements are Shia and Sunni traditions. In both traditions, many different religious and cultural social movements different from each other still exist. In particular, many of the social movements in the Islamic world, even if they emerge with purely secular purposes are fed by religious identity. Political and cultural structure of geography does not allow creating a different opposition and existence language. Therefore, both Turkey as well as in other Islamic lands, no social movement has not only religious or secular. In addition, all social movements are related to the problem fields such as human rights, human dignity, and human will and so on. Like all social movements, they are expressions of the demand for change.

No doubt, some religious movements are willing to transform the world within the framework of certain religious and moral principles. However, mainstream religious movements cannot be explained in this way as they make up a new kind of social movement. They usually demand to (co-)exist in a pluralistic society with respect for their own values and identities. Therefore we argue that today's religious social movements should be addressed through a paradigm that integrates resource mobilization and new social movement paradigms.

In this respect, the example of the Gülen movement which Turkey experienced in the last year, forces us to rethink about religious social movements. Until the 15th of July 2016, the Gülen movement was known as an orthodox, harmless religious movement in Turkey. Even in other religious movements, it was regarded as the only movement that was most compatible with the modern world, limited to education, religion and the civilian space. However, this movement attempted to overthrow the legitimate government with a military coup and take over the state entirely, at a time when it thought it had achieved sufficient power. By cruiser tanks, airplanes, etc. used by soldiers acting in the name of Gülen, the state tried to be controlled. The people who resisted it were fired and 250 people lost their lives. This case brought the concepts of state, citizenship law, civil society and so on. again to the agenda in Turkey. This case shadowed over other social movements seems to be innocent and having no political and economical demands and claiming to exist only on the basis of goodness. Such events imply that no social movements in Turkey and neighboring countries can avoid targeting political and economic power centers. But it would be unfair to claim that all religious, social movements can do the same thing one day.

MOVING AWAY FROM CLASSIC SOCIAL MOVEMENT THEORIES

When reviewing general sociological literature on social movements, two types of perspectives on social movements can be distinguished. In the first strand social movements originate in ideology, whereas in the second strand social movements are explained by structural causes. Both views are underpinned by classical theories which attempt to explain a process before postmodern thought. Also when assessed in terms of religious social movements, it appears that a single theory is not sufficient and that classical theories, whilst explaining the ideological dimension of religious movements, are far from explaining the sociological formation of social religious movements in today's world. Especially in Turkey, a new theoretical approach that combines resource mobilization and new social movement theory may be more explanatory in the new era. Over the last thirty years the development of religious social movements has been strongly intertwined with the distribution of economic resources as well as a continuous struggle to strengthen their positions in the cultural and public sphere in Turkey. In order to come to a better understanding of this social phenomenon, classical and contemporary social movement theories are briefly discussed, with an emphasis on two diverging approaches that explain the emergence of social movements either as a result of a structural crisis or as a social psychological phenomenon.

Resource Mobilization Paradigm

Resource Mobilization Paradigm is a continuation of the theory of collective action (as seen in the example of Smelser) and considers the social movement as either the result of a structural crisis or a common faith / unity of purpose. Yet, it differs from the classical sociologist view, as social movements are no longer treated as pathological movements that are marginal, rootless, temporary, without yielding rational results. The resource mobilization paradigm conveys social movements as ordinary and normal, while discussing their origins along a political and economic dimension: A social movement could either emerge as a product of a marginalized social groups' struggle for political participation, or as an expression of the demand for a reallocation of economic resources. Although resource mobilization theoreticians do not form a homogeneous intellectual integrity, consensus exist on the notion that the actors of collective movements act with an understanding of obtaining economic interests and political recognition (Çayır, 1999, p.20). However, the internal structure of contemporary social movements remains overlooked and the presence of movements not in pursuit of economic and/or political power is far from explained.

Other critics, like Melucci (1985) contend that the resource mobilization paradigm cannot escape dualism in describing social movements as the result of collapse or solidarity. The previously described theories of collective behavior and mass society reside under the umbrella of the collapse perspective, whereas a rather Marxist perspective (on class situations) follows a solidarity perspective by reading social movements as the expression of common interests (Melucci, 1985, pp.790-791). According to Melucci both perspectives have their shortcomings, as collapsing theories neglect conflicts within the collective movement and easily marginalize them and theories of solidarity remain unable to clarify the transition from a specific social situation to a collective movement.

In accordance to the critical readings of Cohen (1985), we like to emphasize the common assumptions of structural functionalist models, collective behavior and mass movement theories that underpin the resource mobilization paradigm:

1. Social movements must be understood from the collective behavior's confrontational perspective;
2. There's no basic difference between institutional and non-institutional collective action;
3. Both of them incorporate conflicts of interest within the framework of institutionalized power relations;
4. Collective action includes the benefits of groups by defending them in a rational way;
5. Aims and complaints are the permanent products of power relations and they are unable to explain the formation of movements;
6. Formation of a movement depends on the changes in resources and opportunities;
7. Success comes true with the recognition of a group as a political actor or an increase of material benefit;
8. Mobility (mobilization) includes bureaucratic organization with large-scale, special-purpose. When analyzing resource mobilization, the objects of analysis of social motion are the collective actors from the group whose benefits are conflicting, not the social movement itself.

Analysis of resource mobilization does not major on the interpretation of ideology or understanding of collective actors itself. The focal point lies at the organizers' rules of the movement on the one hand and political conditions on the other hand (Cohen, 1985, pp. 675-676). Adding a historical angle to this, Charles Tilly has synthesized resource mobilization studies and made invaluable theoretical contributions by examining social movements as the products of both historical and political conditions. Starting with the emergence of social movements in Western Europe and North America at the end of the 18th century (as of 1750), Tilly conveys social movements as political phenomena which comprise peculiar political conflicts and specific historical understandings. He discerns three elements that define the social movements that spread in Western societies during the 19th century and constituted political systems: a) Continuous, organized public initiative which is demanding common right against the targeted authorities; b) Fulfillment of various types of political action. This type of actions come true with the actions such as founding associations and unions, organizing people rallies, the official parades, seizures, demonstrations, petitions, statements to the media; c) Participants demonstrate principles, such as being reasonable, unity, and commitment to themselves or their voters (Tilly, 2008, pp. 16-17). Doing so, Tilly reviews today's conflicts at large and evaluates the emergence of social movements by the entrance of excluded marginal groups on the political stage during a certain political period. He supports the idea that large scaled structural changes affect the collective movement acknowledges that social movements went through various conversions in the 20th century (Çayır, 1999, p.21), but concludes that the mobilization of excluded groups has been provided with modernization.

Globalization, especially evaluated in the context of communication and the flow of capital, has provided flexibility to organizations as well as a wider playing field for social movements. Moreover, the process of democratization has contributed to the development of the social movements by providing formation of intergovernmental and supranational civil society. Tilly notes that all movements which are radical and reformist have adopted campaigns of social movement, which tend to change in parallel with social movement activities and the communication devices available. In many cases, the actions of social movements have been normalized within political culture, as relations between social movements' organizers with local authorities and in particular, authorities providing public peace, have routinized (Tilly, 2008, p.133). Historical experiences of European and American society confirm that political processes, and in particular, the process of democratization, accelerate the formation of social movements. Yet, social movements strongly depend on peoples' sovereignty and those political entrepreneurs who achieve scale, continuity and efficiency. After social movements are constituted in a political environment, model-making, communication and collaboration may allow for support from other environments. However, in terms of structure, staff and claims, social movements remain subject to historical evolution, which means that social movements may disappear in the future or may turn into a very different political entity (Tilly, 2008, pp.64-66).

In addition to a political point of view, the capitalist economy and the institutions of the nation-state develop a logic for collective action. In this context, actors protect their material benefits and they aim for political power so as to control the state and market economy (Cohen, 1985, p. 677). McCarty and Mayer Zald, have therefore explained social movements and their actions through economic factors

(Mc Carty, Zald, 1977, pp. 1212-1241). The organizational structures, including the economic structures of movements, are imperative to yield financial resources. Within this view formal organizations are considered key entities that pass a motion into action, as they transfer their beliefs to potential recipients, who collect resources and personnel in their turn. In this way, the social movement has the opportunity to grow (Çetinkaya, 2008, p. 24).

A different view on social movements in relationship to the capitalist economy is found in anti-system movement theory, which examines the structures of unprecedented anti-system movements of the 19th century, which essentially were political movements. According to Wallerstein, both social movements and national movements with their own members, authorities and specific political aims emerged in opposition to the system at that time. Social movements were characterized by pressure in terms of pressure of bourgeoisie over proletarians, whereas national movements were characterized by pressure of ethnic and national groups over another group. Both types of movements - after solving important inner conflicts - reached the status of official organizations and shared mutual strategies and solutions. Both movements see the basic political structure of modern world as a state and attributes change to something that takes over the power of the state. Particularly in the 1960's and 1970's a shift in the structure of movements is observed as movements with varying origins, historical experiences, ideologies rise in order to gather against social and political events, like Vietnam (Wallerstein, 2004, p.36, 37, 40). Wallerstein, explicitly links this organizational change to political and economic conditions.

According to Touraine, the conversion of the relationship between culture and community has not gained as much research attention as the social structures and historical development of social movements. State has become the main actor of change, whatever its structure and features were during the period of industrialization, while social movements manifest themselves in an area of cultural conflict (Touraine, 1999, p. 44). When evaluated with the structural approach, rather than concentrating on the state, the tendency towards the combination of economic development factors with administrative class initiated. For this reason, the paradigm of source mobilization is maintained inefficient to explain contemporary social movements for as much struggles, statements, aims and participants' socio-economic conditions of contemporary social movements are not meshed with the solutions offered in the source mobilization paradigm. To explain movements with no political aims struggling for identity and recognition in a completely cultural civic area, the paradigm of new social movements has been developed.

Paradigm of New Social Movements

The paradigm of new social movements deals with conflicts not in terms of economic-political subjects but in terms of cultural subjects. The new type of social movements that sprout from Postmodern thought did not have ideals such as changing the world or establishing a paradise on earth nor did they provide hope for the oppressed by overtaking control. It is a method of solution of old type movements that capture the state for saving humanity. Such old movements tried to create a classless exploitative, non-capitalist community by revolution. However, after 20 years of revolutions, conditions had often worsened. In contrast, new social movements rose on a lot richer historical and cultural heritage. For this reason, the purposes of the new social movements are not to seize the state, but to create awareness among a large audience by gathering participants with various socio-economic backgrounds and world views that form a movement around a specific issue. From this point of view, it is important to evaluate new types of movements with a new approach.

According to leading scholar A. Touraine modern social movements cannot be understood as a response to a given situation. Hence, the situation is a product of changing relationships among the actors. Society itself especially is not a given structure but produced over and over again. Social movements are the parts of a model of society that can be defined as the system of social power which is fighting for the control in the cultural field. Here the cultural field is understood to lie at the heart of the most fundamental conflicts. Social movements are confrontational in the social field, but they also encompass conflicted actions directed to the cultural field (Çayır, 1999, p.23). These conflicts can give rise to cultural reforms or breaking of political system. Social movement is a conflict behavior that transforms cultural orientation and historicity field to a form of social organization. This process cannot be absolutely abstracted from class phenomenon. However, the thing which is distinguishing social movement and class is that social movement is action of an actor who is questioning the historicity even though a class can be defined as a situation. According to this, social movement is a confrontational movement of a social class which is defined as an addiction or domination over models of morality and information, cultural investment, historicity format. (Touraine, 1999, p. 48, 50, 51). According to him, modern movements are new movements as they struggle in an area which was opened by the post-industrial society. The struggle is not marked by a market or a state, but is contextualized by a civil society in which the boundaries between private and public are permeated. The dissolving boundaries between the private and public sphere have brought forward new conflicts in terms of public identities (Çayır,1999, p.24), particularly when groups that were previously sentenced to the private area step to the fore (Touraine, 1999). Thus, the new social movements contribute to the expansion of the civil area outside of economic and political structures and colorize the civil area.

Now new social movements have been placed at the crossroads of a public and private sphere, the main purposes of new social movements are situated at the political non-institutional and unforeseen levels against a canvas of of liberal democracy and welfare state practices (Offe, 1985). This is reflected in a continous demand for recognition as political actors in order to affect the society at large, instead of the group itself. Describing new movements by their political impact and achievements, Offe has discussed four movement types. 1) environmental/ecological movements spurred by urban problems; 2) human rights movements, with feminist movements as the most prominent ones protecting the identity of women and others sparring for equal participation regardless age, language, ethnic origin, sexuality or physical ability; 3) pacifist and peace movements; 4) movements aiming for community formation and alternative production and distribution of goods and services. The latter group of new social movements are particularly interesting as they may set any issue from daily life (like body, health or sexual identity) on the agenda by moving it from a particular neighborhood, to the city level and drawing regional or national attention. Although these issues are not new, they are novel with regards to being a ground of opposition and social movement. Moreover the structure of movements have been affected by this novelty. In new movements public and private roles, community (jemaah) and organizations, are fused into new structures; there is especially transitivity between the members' and official leaders' roles. In addition, new participants do not tend to define themselves with traditional political self-codes and previous markers like socio-economic class or ideological lines and political matrices, turnout invalid criteria for the self-identification of the movement and its members. Consequently, it is impossible to explain a new movement by the conflict models that our grounded in class conflict. The conflict is no longer staged by a single class, but a variety of classes, and mounts even non-class elements. This staged conflict has not crystallized from basic economic factors of production and consumption, but overcomes fundamental class differences by universal demands (Offe, 1985, p. 826- 835).

According to Melucci (1985, p. 793-795) new social movements shape conflict areas by strongly drawing upon cultural symbols that are created via opposition messages distributed throughout the system. In this view, new movements have transformative effects by constantly reshaping identities in social interaction. Melucci defines the social movement as a form of collective action, which is driven by solidarity and shakes the roots of a system by addressing a particular issue of conflict. Today's movements are more driven by self-interested than the societal benefits, they ignore policy making and do not fight against power. As in the old social movements, they do not have great ideals and strong hierarchical organizations.

On the contrary, they can be characterized as network organizations, constituted of various daily life inter-related groups. This type of organization networks allow multiple memberships and their actions are short-term and temporary. Nevertheless, individual time allocation to the group and emotional solidarity are desired conditions for participation. The actors mainly contribute by making problems visible, instead of battling for financial goals or participation in the system they fight for symbolic or cultural goals. By altering society they believe their life will also be changed.

Despite the fact that new social movements do not have the explicit purpose to achieve political change, Melucci (1985) also notes that new social movements do affect political institutions by creating a new elite that generates culture. As these movements generate modernity, they bring the demand for innovation forward, and push the system to reform. New social movements with their own structures, problems, and areas of action make up a new social field that goes beyond the dichotomy of traditional state vs. civil society. According to him, a "mid-public sphere" is created which further institutionalizes movements, forwards messages to society and does yield political effects.

New social movements, in terms of Habermas are seen as vision of the future of the unfinished project of modernity. New social movements emerge in late capitalist period against as an opposition to the colonization of the world of life by the system. Movements such as peace, women, environmentalists, LGBTs, and so on. are promising for freedom and democracy. Keeping alive the ideals of modernity such as freedom and democracy may contribute to self-fulfillment of the modernity project (Edwards, 2004, pp.117). As well as these expectations, there are number of criticisms based on the "diversity case" which is consecrated in the new social movements' literature. Accordingly, ethnic, cultural and religious differences and so on. Are being raised as originality and authenticity, combating with social inequality is being devalued. In fact, on the grounds that equality demands eliminate differences, interest is being moved to the "diversity" completely (Blackburn, 1999, pp.5).

New social movements in terms of Western societies can be stated that it is really new in many respects such as social base, quality, organization, problem areas and so on. It is a reality that these movements have oppositional stance to expand the boundaries of freedom and democracy. The past decade that we experienced does not provide grounds for us to be more optimistic about freedom and democracy, and the solution of problems in the world even though new social movements.

War is not finished despite the peace movement. Our world is not peaceful or safer than a hundred years ago. Moreover, selecting specific areas of conflict, the death of more people is followed by millions of people such as an action movie. Similarly, despite a lot of women's movements and feminist movements, patriarchal relations in social, political and economic fields regenerate itself, and the social position of women has not changed. Barricades in front of career of working women in different sectors are not disappeared. The concept of glass ceiling, is used to express the barricades in front of the woman's career. Merely public broadcasting and media sample is sufficient to express the position of women. For example, in the British broadcasting corporation (BBC) which is one of the outstanding institutions of public service broadcasting, the percentage of women executives with decision-making power until the early 1990s, did not even reach 10% (Şanlı,2015,pp.107). For women, the situation is not much different in many sectors. Above all this case is to show the continuity of the patriarchal mentality. Despite the political social achievements of the feminist movements, status of women in public and business life it is still not very good.

One of the most important actors of the new social movements' literature is also environmental movement. Despite the effectively organized environmental movement in many countries, it cannot be claimed that we live cleaner, safer, and risk free social and ecological world. Even experienced in 2011 after the Fukushima nuclear disaster, the world has had to confront the global risk once again. Therefore Beck characterizes today's societies as risk societies. Risk society model is a model of society that raises concerns arising from global threats and risks. In this society, which is essential is to prevent the worst, is not to reach a good thing. The utopia of the risk society is to protect everyone from poisoning (Beck, 2011, pp. 71). In a world where the environmental movement is the most powerful and pervasive, it seems inevitable to live with global catastrophe and the risk.

In the last five years, the human drama on the agenda in the framework of the refugee crisis is a separate issue that requires re-questioning of the basic human values of our civilization. This is an important event as the human tragedy of the Holocaust that necessitated questioning the basic values of our civilization related to human dignity and value. As you know, in the last five years, several thousands of refugees fleeing from the local conflicts in the Middle East and African countries lost their lives while trying to reach Europe. For this purpose, the number of refugees who drowned in the Mediterranean is estimated that almost half a million. The process is still continuing. However, no social movement show interest in this drama or global political attention has been unable to take attention to this tragedy.

When all these events are considered together, it is difficult to claim that the new social movements are successful. However, it is said that the new social movements bring a variety of local and global challenges to the public. It is said that various identity groups have the opportunity to express themselves in public. It should be expressed that this case also expanded the areas of freedom. However, social movements based on criticism regarding the nature and functioning of the global economic and political system such as peace, environment, and equality ideals remain weak and ineffective. These criticisms are also valid to Turkey and Eastern societies.

SOCIAL MOVEMENTS IN TURKEY

The theoretical analysis of the concept of social movement provides a framework to come to a better understanding of the social movements that are emerging in Turkey. Although Turkey has a rich structure with regards to social movements, systemic studies have not been conducted on these movements up to now. The analysis about social life and social movements has been total and reductionist. Turkey has a social structure, inherited from an empire, which attempted to modernize through prestigious ones. Further, Turkey has the structure of a society and a state that prohibits all identities and social movements rejecting the project of modernization. Different ethnic, religious, cultural and social movements did not obey to modernization projects for a very long time, and continued their resistance as underground organizations. It is possible to explain the social movements from this period by structural factors as most of them believed in transformative power of the state and established strategies for seizing the state. However, after the 1980s public and cultural space have been rediscovered with the transition to a more liberal economy, the end of the cold war and increasing economic well-being, public and cultural space. In general four phases can be discerned in the development of social movements in Turkey. First the pre-republic period, followed by a period of transition from the proclamation of the republic to a multi-party period (1923-1950). The third multi-party era (1950-1980) is less relevant to the development of social movements, although it preceded the fourth period in which social movements in modern Turkey flourished from the 1980s onwards. In our analysis of each period, the imperial period's structure of society, society and state relations and the infrastructure of social motion are given the required attention.

Early Republican Period (1923-1950)

Between 1923 and 1950 Turkey was marked by a very poor economic situation, which paved the way for strongly rooted religious movements that succeeded to mobilize citizens for the sake of religious and moral values (Ağartan, Choi, Hyunh, 2008, p.59). At this time, the administration of the single party provided a strong state sovereignty and no social differences were allowed. In fact, the whole republican history can be read from the general contradiction between a bureaucratic elite favoring modernization versus a community rejecting this project, with even the most exclusive state elite imposing modernity through an authoritarian mentality. Rooted in its unique past, Turkey's state tradition actually further impeded the development of a civil society and social movements. Attempts to legitimize the newly established nation-state developed in conjunction with efforts to suppress the public visibility of cultural differences and caused repressive acts (Sanlı, 2003, p.13), further homogenizing community. All kinds of media, associations, social movements, parties and social organizations that did not comply with the new understanding of society were forbidden and public social sphere was made completely unitary. During this period political power was unlimited and absolutely dominated all political, cultural, and economic areas. By the beginning of the 1930's only the "Turkish Women's Union", "Masonic Lodges", and some other organizations seemed to survive. But only a few years later these were closed down as well, so all social institutions and actors bared the official ideology (Çaha, 1999, p.62) in order to modernize and create a modern society.

The process of modernization which was realized through authoritarian mentality meant to break jemaah-structure and to homogenize society through citizenship. However, the Republic could not remove formerly installed hierarchical structures from the Ottoman period. Despite claims of modernity the hierarchy in state-jemaah relationships was made permanent. To the satisfaction of Alevi communities and minority groups only the identity based on citizenship could survive the destruction of social structures. However, the door remained locked, as the relationship between the state and society and internal governmental relationships remained unchanged to the Ottoman period and patriarchal mentality continued (Mahçupyan, 1998, p.44). Equality among citizens did not mean anything in terms of political statement and participation in decision making.

During the multi-party era, continuous communitarianism and strengthening patriarchy put communal demands on the agenda again, while secularism of the state erased religious demands from public life. Consequently, the Muslim community was alienated from politics, and a new breed of self-dependent congregation based on ideological background was legitimized (Mahçupyan,1998, p.44). The relationship between religion and the history of Turkish modernization is paradoxical: On one hand, religious demands and Islam were banned from public sphere, on the other hand institutional control over religion was obtained. Sunni-Islam tradition was taken under state control and was supported secretly. So it did not approach state power but served the framework of ideals of religious discourse, education, modernization and rationalization (Göle, 2000, p.22). At the end of modernization, religious political power did not withdraw to its own area. Some of the newly built institutions had embraced a religious concept in line with the state's modernization project and were supported by the state, with a few exerting political power. In the end, Islam was never privatized as it was in the west. The clergymen who accepted state secularism were allowed to take new positions in the religious institutions (Davison, 2002, p.254).

De facto, the public sphere's homogenization was political, religious or cultural from time to time, but was always fostered by the politicization of sub-identities. A novel upper class, which held state power, emerged and considered itself as the owner and protector of the Republic. While the wealth of the region concentrated around the governing elite, conflicts concentrated in the poor and marginalized segments of population and affected religious and Kurdish parts of society. The state's elite's effort to impose their ideological identity created resistance to Islamic concepts and institutions, and constitutes the essence of the political history of the Republican era (Yavuz, 2003, p.164).

In general, the years between 1923 and 1950, depending on the efforts of uniformity of social life, a strong authoritarian state did not allow difference and during this period social movements were excluded from the public sphere. Ethnic and religious movements, embodying difference, continued to exist as underground organizations. None of the actions, employing political, social, or cultural identity, were permitted until a light change occurred in the 1950s.

First Trial of Democracy

It was the period between 1950 and 1980 in Turkey that social movements gradually entered the public sphere, but remained under state control. From time to time the state even re-established political life by military coups. Especially after World War II, global politics increasingly affected Turkey. The defeat of totalitarian regimes in Germany, Italy and Japan, and the triumph of Western democracies inevitably affected Turkey which was always responsive to the political climate in the West. In the post war context, Turkey was increasingly isolated and Stalin's hostility forced Turkey's government to lean against Western powers. In this position, the state had to listen to criticism either internally expressed or from opposition (Ahmad, 1996, p.23). Although the state paradigm did not end and it was to early to speak of a spring air in terms of participation of civil society and the entry of social movements into public life, the multi-party system certainly conveys a turning point in Turkish political life.

With multi-party, people expressed their political preferences for the first time, and voted against the statist tradition which had lasted for hundreds of years. The conception of the state as a father and centralized control was rejected by society, while top-down imposed reforms aimed to remove possible obstacles for capitalism. This progress constitutes the beginning of a transformation from bureaucracy dominated capitalism into the market dominated capitalism. Bureaucracy ruled the country during the empire, even during the 19th century reformist movements. When the bureaucracy started out to provide the transformation of the social structure, it had allied with the bourgeoisie which was chosen and developed by itself (Keyder, 1999, p.103). However, this alliance was completely under the control of the bureaucracy and (not surprisingly) both sides didn't trust each other very much. During this alliance, the bourgeoisie, giving up civil society rights and staying within the boundaries set by the state, achieved substantial gains. However, with multi party life, several economic and social changes initiated the beginning of the end of this alliance.

Social mobility started with migrations from the rural to the urban: Between 1950 and 1960 the urban population in Turkey increased from 19% to 26%, with a rise up to 75% in the four major cities of Turkey. The economic vitality of cities increased employment opportunities and was clearly manifested by an increase of shanty houses and a cultural upheaval in the cities. The traditional city with economic and cultural superiority of the ruling class vanished. This spatial mobility formed the actual start of integration by confronting the cultures of center and the periphery. This situation represents a new era that constitutes the beginning of the articulation to the traditional bourgeoisie. The most influential industrialists and businessmen of the 1980s and 2000s went into professional life for the first time during these periods and mostly obtained their savings in this period (Keyder, 1999, p.111, 112). This period can be regarded as the beginning of a transformation in terms of both economic and social movements.

With this period, both the traditional social solidarity organizations in Ottomans and some newly established civil society elements began to emerge. Closed and prohibited religious communities, sects, foundations, *madrassas*, as well as political parties, associations, labor movements, women's movements, political ideologies and even the media prospered for a while (Çaha, 1999, p.77). As the public sphere was less developed only private activities were not associated with the state. Gaps were created in the areas where the state gave up auditing on behalf of economic freedom, and these gaps were rather filled with individual expansionism than with building blocks for strong civil society. Since the civil society didn't flourish rapidly, the remaining gaps offered a chance for state penetration on behalf of the protection of political stability. In 1960 even a military coup took place, although that did not last very long (Keyder, 1999, p.114). The real development of both the bourgeoisie and civil society in Turkey took off in the 1970s.

Indeed, after the foundation of the Turkish Industry and Business Association (TÜSİAD) in 1971 a bourgeoisie community begin to emerge, which initially criticized political power from an economic perspective in the 1970s, but evolved to both an economic and political player in the 1980s. They yielded success with several alternative projects in monetary policy, education and other domains when the electoral system shifted to a constitutional system (Yilmaz, 2006, p.202). This was not just a transformation in terms of economic organization, but constituted a differentiation in religious, cultural and political domains of the public sphere.

Period of Political and Economic Liberalization

Turkey after the1980s, witnessed the emergence of many different social movements which struggled for their existence in the public sphere. Both specific structural conditions in Turkey and more general political trends after the 1950s fed this process. The end of the Cold War vanished most security politics on external enemies, while more liberal policies were implemented. Consequently, the underground processes of social identity construction started to surface with the rise of new interests and pressure groups. At this time debates centered on subjects like air pollution, health, tourism, environment, human rights, ethnic identity, religious rights and women's rights affecting particular parts of society and disclosed many social differences. More uniting subjects of concern, such as modernization, national identity, national solidarity were hardly part of public discourse. In other words, social differences were magnified rather than considering issues related to the protection of the state (Çaha, 1999, p.77). Thus, a new social movement period started in which different identity, politics, economic interests and rights and demands were staged on the social platform. And the public sphere became a place where Kurds, Alevis, and different religious movements contended.

So, until the 1970s the development of social movements in Turkey can be explained as a phenomenon with structural reasons. Until then the absolute power of the authoritarian understanding of the state was at stake. Almost all social movements emerged within this process, opposition movements and movements established to achieve politics targeted to the state and sought out the possibility of achieving their political projects by seizing the power of the state. We can evaluate this kind of movement as an example for a social movement in the old period. These movements, have survived as secret organizations until the present period. After 1980, almost all of these movements have engaged in fighting an identity in the public sphere.

Changing world conditions and Turkey's socio-economic conditions became effective on these social movements emerging from underground and retaking strategies. After the 1980s, new movements which developed strategies based on identity politics such as women's movements, environmental movements, human rights started to appear. Struggles started to move slowly to cultural areas.

Hence, social movements and organizations, which the state both deemed legitimate and considered dangerous, entered in the opened door together with enlivening of liberal approaches with economics and politics in the 1980s. The intensification of economic, social, political subjects met appropriate opportunities which would produce the massive organization. So a new public area was formed (Sanlı, 2003, p.15). With the participation of new actors to the newly formed public area, a new way to express social differences was found. A lot of commentators connected the development of social change in these years to the evolution of economic and political liberalism. In fact, there had been an economic structure that imposed the state-mandated traditional bourgeoisie to operate within confined boundaries. However, after the 1980s, there were new participations into the job market and the state's economic and political policies started to be criticized by this newly shaped bourgeoisie. Although the relations with the state still kept its traditional style, critical approaches developed via some associations based on voluntary membership, and the need for the state decreased (Buğra, 2010, p.53). Besides, through the slum culture that occurred in large cities as the result of a large wave of immigration in the 1950s, various cultural, economic social solidarity organizations emerged. Even in the 1990s, important holdings and entrepreneurial organizations in global business are rooted in the small manufacturing and service sector activities that entered the Turkish market back then.

Thus, the newly formed bourgeoisie developed independently from both traditional bourgeoisie and bureaucracy. The 1990s became the year of emergence of holdings, trade companies, organizations of businessmen, alternative banks in many places in Anatolia and they developed very fast. Until the years of 1980s, there were businessman organizations which represent the traditional bourgeoisie assisted by bureaucracy including just TUSIAD and TISK (Turkish Confederation of Employers). After 1980, many organizations appeared such as MUSIAD (Independent Industrialists ' and Businessmen's Association), AGIAD (the Association of Young Businessmen of Anatolia), and ASKON (Anatolian Lions Businessmen's Association) (Köroğlu, 2012, p.91). All of these developments provided the formation and strengthening of urban, educated new middle classes. This ambient also provided the participation of many different social structures having difficulty to translate their demands into political speech.

In 1990s, Turkey was a witness of long-term actions of social movements such as the civil constitution, The Citizens' Initiative for Bright, Bergama Environmental Movement, The Black Sea Environmental Action Network, The Saturday Mothers, Victims of Torture, feminist movements that put violence against women and women's education on the agenda. In addition, Ethnic Kurdish movements and dozens of Alevi associations were arising and began to compete. Within this social mobility, there is a special position of religious movements.

Some of the religious movements are historically internalized by the Republic of Turkey as structural elements of the Sunni school, forming the largest part of congregates of The Ottomans and being suppressed by the modernization movement. However, this frustration could not be reached to the point of destroying and eradicating. Traditional religious movements withdrew their activities to secret underground bases until the 1980s. After surfacing, conflicting and tense relations between Turkish modernization and religious movements have fuelled a more radical stance towards the state. There have always been in tension on the control of state-owned areas among the modernizing elite with traditional Islamic perspectives. In terms of the state, religion is understood as a system that is limited to the conscience of the individual, the fulfillment of the traditional ritual merely. In contrast, the desire of Islamic movements' transforming society by seizing the state has formed end point (Kırık, 2005, p.127). However, after the 1980s, religious social movements have experienced a transformation within themselves. As a result of this transformation, they gave up the target of seizing the state and focused on participation in the public sphere and dominating the cultural field. The next process, has been a process of struggle for identity.

Seeing Islamic social movements, as it was not one homogeneous movement, as aiming for the state has been a classic reflex of authoritarian states. As a result, Islamic social movements in Turkey have hardly been modernized before the 1990s while they were detected as pathological movements aiming for the founding of a sharia state by seizing the state insidiously. Consequently, like many other traditional groups, religious movements have mostly been considered as a threat to the public sphere and they have been disadvantaged in terms of education, economic distribution and urbanization for a long time (Mardin, 2010, p.144).

When evaluated in general, religious movements in Turkey and fundamentalist religious movements in the general sense of piety are barely present. All of the religious and social movements have had criticism against to the system. Most of the religious movements' criticism related to the system has been merely non-political and opposed to moral degradation, the spread of evil, or children moving away from religious values, etc. (Başer, 2000, p.150). The vast majority of the Islamic social movements emerging in the Republican period, predominantly expressed demands for religious freedom rather than seizing the state. However, power and power relations are a returning axis at the center of the state, just because traditional religious movements had, like any social movement, no other way of channeling their opposition than through the state.

Based on this context, the larger religious movements in Turkey (attracting wide social participation), can be divided into a few categories. The first and largest category consists of Sufi movements, which transferred the elements of sects from the Ottoman to the Republic of Turkey. These movements represent a religious concept that emphasizes personal experiences and mystical aspects of religion (Atacan, 1990, p.23). Despite losses and prohibits in the sense of economic and institutional and religious rituals, the Sufi movements did not target the state, but focused mostly on worship and moral issues. They possessed a moral vision of society based on the individual's physical and mental training as mystical.

A second category is made up of the very active Nur movement and Süleymanizm movement. Nur has had many branches and was born in response to positivist approaches about faith in recovery and belief. Whereas, Suleymanizm has emerged as another important religious movement in response to difficulties on the point of raising the cleric, which was experienced in the first years of the Republic (Çakır, 1990, p.83, 130). These movements are centered on unity of belief. There is a sense of unity among them. They apply similar religious rituals. Together with these movements, radical movements that considered the state as the source of all problems have been in place too driven by the assumption that they may be able to solve both their own problems and the problems of the community by seizing the state. In that sense, radical movements are quite similar to the revolutionary movements of the 18th and the 19th century. Seizure of the state, as a golden age utopia for society, has fostered the main motivations of such movements. However, in the 1980s, especially after the 1990s, these movements entered in a serious transformation period exceeding their closed nature and sliding their struggles for the public, the cultural sphere.

CONCLUSION

The case of social movements represents a dynamic process. Constant new problems and reactions to them reveal new modes of social movement and action. In terms of the western world, the phenomenon of the new social movement is a style of structuring a service society that emerges after the heavy industrial economy. The action style, organizational structure, member profile and problem areas of the new period movements are quite different. On the other hand, the agenda and expectations of new social movements in Turkey and neighboring countries are different. In Turkey and neighboring countries, religion is an important element in terms of acquiring societal base and mobilizing the masses. For this reason, social movements have a positive or negative relationship with the phenomenon of religion. Social movements that have emerged in order to promote a completely religious identity and a religious lifestyle can also focus on different problems over time. Sometimes they can lose their own agenda and turn into a completely political organizations, even commercial companies. Gülen's experience of Turkey is a typical example of the radical transformation of a religious movement ending with a terrorist action. It is understood that the social movements that emerged after the Republic in particular do not yet develop a relationship based on trust with the state. In this respect, there is no difference between the secular or religious social movements.

The phenomenon of the new social movements in Turkey and in neighboring countries seems to vary from Western countries. Social movements in the West have become widespread with the development of community services, their participants are mostly secular, educated, and from upper and middle-class. Social movements in Turkey are similar to these movements. However, Turkey's social experience and economic structure provide more opportunities to the development of social movements. Therefore, religious social movements have a special place in the social relations system of Turkey. In addition, religious social movements not only bring to the religious themes to the agenda but also themes such as environment, peace, women and so on. For both Turkey and surrounding countries, it is still very difficult to establish universal and intellectual opposition language except religious literature. Therefore, whatever the problem area, all social movements use religious literature in certain proportions. In Islamic countries except Turkey, for they are on the bottom rung of the social hierarchy of needs, it is impossible to see movements such as women, the environment, and peace and so on. New social movements are fantasy for people who are concerned about the safety of life, and starving. In this respect, Turkey has a special position.

When generally evaluating the development of social movements in Turkey, we can conclude that until the 1980s, religious movements acted either as a mere-faith movement within closed organizational structures, or with the aim and strategies of state seizure. After the 1980s this changed, as religious movements entered in the market and economic sphere, participated in the intellectual sphere and at the same time they entered modern social and political life. On the one hand they dealt with educational problems by creating educational institutions, on the other hand they increased their economic prosperity by entering the market through holdings, Islamic banks and business associations. As a result, a bourgeois class was created that took significant shares of economic prosperity as well as an intellectual role by criticizing Western modernism in order to become recognized as a well-educated nation, instead of being represented as the other through representations of peasantry, ignorance, poverty, and being uncultivated. In addition to faith problems, religious movements drew attention to issues regarding inclusion of women, family, and social service needs, social differences, the environment, civil rights and freedoms, democracy, etc.. Parallel to these developments, it would be wrong to evaluate Turkey's religious movements as a purely religious movement or to see them as temporary reactive movements that emerge from a specific situation or struggle for economic prosperity.

Although religious movements or people who have obvious identity of religiosity have been contesting for larger shares from the increasing economic prosperity, this has not been their only fight. Apart from this, they also demonstrated an intellectual position related to all kinds of social and political problems. And they have mainly tried to secure their position within the public and cultural spheres by their own discourses and claims for sovereignty. In other words, they bring forward identity politics.

In this regard, to make sense of religious movements in Turkey, we think that resource mobilization and new social movements' paradigms should be addressed as a theoretical whole. Of course the desire to gain economic resources and structural changes affected these movements, but today it is not possible to demarcate Turkish religious movements solely by the transformations they went through and their current social position. Religious movements are both similar and different to new social movements in several respects.

These religious movements convey the same critical sensitivity as other Western contemporary movements in attitudes toward enlightenment and modernity. Their responses depend on their identity as well as issues related to religion, ethnicity, race, sex, or nature that have been pressed by modernism. They follow identity politics expressing local identities rejecting the modern project. Paradoxically, they revived by the politics of identity, particularism and localism against the canvas of abstract universalism and a tendency towards uniformity. From this perspective, religious movements resemble feminism. Religious movements have also defended the difference of Islam to problematize the universality of civilization. Besides, unlike structural political tradition, a cultural understanding of religion has put great emphasis on new movements (Göle, 1998, p.40, 136). New religious movements, are transforming religious definitions, form new actors and new autonomous areas and also question current cultural programs of ego and the modernity. Consequently, lifestyle is now expressed as an integral part of the cultural sphere due to the rise of new religious movements (Göle, 2000, p.30, 31). By these transformations new religious social movements distinguish themselves from the classic class movements or benefit source distribution movements.

In sum, being present in the cultural and economic sphere is crucial for new period religious movements. For these movements targeting the state makes no sense and has no charm. By means of the autonomous public spheres they formed, they seem to have obtained facilities and places that will enable them to live according to their own identity, differences, and values. Through religion new movements have attempted to give meaning to complexity in everyday life, doing so they offer individuals a way to make sense of the world. It can be predicted that both the secular and religious social movements of the future in Turkey, will turn into movements that revive issues of identity and difference, next to environmental and local problems.

REFERENCES

Ağartan, T., Choi, W. Y., & Hyunh, T. (2008). *Kapitalist Dünyanın Dönüşümü:1750-1850.Toplumsal Hareketler 1750-2005 Dipten Gelen Dalgalar* (D. Keskin, Trans.). İstanbul: Versus Yayınları.

Ahmad, F. (1996). *Demokrasi Sürecinde Türkiye.1945-1980.* İstanbul: Hil Yayınları.

Atacan, F. (1990). *Sosyal Değişme ve Tarikat Cerrahiler.* İstanbul: Hil Yayınları.

Baratov, K. (2005). *1980'li Yıllarda Türkiye'de Sosyal Sınıflar ve Bölüşüm.* İstanbul: İmge Kitabevi.

Başer, A. (2000). *Ortadoğu'da Modernleşme ve İslami Hareketler.* İstanbul: Yöneliş Yayınları.

Buğra, A. (2010). *Devlet ve İş Adamları* (F. Adaman, Trans.). İstanbul: İletişim Yayınları.

Çaha, Ö. (1999). *Sivil Toplum, Aydınlar ve Demokrasi.* İstanbul: İz Yayınları.

Çakır, R. (1990). *Ayet ve Slogan.* İstanbul: Metis Yayınları.

Çayır, K. (1999). *Yeni Sosyal Hareketler.* İstanbul: Kaknüs Yayınları.

Çetinkaya, Y. D. (2008). Tarih ve Kuram Arasında Toplumsal Hareketler. In Y. D. Çetinkaya (Ed.), *Toplumsal Hareketler* (pp. 15–61). İstanbul: İletişim Yayınları.

Cohen, L. J. (1985). Strategy or Identity: New Theoretical Paradigms and Contemporary Social Movements. *Social Research, 52*(4), 663–716.

Davison, A. (2002). *Türkiye'de Sekülerizm ve Modernlik* (T. Birkan, Trans.). İstanbul: İletişim Yayınları.

Göle, N. (1998). *Modern Mahrem.* İstanbul: Metis Yayınları.

Göle, N. (2000). *Melez Desenler.* İstanbul: Metis Yayınları.

Keyder, Ç. (1999). *Türkiye'de Devlet ve Sınıflar.* İstanbul: İletişim Yayınları.

Kırık, H. (2005). *Kamusal Alan ve Demokrasi.* İstanbul: Salyangoz Yayınları.

Köroğlu, M. A. (2012). *Özerk Kamusal Alanların Oluşumunda Dinin Rolü* (Unpublished doctoral dissertation). Necmettin Erbakan University, Konya, Turkey.

Mahçupyan, E. (1998). Osmanlı'dan Günümüze Parçalı Kamusal Alan ve Siyaset. *Doğu- Batı, 5*, 22-47.

Mardin, Ş. (2010). *Din ve İdeoloji.* İstanbul: İletişim Yayınları.

Marshall, G. (1999). *Sosyoloji Sözlüğü* (O. Akınbay & D. Kömürcü, Trans.). Ankara: Bilim ve Sanat Yayınları.

McCarty, J., & Zald, M. (1977). Resource Mobilization and Social Movements: A Partial Theory. *American Journal of Sociology*, *82*(6), 1212–1241. doi:10.1086/226464

Melucci, A. (1985). The Symbolic Challenge of Contemporary Movements. *Social Research*, *52*(4), 789–816.

Offe, C. (1985). New Social Movements: Challenging the Boundaries of Institutional Politics. *Social Research*, *52*(4), 817–868.

Sanlı, L. (2003). Türkiye'de Toplumsal Hareketler. In L. Sanlı (Ed.), *Toplumsal Hareketler Konuşuyor* (pp. 9–32). İstanbul: Alan Yayınları.

Tilly, C. (2008). *Toplumsal Hareketler* (O. Düz, Trans.). İstanbul: Babil Yayınevi.

Touraine, A. (1999). *Toplumdan Toplumsal Harekete. Toplumsal Hareketler* (K. Çayır, Trans.). İstanbul: Kaknüs Yayınları.

Wallerstein, I., Hopkins, T. K., & Arrighi, G. (2004). *Sistem Karşıtı Hareketler* (C. Kanat, B. Somay, & S. Sökmen, Trans.). İstanbul: Metis Yayınları.

Yavuz, M. H. (2003). Türkiye'de Çoğulculuk ve Kamusal Alanın Kullanımı. In EDAM (Ed.), Sivil Bir Kamusal Alan (pp. 163-192). İstanbul: Kaknüs Yayınları.

Yılmaz, H. (2006). Kamu, Kamu Otoritesi ve Devlet; Habermas'ın Işığında Türkiye'yi Düşünmek. In A. Karadağ (Ed.), *Kamusal Alan ve Türkiye* (pp. 192–209). İstanbul: Asil Yayınları.

Chapter 4
Remediating Interaction:
Towards a Philosophy of Human-Computer Relationship

Roger Gacula Pineda
Emopowerment Oy, Finland

ABSTRACT

The concept of interaction is foundational in technology interface design with its presuppositions being taken for granted. But the interaction metaphor has become ambiguous to the extent that its application to interface design contributes to misalignments between peoples' expected and actual experience with computer-enhanced actions. This chapter re-examines the presuppositions governing human-computer interaction with the motivation of strengthening weaknesses in their foundational concepts, and contributing a theoretical framework to designing for artistic as well as mundane experience. It argues for abandoning the interaction metaphor to refocus design discourse toward the intermediation and mediation roles of technology interfaces. Remediation (i.e., representation of one medium in another) is proposed as a conceptual model that more precisely describes the human-to-computer actions.

INTRODUCTION

"Art calls for technology. An emergent technology can produce leaps into both science and art" (Risset, 2003). The rapid integration of computing technology in everyday life, thus leaving its mark on human thought, emotion and action, attests to the leaps Jean-Claude Risset envisions. Such historic velocity of change

DOI: 10.4018/978-1-5225-5023-5.ch004

in the human-computer relationship, or 'interaction', however, leaves experts and non-experts little time to reflect on the foundational principles of this relationship. This essay examines the nature of human-computer interaction. The motivation is twofold: first, to see whether there are any cracks in the foundation of the *interaction concept* as it is used in computer interface design, and thus consider whether these should be repaired or replaced; secondly, with an eye toward developing theory that serves as a framework for research and design of technology, contribute to revisions of current descriptions of the relationship between people and computers, concepts which function as models of human embodied cognitive processes, and models applied to design and engineering of computer interfaces, impacting people's everyday experience, culture and art. These purposes are realized with and through an interpretive discourse and a sketch of a prescriptive approach to methodology.

The author's strategy for decomposing presuppositions is grounded on 'foundational analysis' (Saariluoma, 1997), but deviates from the framework by using methods of the phenomenological tradition. The concepts employed in the critique can be applied to technology in general, but the following treatise is limited to the domain of computer technology, and specifically to the concepts used in design of computer interfaces. Throughout this work, the term 'ordinary user' denotes people without expert knowledge of computer technology, in contrast to the term 'expert', which is used interchangeably with 'designers and engineers'.

Two themes emerge in the following sections. First, interaction as a conceptual metaphor denoting what happens during human usage of computers foregrounds illusory aspects of the intermediation, mediation and operational roles assigned by experts to computers. The author contends 'interaction' is a misplaced label. This misplacement leads designers and engineers to rely on a partial view of design discourse. Modeling and labeling discrete events as interaction, without foundational analysis, contribute to the problem of misalignment between ordinary users' expected and actual experience derived from computer usage. Second, there is a need for a philosophy of human-to-computer relationship that reframes the modeling process in order to assist experts to refocus the design discourse towards the intermediation and mediation roles of the computer interface. To this end, the author proposes using the concept of 'remediation' as a more accurate description of constitutive events in the realm of human usage of computing devices, to understand them well enough to make decisions on interface design.

WHERE THE INTERACTION IS NOT

One cannot deny the contribution of metaphor usage to modeling technical aspects of the operating principles used in contemporary digital computers: it has triggered the evolution of computer interfaces to its current iteration. Metaphor usage, however, has also become an object for further analysis (Marcus, 1994), related to studies of resultant problems encountered by ordinary users of computers. Researchers are aware of the need for empirical evaluations of metaphors, as shown by the work of e.g. de Castro Salgado et al. (2011), but the literature on computer technology design provides few reflective analysis of interaction, such as the assessment by Keeler and Denning (1991). Precedent critiques by Dreyfus (1992), and by Winograd and Flores (1988) only lightly touch upon the issue of interaction *per se*. Works expounding a philosophy of interaction relevant to design of computer interfaces (e.g. Dourish, 2001; Janlert & Stolterman, 2017; Svanaes, 2014) are even fewer, and these precedents do not fully describe the gestalt of interaction. To date, partial answers based on analysis of its components have enabled development of models used in design and engineering of user interfaces. They have helped in realizing the technical aspects. There remain, however, questions regarding human factors in computer usage. Partial answers to them are proving to be inadequate, and consequences manifest in misalignments between people and technology. What is interaction? Where does it happen? How is interaction with a computer experienced? What are the presuppositions and the contingencies that must exist in order for something to be perceived as interaction gestalts? These are the foundational issues which remain partially understood.

Critical deconstruction of the term has already been remarkably absent in seminal literature such as in the work of Card et al. (1983) and Norman (2002). The interaction metaphor has become one of those taken for granted concepts. Practices which change perception of the interface and whose origin and means for *becoming* something have been forgotten, are now taken for granted as intuitive cultural perception (Husserl, 1970; Polanyi, 1958). There seems to be a tendency to jump directly into the derivatives of action, without an equal amount of effort in describing the phenomenon. Descriptive and prescriptive aspects of technical actions assigned to computer interfaces tend to focus on the constituent parts of interaction without deep critical reflection on the concept of interaction *per se,* or its constitutive presuppositions.

If one insist on a definition explicitly distinguishing between reciprocal influence that happens between persons, and reciprocity in events between things (see e.g. Oxford English Dictionary), one immediately confront problems of equivalence and compatibility. The ontology of interaction has to confront this problem. Dealing with this type of ontological issues dates back to the ancients, demonstrated in Aristotle's distinction between natural things and artifacts (Aristotle, *Physics* II.1, 192b8-193b20, trans. Cornford & Wicksteed). From what level of observation could one classify events as interaction that is not restricted to a simple feedback mechanism? Developing a nuanced description of the construct proves to be a tricky problem.

Interaction as a *scientific concept* (Carnap, 1966) is primarily treated as a classificatory concept—i.e., a taxonomy concept (e.g., memory, attention), which is used to develop and classify quantitative concepts (e.g., time to task completion), while skipping or ignoring development of comparative concepts—i.e., qualitative descriptions of concepts such as utility and usability, which have been deemed as preconditions to developing measurable concepts. One of the results of this state of affairs is that designers and engineers often miss crucial attributes, such as 'cultural responsiveness' (Clemmensen & Roese, 2010; Paterson, Winschiers-Theophilus, Dunne, Schinzel, & Underhill, 2011; Sivaji, Nielsen, & Clemmensen, 2017), which are relevant to their work.

Equally problematic is that 'human-computer interaction', as a linguistic representation of the relation between people and machines, misleads ordinary users to presuppose intelligence and intentionality on the part of the computer. Characterizing the relation between people's movements and movements of computers as interaction could be considered as psychological priming, to the extent that people tacitly assume presence of a living being as the agent behind a technical object. Observed behavior whereby people expect human-like qualities from technical devices and thus treat them as if the devices are human (Reeves & Nass, 1996; Turkle, 2011) is illustrative of the type of resultant behavior from such priming. This contemporary phenomenon, however, is not new, for it can trace its lineage to the phenomenon of projecting agency—human and superhuman—to inanimate objects, such as the practice in medieval European culture of venerating reliquaries and sacral images (Belting, 1994; Hahn, 2010).

In observing interaction between humans, one can infer preconditions for this event: sentience, intelligence, and intentionality. There are certainly more preconditions; however, the present focus will be limited to these three. To a certain extent these preconditions are also satisfied when one observes interaction between humans and members of other animal species. Observations of human actions relative to an environment require different conceptualizations of interaction with regard to its preconditions. Does our inference for the presence of sentience, intelligence, and intentionality still applies? This question underscores the reminder to be "... alert to the idea that agency is an attributed property in systems; that is, agency is something to be discursively granted (or withheld) in our accounts of the operation of the world and our place within it" (Dourish, 2017).

One approach is to ignore conceptual preconditions. One can apprehend this approach in descriptions of interactions between entities as observed at the atomic and cellular levels, i.e. inferred actions which manifest between these entities. Even a non-expert could reasonably accept the notion that interaction amongst cells is not contingent on sentience, intentionality, and intelligence. Such recognition of agency has been extended to artifacts and environments, as participants in collective action, a construct within a framework for analyzing and describing the dynamic of social relations, with the view that "...we have to take non-humans into account only as long as they are rendered commensurable with social ties and also to accept, an instant later, their fundamental incommensurability" (Latour, 2005, p. 78). One, however, should distinguish between *causal descriptions*, using language, of dynamic processes as these descriptions are used for communications between people, and the *actual physical manifestations* of these dynamic processes. For example, at the cellular level, it is also accurate to say that the stimulus-response process is a series of physical perturbations structurally perceived by an entity in relation to a given environment as well as in relation to other entity or entities. Perturbations are set off by physical energy oscillations, i.e. movements at the quantum level of analysis (Maturana & Varela, 1987). Interaction is a label assigned to a causal description. The actual manifestations within a human agent and a non-human agent respectively, and between the two agents dynamically, could be something else. One can create an artifice of interaction without claiming it a simulation of cognitive processes (Rowe, 1993, p. 96).

From Homo Faber to Homo Communicationis

The concept of intelligence tends to complicate analysis of interaction. With some exceptions, as in analyses of using computer programs in interactive artwork and live music performance (see Dixon, 2017; Lippe, 2002), the ordinary user's ignorance of the artifice (of sentience, intentionality, and intelligence) has been largely overlooked in design and engineering discourse, beginning with the publication of the Turing test (Turing, 1950) all the way to contemporary work in robotics. The success of something like a Turing test is contingent upon keeping the human participant unaware of the machine's presence and role. The issues at stake have practical implications vis-à-vis design, as well as philosophical and ethical dimensions.

Conceptualizations of human-computer interaction began to appear in computer science literature notably during the 1970s, e.g. in the works of Engelbart (2001/1962), and Kay and Goldberg (1977). Prior to the introduction of terminal devices (i.e., dumb terminals) as interface to mainframe and minicomputers, and consequent development and mass production of personal computers, the user of a computer was also the designer-engineer as well as the operator of the device, which was customized for a particular task. The received view changed with the advent of compartmentalization of human roles—i.e., engineers, programmers, operators, and ordinary users. Card et al.'s (1983) seminal work on the psychology of human-computer interaction contributed to the foundation of the received view: they took away from ordinary users the role of operator of machine, and assigned to them the role of communicator in dialogue with a machine. In effect this event helped transform the computer user from *homo faber*, the artificer whose creative labor produced the instrument of artifice, to *homo communicationis*—stripped of artistic license, a laborer talking to a ghost. It is important to note that early users of computers created and operated them in order to automate the manipulation of symbols. It is an act of off-loading the cognitive load of calculations, which supported the task of solving problems, primarily of military nature. In this context, the artificer's relation to the artifact can be labeled as manipulation of objects, but symbolic manipulation is not interaction.

Conceptualizing the human-machine relation (artificer-artifact) as a dialogue came at a later stage in the development of human-computer interfaces. Yet the artificer (designer-engineer) may not have been aware that the dialogue is necessarily illusory. It would be more accurate to label the communication as a *monologue* scripted and performed by the artificer. That is to say, the designer-engineer uses linguistic means to create machine-comprehensible linguistic forms that trigger chains of events, which transform electrons from one state to another. This production process and resultant product are derivatives of the monologue. They are components of a *human* monologue.

The dyadic format of the dialogue metaphor is useful in modeling and programming the action-reaction sequences that must be represented at the interface. From a user-centric design perspective, modeling events as interaction allows the designer-engineer to step into the role of the computer-intermediary and conceptualize trade-offs that must be considered in the interest of balance between affordances—cf. 'signifiers' (Norman, 2010)—and constraints, which are operationalized at the interface. Affordances and constraints shape the selectivity of the user's attention, therefore structuring the user's experience. It is useful in designing feedback loops, and in deciding what part or parts of the complex whole must be hidden from the ordinary user. The principles of 'direct-manipulation interfaces' (Hutchins, Hollan & Norman, 1985; Shneiderman & Plaisant, 2005) embody these design goals, and concepts such as virtual reality, augmented reality, and tangible user interfaces extend these principles. The masking of complexity, however, often leaves the ordinary user dumbfounded when the computer itself becomes the focal point of attention due to some breakdown.

In contrast to the user-centric approach to analysis and modeling, an artifact-centric methodology can be deployed to deconstruct the interaction model and to guide the modeling processes back to the interface of computational artifacts. This approach implicitly suggests metaphorical attribution of agency to artifacts, and assigns anthropomorphic independence to computers to the extent that machines can urge or force people to interact. Janlert and Stolterman's (2017) notions of 'interactability, recognizable agency, receptivity, predictability, and pace' enable differentiation of the preconditions to machine-to-human actions, as well as analysis of computers as a proxy of human artifice, but tend to further mask the human intelligence and intentionality behind the artifice.

Interpreting a series of spatiotemporal events as interaction is dependent on human gestalt and semiotic abilities. "...[A]ny pragmatically useful analysis of artifacts is inseparable from an analysis of the values and experiences of those who use them or feel their effects, or indeed from the values and experiences of those who design them" (McCarthy & Wright, 2017). The interaction metaphor was indeed relevant to the modeling processes when the user was also the artificer, but the metaphor's relevance to the contemporary ordinary user has become ambiguous because the artificer now remains in the background, and the expert's knowledge of the artifice is not readily accessible. There is no direct spatiotemporal contact. Instead the ordinary user is left with a *proxy* of the artificer in the form of an interface. From the ordinary user's perspective, the knowledge is lost. The metaphor has been useful in reifying the mental contents of the artificer through and with a computer, but the metaphor comes with several tacit presuppositions which are opaque to the ordinary user yet nevertheless influence his or her perception.

Keeping a Logical Account

Conceptual postulates used in human-computer interaction revolve around human activities. Thus the study of events are framed by human interactions within a world which include material things. As soon as one acknowledges that interaction need to be considered in the context of physiocognitive actions, one faces another problem: concepts are culturally laden. If all entities "enter our conception only as *cultural posits* [emphasis added]" (Quine, 1953, p. 44), then conceptual postulates cannot be assumed invariant. Though discussion of the intertwined relationship of culture and technology is beyond the scope of this chapter, and is discussed in detail elsewhere (see Pineda, 2014), it is useful in keeping a logical account of the human-computer relationship to note that computer technology is a socio-cultural phenomenon. Broadly speaking, "...the cultural role of commodities...cannot ultimately be divorced from questions of technology, production and trade" (Appadurai, 1986, p. 35). For experts and ordinary users alike, tacit criteria for expected experience of interaction include culture dependent-independent ontology, and culture dependent-independent epistemology.

Experts of representationist-computationalist inclination often describe cognitive processes in terms of the information processing framework. The notion of information as stimulus is a conceptual presupposition in this context. The supposition is committed to a theory underlying *stimulus* as a component in causal description of the action-reaction chain. As such, theory ladenness is not a problem. One, however, has to keep a running account of this status. Stating that attention is a process by which information enters working memory and manifest as consciousness in the human mind (Welsh et al., 2012) has different implications in contrast to stating that attention is a process linguistically describing an entity's spatiotemporal movements in response to perturbations, as heuristic natural drift toward structural coupling that preserves the entity's physical structure.

The previous enunciation of the problematic notion of interaction leads to a hypothesis regarding its location: interaction is in the mind and not at the interface. More precisely, the *experience* of interaction is located in the apperceptive active human being who possesses intelligence, intentionality, and capability to transparently use instruments during (inter)actions in the world. Risset (1996) says,

"[I]n the quest for the sonic resources available from the computer, one must explore the psychoacoustic [emphasis in the original] relation - the relation between the physical structure of the sounds and their perceived effects: synthesis programs require the precise and exhaustive specification of the physical parameters of the desired sound, but what counts in music is the auditory experience [emphasis added], not the physical structure".

Interaction denotes a description of what happens between people, and people's actions with things in relation to the world. It is a connection between action and experience. Articulation is difficult because we are dealing in language. One becomes aware of thoughts as they take place in language. Language is therefore one's launchpad, cognitive instrument, and stumbling block (Maturana & Varela, 1987). Keeping a logical account of and articulating descriptive and prescriptive aspects of interaction entails a process of minding the ontological gaps that one experiences as an apperceptive being.

MINDING THE GAP: A PHILOSOPHY OF HUMAN-COMPUTER REMEDIATION

The current conceptualization of interaction fails to adequately model phenomena such as semiotic and perceptual transparency (Keeler, 1990), intentionality (Merleau-Ponty, 1962) and intelligence (e.g. Dreyfus, 1992). Transparency, intentionality, and intelligence as applied to technology and to computers in particular, are not isomorphically perceptual; they are interpretive. An interaction model useful to design and engineering of interaction using computers is contingent upon distinctions between social interaction and other abstractions of the concept. Adequately dealing with intentionality, from a design perspective, presents myriad design problems, which are amplified particularly in contemporary design of technology assisted activities, ranging from computer-assisted live performance art (e.g. interactive music systems, live interactive performance art) to brain-to-computer interface and robotics. Whilst preceding critics of computer technology have already flagged these issues, there can be a renewed sense of urgency in developing the underlying philosophy. The author proposes a philosophy of human-computer remediation.

Organization of Interaction

If interaction is in the embodied mind of the observer, not at the computer interface, how does one go about classifying myriad observable events as interaction? As a being predisposed to categorizing events and objects of the world, one could use the notion of *organization*, which denotes the observed relations between the parts of the whole (relations between events). What relations must be present in order for interaction to exist? What are the attributes of the relations? How do the relations work? This deconstruction of interaction's organization is of cognitive relevance to the process of apprehending a more nuanced interpretation of active events.

Interaction has been characterized as a dyadic relation in terms of changes of state (Harel, 1987), as information processing (Card, Newell, & Moran, 1983), as usage of tools (Heidegger, 1977) as perception-action (Merleau-Ponty, 1962), as theatre (Laurel, 1993), and as 'feel' dimension of experience (Svanaes, 2014). Definitions like interaction as "...interplay between the state of the environment and the actions of agents (which are usually people), where the actions change the state of the world, ... can be intentional or non-intentional" (Norman, 2014), bring us closer to a gestalt of interaction, but the ontological gaps still have to be narrowed and bridged.

In Svanaes' (2014) example of usage of a switch, the switch is labeled as an interactive object. If one takes the position of an observer of a scene, as if a spectator in a theatre, wherein an actress is putting her hand on a light switch with a consequent result of a light bulb turning on, would one call the series of events interaction? Then if one adapts the position of the actress, would one experience the series of events as interaction? If the light bulb does not go on, how does one know that the event is (un)intentional? Placing interaction at the switch is partially helpful in conceptualizing the action. The human action upon the switch triggers action-reaction chains resulting in a light turning on (a change of state for the light bulb). But which component affected a change of state in the human? The light turning on or the switch? Missing in the picture: the switch and light bulb are components of a whole electrical system, i.e., the gestalt of electricity and possible applications of this integrated wholeness to bring to fruition objects of human desires.

An alternative interpretation of the vignette is that the actress integrates the electrical system to her structure, her being, in order to amplify or extend perceptual powers (e.g. to better see the texts of her book), and at the same time preserve the integrity of her being (e.g. not get electrocuted). This seems counterintuitive until one assigns the role of mediator to the switch, as well as switch one's observational-interpretive level to the biological roots of existence, and use the notion of 'structural coupling' (Maturana & Varela, 1987). This has several implications for computer interface design: first, structural coupling, i.e. the interaction, must allow the human entity to preserve its wholeness. This is fairly obvious in the context of electrical switches, or anything electrical, for if using a switch consequently results in literally integrating the actress into the electrical system, this could possibly result in death from the experience of electric shock. It is not entirely obvious in all the aspects of the computer interface, but preservation of physiocognitive integrity (i.e. wholeness of experience) is foregrounded in the context of tangible user interface, interactive performance art, brain-to-computer interface, and robotics. Second, the computer should transparently mimic a dialogue, through designed feedback loops represented at the interface, in order to help the actress react to breakdowns. Third, the semantics used at the interface must be aligned with the way humans make meanings through multiple modalities. Application of interactive music systems to live music performance is exemplary of the principles discussed thus far. As in the early days of computer usage, the artificer can have multiple roles—as designer, programmer, engineer, operator, composer, conductor, performer; the designed solution, i.e. the artifice-artifact that projects sonic events, strives to mimic the relationships between musicians in real-time performance. Powered by algorithm informed by music theory and thus capable of actions intelligible to musicians, the artifice-artifact is integrated into the gestalt of performance through sonic manipulation by human participants.

The above implications are in turn loaded with their respective implications on the role of the expert. Does the notion of structural coupling adequately aid in keeping account of the different levels of observation relevant to the design of computer interfaces? For as soon as one take this notion into use, one have to further distinguish between social and non-social coupling. A human being might cry out in fury and despair against nature (non-social coupling) upon looking at the wreckage of his or her home as aftermath of a tsunami. He or she expects a response from the insurance agent evaluating the damages (social coupling) but does not expect a response from nature. During the communication with the insurance agent, this distraught person might verbally note a lack of empathy on the part of the agent. These appraisal, categorization and consequent articulation in language are in effect brief disruptions in the flow of interaction. These disruptions trigger a conscious awareness of interaction—cf. Foucault's (1970) discussion of discontinuity.

A dramaturg might cause an actor to rile against a tempest. The dramaturg can cause nature to respond, a *deus ex machina*. An observer sitting in the theatre audience apprehends the artifice and comprehends the meaning of the exchange between the actor and nature, but he or she does not classify these observed events as real interaction. The act of classifying in language and in context requires conscious reflection. The computer designer and engineer, like the dramaturg, create the illusion of interaction. The ordinary user of a computer, like the observer in the theatre audience perceives, understands and might in reflection classify the observed event as interaction, but is unable to immediately apprehend the artifice. An implication for design: making distinctions between structural coupling events taking place within and between humans, within and between computers, and between humans and computers is incumbent in interface design. In design of computer-mediated music interactivity, for instance, this principle foregrounds the difference between modeling performers or instruments, and modeling the relationship between musicians during performance (Lippe, 2002). Triggering transparency helps ordinary users to focus attention on experiences within and with the world. Feedbacks are necessary for appraisal of any discontinuity to the moment-to-moment reciprocal transformative exchange, which is necessary for avoiding disintegration of the experience. Note also that the semantics and semiotics constituting human structural coupling require strong cultural agreement. A semiotic system is also a cultural system; therefore, one could infer variance in the agreed meaning of interaction. Considering these variances, how does one account for and mimic intelligence and (un)intentionality at the interface?

Human-to-human social coupling is contingent upon knowledge shared by a community of people; that is knowledge about the world, a common level of perspective among multiple veridical perspectives of reality. People use this knowledge as part of their daily experience. In computer systems, knowledge is represented and stored in forms—i.e., binary electronic signals, machine language, etc, which are unintelligible to ordinary users. These forms must be translated and represented at several levels of interfaces all the way to the physical touch point with the ordinary user. How does one ensure that nothing gets lost in translation?

The attractiveness of interaction as a metaphor is understandable in view of the expert's desire to model the chain of events which have to be reified, given the limitations of computer technology and its instrumentation, because one can then use the dyadic action/reaction movements that can be characterized as a form of communication. One has to remember, however, that the resulting reification through instrumentation represents an idealized situation which is conceived within the boundaries of an idealized environment, i.e., controlled experiments. One cannot expect the resulting functions assigned to the instrument to perfectly map to the uncontrolled environment of everyday existence in the world. The communication is not a dialogue. It is rather a user's internal monologue in which he or she conceives the message and the meaning, while manipulating an object constituting representations of 'statements of probability' (Polanyi, 1958).

Computers are configured to make mathematical and logical deductions. In contrast, excepting people who are trained in scientific perception, ordinary users usually do not think, feel, and behave according to neither logical mathematical deduction nor induction. Automatization of tasks and events as manifest in computers is discrete and rectilinear in movement. Human cognitive processes, e.g. semiotic process, are analog and circular. In this sense, one could characterize interaction, as currently conceived for computers, as rectilinear movement which is tangential to or circumscribed by circular motion—cf. Hintikka and Remes' (1974) discussion of geometer's analysis and synthesis process.

One of the functions of the computer's interface is to make two different things compatible. A digital computer is configured to move in formalized discrete rectilinear fashion. The human configuration's attributes are characteristically analog and non-linear. It is difficult to make computer interaction and human interaction compatible because of differences in the degree of restraints put upon them by their respective structures and configurations. A computer's hardware and software structures restrain their formalized movements to a closed system. In contrast, human embodied actions, i.e., perceptual and motile, do not conform to formalized rules. Their continuous movement is characteristically non-linear and happens within an open system.

Going back to the actress and taking up her position, it becomes more comprehensible that labeling observed patterns as interaction requires conscious reflective thought in language. Articulation of thought can be tacit or explicit (Polanyi, 1958). People not trained in scientific perception will not explicitly articulate observed correlations as interaction. Perception has to be also understood as cultural or contextual perception (Ihde, 1991). Positive interaction is reciprocally transformative action that results in preservation of integrity of human wholeness and therefore of wellbeing. Conversely, negative interaction is transformative action that results in partial or full disruption of wellbeing. Feedback is foundational in creating the illusion of interaction as well as allowing the possibility for interaction.

The most relevant interaction is interaction between living beings. That is to say, interaction between people is foundational; manipulation of instruments used during the interaction is a derivative. People's actions with a computer interface are manipulations of symbols in order to integrate the instrument in his or her structural coupling with the world. Symbol manipulation is action with and through an intermediary (a computer is assigned the role of mediator and executor of commands). A computer has to be transformed as an extension of the living being. In order for the computer to become an extension, it must withdraw to the background. In order for it to withdraw to the background, an artifice must trigger transparency. This interpretation becomes more comprehensible in the context of art installations and performance in virtual world settings such as *Second Life*, in which the creative process itself is mediated through an avatar (Doyle, 2014). It is also reflected in the context of computer-assisted prosthetics—see e.g. work at the Applied Physics Laboratory in John Hopkins University (Bridges, Para, & Mashner, 2011); muscle actions are used to manipulate robotic limbs through an embedded computer interface. Empirical work in robotics, however, has not progressed enough in order to deal with the issue of (un)intentionality. A detailed description of intentionality and intelligence is beyond the scope of this essay, but these issues has to be flagged.

Intelligence and intentionality are preconditions of interaction among humans; they are embodied conditions. Human interaction can thus be distinguished from that of other animals. It can be distinguished from that of machines. Within the context of doing a task with a computer, an observer apprehends interdependent relations between manipulation events triggered by people and operational events of a computer. Still, this observed interdependence must be framed by a human desire and intention to move from one state of being to another. The chain of events, which constitute movement between states of being, while interdependent, cannot be fully described as interaction because intentional reciprocity is missing. The possible actions of two independent self-contained entities, a human and a machine, are interdependent to the extent that the human agent must trigger the chain of actions. Of course one can say for instance that interactive music systems can simultaneously and interchangeably assume the role of performer, composer, conductor (Lippe, 2002), and thus act indepently, or that life sustaining machines like the ones used in hospitals have the active role in triggering certain functions of the human body to keep a comatose person alive. However, one must also remember that a human agent engages the machine with intentional action to start the triggering event.

Design of Interaction Between People

To create a material artifact that embodies linguistic representations of somatic proprioceptive experiences and mental reflections (e.g. texts, images, sounds), one has the option of acting, i.e. with one's mind and limbs to produce the artifact, or ask another person to do it. Secretaries and stenographers taking dictation are modern version of the ancient scribes; contemporary artists—e.g. dramaturgs, composers and musicians—extend the practices of their ancient lineage. In both cases, there is a need for an instrument (technology) to accomplish the creative act. The point here is that the collaborative relationship between the commissioner and executor of the artifact constitutes human-to-human interaction, a relationship that meets the preconditions of intentionality and intelligence. The relationship between the mental processes of the lone artificer may still be labeled as interaction to the extent that it is a discursive monologue, which also meets the preconditions. Note, however, that there is no attempt to label the human manipulation of the instrument as interaction. But since the diffusion of personal computers, the received view of perception of these series of events is now based on the substitution of interaction, which presupposes intentionality and intelligence only for humans, for manipulation-operation.

The author have thus far asserted that humans integrate technology into their actions in order to amplify or extend perceptual powers—e.g. via off-loading of cognitive and physical load, to serve as extensions to creative and observational capabilities, to amplify subjective experience; in other words, to transcend constraints and discontent. The fundamental value of the computer lies in its function of enabling people to transfer some of the cognitive and physiological load to machines. This off-loading act frees people to realize a portion of *vita activa* (Arendt, 1959), i.e., to be released from certain form of labor.

Experts desire to understand human interactions within the world among materials, including machines and others, to derive a model of movement which is reified and represented by a material interface. The designer and engineer of an interface target an ideal state, i.e., intuitive, easy-to-use, etc., that exploit the human powers for learning. In this context, modeling the events taking place during usage, as interaction is partially useful. It is partial because the model constitutes problematic limitations. Mental images, for instance, lend themselves easily for reification. Whereas embodied phenomena, e.g. feelings and emotions, are more difficult to reify and represent at the interface. Current work on developing a emotional and gestural vocabulary—e.g. Wodehouse & Marks (2013)—for designing gesture-based interaction that mimic commonly used movements and emotions associated with them attest to this complexity.

In summary, the objections to the use of interaction as a metaphor in computer technology are: (1) interaction among humans connotes intentionality and intelligence. For contrast, one's observations of relationships of events occuring in the natural world, at the quantum level and level of universes are also metaphorically labeled interaction, but it is accepted that connotations of intentionality and intelligence are also metaphoric. The same cannot be said for usage of interaction in computer technology parlance. The construct tends to foreground something that has already withdrawn, i.e. the programmer(s) emblematically involved in human-to-human communication with ordinary users. In human-to-computer actions, the semiotic process is participated by a machine, which represents another human by proxy. In this sense the machine mediates the mediator, a process of remediation. (2) The interaction metaphor fails to acknowledge a distinction between ordinary perception and scientific perception. Note both objections are concerned with incompatibilities in worldview, and both have tacit dimensions of power dynamics.

It is difficult for interaction designers to fully capture the nuances in ordinary users' requirements without minding the gap between perception and descriptions of the actual physical manifestations of phenomena. Interaction as a label for what happens during human usage of computing devices leads to the proverbial mistaking the map for the terrain. The notion of interaction design can be reframed as primarily design of interaction between people, and between people and the world. This implies that interaction design experts should cultivate conscious awareness of cultural variability in the semiotic component of perceptual processes, as well as of the notion of 'emergence' (Seevinck, 2017) in human actions with other beings, including technology artifacts, and the environment. Predicting emergent actions during the design process can be foregrounded earlier in the process and can be inspired, for instance, by characteristics of improvisation practiced by musicians and actors. Moreover, reframing necessitates a switch in focus on how humans mediate and remediate the semantics and semiotics of structural coupling. This approach explicitly acknowledges the inevitable inaccuracy of conclusions derived from semiotic analysis.

The author submits that the notion of 'remediation' (Bolter & Grusin, 1999), i.e. the representation of one medium in another medium contributes clarity to the description of events. Modeling human-computer events as remediation has the advantage of switching the focal point of attention back to analysis of both analogue and discrete human physiocognitive movements in manipulating artifacts. Anyone who has played the game in which one whispers a statement to an ear of a participant who in turn whispers it to the next person in a chain of participants until audibly articulated by the one at the end of the chain has experienced what could happen during the process of remediation. Foregrounding the complexities of action through a proxy can help experts to keep a logical account of where interaction takes place. The notion of remediation applied to human-to-technology events can inspire interaction design: "Remediating such events via critical and creative intervention, the creative media project we have in mind has the potential to become a new incarnation of the age-old 'theatre-within-the theatre' device, whose actors are also at the same time critics" (Kember & Zylinska, 2009, p. 27). Given the motivation to design computer interfaces that reduce misalignments in semiotic and perceptual processes, the author proposes that the metaphor of interaction be abandoned and replaced with the concept of remediation, to better model the human-computer interface.

For the interaction designer, it is proposed that phenomena be examined through the lens of culture to determine whether what one sees fits what one has seen before. Through this process one can develop methods for a new way of 'seeing as' (Wittgenstein, 1958) to revisit prior conceptions of phenomena. Semiotic processes of the human mind constitute the immaterial condition for perception of (inter)action. The computer is part of the material condition for this perception in the sense that it embodies human understanding of these processes. Taken together, they serve to operationalize and realize tasks, as means to desired ends.

Deployment of design processes informed by a theory of remediation implies the notion of interaction design can foreground the primary task of modeling relations between people. The secondary task is to represent this understanding of human actions and concepts of the range of probable responses via a proxy, i.e. the computer interface. Inspiration for operationalizing design principles reflecting the gestalt of human interaction can be obtained from extant work in performance art. Current empirical work in applying bio-sensing technology to interactive performance (e.g. Rostami, McMillan, Segura, Rossitto, & Barkhuus, 2017) exemplifies the effort to reify human semiotic apperception processes through computer interfaces while concurrently, by design, providing the non-expert, the spectator, more possibilities for awareness of the artifice. Rostami et al's concepts of 'temporality of input, autonomy and control, and visibility of input' contribute an extension to the concepts of affordances and constraints, and assist in refocusing experts and non-experts' attention back to the realm of manipulation of symbols.

A theory of human-computer remediation giving primacy to wholeness of human experience is fundamentally contrastive to and mutually exclusive of an artifact-centric theory of interaction. From the point-of-view of design practice, however, mutual exclusivity in terms of methodology is more ambiguous. Work by Oogies and Wakkary (2017) with design videos to address aspects of designing for technological mediation illustrates how designers can criss-cross between poles of conceptual primacy in foregrounding or backgrounding human experience and relations, materiality, and non-human agency to derive design-relevant attributes, gain a wider view of the relationships between participants in domains of action, and speculate on emergent activities.

The notion of emergent action resonates with a theory of remediation, and serves as a reminder of underlying compatibility issues between formalized, digital technology-enhanced actions and natural, analog human actions. And while acknowledging the inherent complexity of accurately predicting emergent, ad hoc actions, designers can apply knowledge of people's 'rule following actions' (Leikas, 2009, p. 117) to make suppositions about the range of possible actions and reactions during manipulations of technology-enhanced artifacts. Application of design practice foregrounding emergence and remediation can be observed particularly in current computer technology-enhanced live performance, inclusive of improvisation (Barkhuus & Rossitto, 2016; Morris, 2013; Salter, 2009; Visi, Coorevits, Schramm, & Miranda, 2017); usage of digital games and game-like interactions for sacred experience (e.g. Sliwinski, Katsikitis & Jones, 2017); converging artistic and sacred experiences through technology integration (Shaw, Gromala, & Seay, 2007; Vidyarthi, Riecke, & Gromala, 2012); 'designing for agency': integrating computational artifacts into art therapy involving people with dementia (Lazar, Edasis, & Piper, 2017; Morrissey, Wood, Green, Pantidi & McCarthy, 2016); research and application of musical and non-musical 'entrainment' to human inter(action) within and with technology-enhanced artificial environments (Mandanici, Rodà, & Canazza, 2017).

Regardless of the theoretical basis chosen for reifying concepts and human actions at the interface, the necessarily partial view of the interaction gestalt has to be flagged early enough to mitigate negative consequences of misalignment between people and technology. Towards this end, future theoretical and empirical work in interaction design for artistic, sacred and profane experiences has to integrate the processual views of design, production, and commoditization of artifacts.

CONCLUSION

At this moment in human history, interaction between humans and computers is ultimately restrained by the limitations of human artifice: people make things according to what is already known within the confines of the material universe. Even things produced based on fantasies contain components of current scientific perceptions. That is to say, the limitations of machines are ultimately limitations of human minds existing within the boundaries of material existence.

In the course of a few decades, designers and engineers have ceased to view the relation between humans and computers as operational. It is now seen as interactive. This shift in paradigm has affected the cultural perception of interaction designers and is being absorbed by other cultures, e.g., culture of marketers. One should note the proliferation in advertising and product labeling of the terms 'interactive', 'smart' or 'intelligent' for indications of how these terms denoting scientific perception is being transformed into intuitive cultural perception.

Designers and engineers can take advantage of the human ability to apprehend meaning at the interface. When done well, the machine withdraws to the background and remains transparent. This event allows focus on the task at hand. Done poorly, the machine comes to the foreground of perception but becomes opaque. Ordinary users are left to suffer through a discontinuity. They become conscious of the differences between their way and the computer's way of thinking and behaving during the state of discontinuity; that is to say, when something breaks down. This situation becomes problematic when the ordinary user realizes he or she is left to deal with the incompatibilities, and is clueless about the techniques of how to transcend the situation. In most circumstances the chain of events lead to feelings of irritability, but in some others, the situation becomes a matter of dysfunction, of life or death.

Developing a philosophy of interaction, in the context of art, culture, computer science and designing interfaces, leaves us stuck with issues of mind and matter. Descriptions of interaction based on models of human psychology are inadequate. Equally inadequate are descriptions of interaction within the boundaries of electronically discrete movements of the computer. One needs yet another conceptual interface between the two domains. As artificer and ordinary user one has to remember interaction is not at the interface but in the mind.

ACKNOWLEDGMENT

The author thanks José Filipe Silva of University of Helsinki and the anonymous reviewers for their comments, and suggestions to earlier version of this work.

REFERENCES

A. Appadurai (Ed.). (1986). *The social life of things: Commodities in cultural perspective*. Cambridge, UK: Cambridge University Press. doi:10.1017/CBO9780511819582

Arendt, H. (1959). *The human condition* (3rd ed.). Chicago, IL: The University of Chicago Press.

Barkhuus, L., & Rossitto, C. (2016). Acting with technology: rehearsing for mixed-media live performances. In *Proceedings of the 2016 CHI Conference on Human Factors in Computing Systems* (pp. 864-875). ACM. doi:10.1145/2858036.2858344

Belting, H. (1994). *Likeness and presence: A history of the image before the era of art*. Chicago, IL: The University of Chicago Press.

Bolter, J. D., & Grusin, R. (1999). *Remediation: Understanding new media*. Cambridge, MA: MIT Press.

Bridges, M. M., Para, M. P., & Mashner, M. J. (2011). Control system architecture for the modular prosthetic limb. *Johns Hopkins APL Technical Digest*, *30*(3), 217–222.

Card, S. K., Newell, A., & Moran, T. P. (1983). *The psychology of human computer interaction*. Hillsdale, NJ: Lawrence Erlbaum.

Carnap, R. (1966). *Philosophical foundations of physics: An introduction to the philosophy of science*. New York, NY: Basic Books.

Clemmensen, T., & Roese, K. (2010). *An overview of a decade of journal publications about culture and human-computer interaction* (HCI). In D. Katre, R. Orngreen, P. Yammiyavar, & T. Clemmensen (Eds.), *Human Work Interaction Design: Usability in Social, Cultural and Organizational Contexts* (pp. 98–112). Boston, MA: Springer; doi:10.1007/978-3-642-11762-6_9

de Castro Salgado, L., de Souza, C., & Leitão, C. (2011). Using metaphors to explore cultural perspectives in cross-cultural design. In P. Rau (Ed.), *Internationalization, design and global development* (pp. 94–103). Berlin: Springer; doi:10.1007/978-3-642-21660-2_11

Dixon, S. (2017). Cybernetic-existentialism in interactive performance: Strangers, being-for-others and autopoiesis. *International Journal of Performance Arts and Digital Media*, 1–22. doi:10.1080/14794713.2017.1301173

Dourish, P. (2001). *Where the action is: The foundations of embodied interaction*. Cambridge, MA: MIT Press.

Dourish, P. (2017). Commentary: Interactivity as cultural. *Human-Computer Interaction*, *32*(3), 153–154. doi:10.1080/07370024.2016.1257388

Doyle, D. (2014). Phenomenologies of practice: The artist in virtual Worlds. *International Journal of Art, Culture and Design Technologies*, *4*(1), 1–12. doi:10.4018/ijacdt.2014010101

Dreyfus, H. L. (1992). *What computers still can't do: A critique of artificial reason*. Cambridge, MA: MIT Press.

Engelbart, D. C. (2001). Augmenting human intellect: A conceptual framework. In R. Packer & K. Jordan (Eds.), *Multimedia: From Wagner to virtual reality* (pp. 64–90). New York, NY: WW Norton & Company. (Original work published 1962)

Foucault, M. (1970). *The order of things: An archaeology of the human sciences*. London, UK: Tavistock publications.

Hahn, C. (2010). What do reliquaries do for relics? *Numen*, *57*(3-4), 284–316. doi:10.1163/156852710X501324

Harel, D. (1987). Statecharts: A visual formalism for complex systems. *Science of Computer Programming*, *8*(3), 231–274. doi:10.1016/0167-6423(87)90035-9

Heidegger, M. (1977). *The question concerning technology and other essays*. New York: Harper & Row.

Hintikka, J., & Remes, U. (1974). *The method of analysis: Its geometrical origin and its general significance*. Dordrecht: D. Reidel Publishing Company; doi:10.1007/978-94-010-2296-5

Husserl, E. (1970). *The crisis of European sciences and transcendental phenomenology: An introduction to phenomenological philosophy* (4th ed.). Evanston, IL: Northwestern University Press.

Hutchins, E. L., Hollan, J. D., & Norman, D. A. (1985). Direct manipulation interfaces. *Human-Computer Interaction, 1*(4), 311–338. doi:10.1207/s15327051hci0104_2

Ihde, D. (1991). *Instrumental realism: The interface between philosophy of science and philosophy of technology*. Bloomington, IN: Indiana University Press.

Janlert, L. E., & Stolterman, E. (2017). The meaning of interactivity: Some proposals for definitions and measures. *Human-Computer Interaction, 32*(3), 103–138. doi: 10.1080/07370024.2016.1226139

Kay, A., & Goldberg, A. (1977). Personal dynamic media. *Computer, 10*(3), 31–41. doi:10.1109/C-M.1977.217672

Keeler, M. A. (1990). Investigating transparency in the conditions of mediation from a semeiotic view. *Semiotica, 82*(1-2), 15–42. doi:10.1515/semi.1990.82.1-2.15

Keeler, M. A., & Denning, S. M. (1991). The challenge of interface design for communication theory: From interaction metaphor to contexts of discovery. *Interacting with Computers, 3*(3), 283–301. doi:10.1016/0953-5438(91)90018-W

Kember, S., & Zylinska, J. (2009). Performance, invention, critique. In M. Chatzichristodoulou, J. Jefferies, & R. Zerihan (Eds.), *Interfaces of performance* (pp. 27–43). Farnham, UK: Ashgate.

Latour, B. (2005). *Reassembling the social: An introduction to actor-network-theory*. Oxford, UK: Oxford University Press.

Laurel, B. (1993). *Computers as theatre*. Reading, MA: Addison-Wesley.

Lazar, A., Edasis, C., & Piper, A. M. (2017). Supporting people with dementia in digital social sharing. In *Proceedings of the 2017 CHI Conference on Human Factors in Computing Systems* (pp. 2149-2162). ACM. doi:10.1145/3025453.3025586

Leikas, J. (2009). *Life-based design. A holistic approach to designing human-technology interaction*. Espoo, Finland: VTT Publications.

Lippe, C. (2002). Real-time interaction among composers, performers, and computer systems. *Information Processing Society of Japan SIG Notes, 2002*(123), 1-6.

Mandanici, M., Rodà, A., & Canazza, S. (2017). Bodily interactions in motion-based music applications. *Human Technology, 13*(1), 82–108. doi:10.17011/ht/urn.201705272519

Marcus, A. (1994). Managing metaphors for advanced user interfaces. In *AVI '94 Proceedings of the Workshop on Advanced Visual Interfaces* (pp. 12-18). New York, NY: ACM. doi:10.1145/192309.192317

Maturana, H., & Varela, F. (1987). *The tree of life*. Boston, MA: Shambala.

McCarthy, J., & Wright, P. (2017). Commentary: Making interactivity meaningful for contemporary HCI. *Human-Computer Interaction, 32*(3), 148–152. doi:10.1080/07370024.2016.1254048

Merleau-Ponty, M. (1962). *Phenomenology of perception* [Phénoménologie de la perception]. (C. Smith, Trans.). London, UK: Routledge.

Morris, J. M. (2013). Collaborating with machines: Hybrid performances allow a different perspective on generative art. In *Proceedings of the 16th Generative Art Conference* (pp. 114-125). Retrieved from http://www.generativeart.com/ga2013xWEB/proceedings1/11.pdf

Morrissey, K., Wood, G., Green, D., Pantidi, N., & McCarthy, J. (2016). "I'm a rambler, I'm a gambler, I'm a long way from home": The place of props, music, and design in dementia care. In *Proceedings of the 2016 ACM Conference on Designing Interactive Systems* (pp. 1008-1020). ACM. doi:10.1145/2901790.2901798

Norman, D. A. (2002). *The design of everyday things*. New York, NY: Basic Books.

Norman, D. A. (2010). *Living with complexity*. Cambridge, MA: MIT Press.

Norman, D. A. (2014). Commentary on: Svanaes, Dag (2014): Philosophy of interaction. In M. Soegaard & R. F. Dam (Eds.), The Encyclopedia of Human-Computer Interaction (2nd ed.). Aarhus, Denmark: The Interaction Design Foundation. Retrieved from https://www.interaction-design.org/encyclopedia/philosophy_of_interaction.html

Oogjes, D., & Wakkary, R. (2017). Videos of things: Speculating on, anticipating and synthesizing technological mediations. In *Proceedings of the 2017 CHI Conference on Human Factors in Computing Systems* (pp. 4489-4500). ACM. doi:10.1145/3025453.3025748

Paterson, B., Winschiers-Theophilus, H., Dunne, T. T., Schinzel, B., & Underhill, L. G. (2011). Interpretation of a cross-cultural usability evaluation: A case study based on a hypermedia system for rare species management in Namibia. *Interacting with Computers, 23*(3), 239–246. doi:10.1016/j.intcom.2011.03.002

Pineda, R. G. (2014). *Technology in culture: A theoretical discourse on convergence in human-technology interaction* (Unpublished doctoral dissertation). University of Jyväskylä, Jyväskylä, Finland.

Polanyi, M. (1958). *Personal knowledge: Towards a post-critical philosophy.* London, UK: Routledge.

Quine, W. V. O. (1953). *From a logical point of view: Nine logico-philosophical essays.* Cambridge, MA: Harvard University Press.

Reeves, B., & Nass, C. (1996). *The media equation: How people treat computers, television, and new media like real people and places.* Stanford, CA: CSLI Publications.

Risset, J. C. (1996). Real-world sounds and simulacra in my computer music. *Contemporary Music Review, 15*(1-2), 29–47. doi:10.1080/07494469600640341

Risset, J. C. (2003). The perception of musical sound. Computer music: Why? *France-University of Texas Institute for Interdisciplinary Studies.* University of Texas at Austin. Retrieved from http://liberalarts.utexas.edu/france-ut/_files/pdf/resources/risset.pdf

Rostami, A., McMillan, D., Segura, E. M., Rossitto, C., & Barkhuus, L. (2017). Bio-sensed and embodied participation in interactive performance. *Proceedings of the TEI*, 17. doi:10.1145/3024969.3024998

Rowe, R. (1993). *Interactive music systems: Machine listening and composing.* Cambridge, MA: MIT Press.

Saariluoma, P. (1997). *Foundational analysis: Presuppositions in experimental psychology.* London, UK: Routledge.

Salter, C. (2009). Environments, interactions and beings: The ecology of perfomativity and technics. In M. Chatzichristodoulou, J. Jefferies, & R. Zerihan (Eds.), *Interfaces of performance* (pp. 44–62). Farnham, UK: Ashgate.

Seevinck, J. (2017). Interaction in art and computing. In *Emergence in Interactive Art* (pp. 31–46). Springer International Publishing. doi:10.1007/978-3-319-45201-2_3

Shaw, C. D., Gromala, D., & Seay, A. F. (2007). The meditation chamber: Enacting autonomic senses. *Proceedings of ENACTIVE/07 4th International Conference on Enactive Interfaces.*

Shneiderman, B., & Plaisant, C. (2005). Designing the user interface: Strategies for effective human-computer interaction (4th int. ed.). Boston, MA: Addison-Wesley.

Sivaji, A., Nielsen, S. F., & Clemmensen, T. (2017). A textual feedback tool for empowering participants in usability and UX evaluations. *International Journal of Human-Computer Interaction*, *33*(5), 357–370. doi:10.1080/10447318.2016.1243928

Sliwinski, J., Katsikitis, M., & Jones, C. M. (2017). A Review of Interactive Technologies as Support Tools for the Cultivation of Mindfulness. *Mindfulness*, 1–10. doi:10.1007/s12671-017-0698-x

Svanaes, D. (2014). Philosophy of Interaction. In M. Soegaard & R. F. Dam (Eds.), The Encyclopedia of Human-Computer Interaction (2nd ed.). Aarhus, Denmark: The Interaction Design Foundation. Retrieved from https://www.interaction-design.org/encyclopedia/philosophy_of_interaction.html

Turing, A. M. (1950). Computing machinery and intelligence. *Mind, 59*(236), 433-460. doi:/LIX.236.43310.1093/mind

Turkle, S. (2011). *Alone together: Why we expect more from technology and less from each other*. New York, NY: Basic Books.

Vidyarthi, J., Riecke, B. E., & Gromala, D. (2012). Sonic Cradle: designing for an immersive experience of meditation by connecting respiration to music. In *Proceedings of the designing interactive systems conference* (pp. 408-417). ACM. doi:10.1145/2317956.2318017

Visi, F., Coorevits, E., Schramm, R., & Miranda, E. R. (2017). Musical instruments, body movement, space, and motion data: Music as an emergent multimodal choreography. *Human Technology*, *13*(1), 58–81. doi:10.17011/ht/urn.201705272518

Welsh, T. N., Chandrasekharan, S., Ray, M., Neyedli, H., Chua, R., & Weeks, D. J. (2012). Perceptual-motor interaction: Some implications for human-computer interaction. In J. A. Jacko (Ed.), *The human-computer interaction handbook: Fundamentals, evolving technologies, and emerging applications* (3rd ed.; pp. 3–20). Boca Raton, FL: CRC Press; doi:10.1201/b11963-3

Winograd, T., & Flores, F. (1988). *Understanding computers and cognition: A new foundation for design*. Reading, MA: Addison-Wesley.

Wittgenstein, L. (1958). *Philosophical investigations* [Philosophische Untersuchungen] (G. E. M. Anscombe, Trans.). Oxford, UK: Blackwell.

Wodehouse, A., & Marks, J. (2013). Gestural product interaction. *International Journal of Art, Culture and Design Technologies*, *3*(2), 1–13. doi:10.4018/ijacdt.2013070101

Chapter 5
Experience-Centred Design and the Role of Computer-Aided Tools in the Creative Process

Folasayo Enoch Olalere
Vaal University of Technology, South Africa

ABSTRACT

In this chapter, the author examines the levels of a cognitive process (visceral, behavioural, and reflection design) in experience-centred design and their applicability in developing effective products. The study further explores the role of computer-aided tools (CAD) in engaging users as the fundamental participants in the creative process. This was achieved by developing two products digitally (CAD models) based on Norman's three design levels, and evaluated using a semantic differential scale to test the emotive response towards the designs. The findings show that CAD is a viable tool for gaining insight into users' perceptions towards a design idea. Also, the results revealed that the process of supporting or sustaining cultural values through design also enhances the affective quality (reflective level) of the design.

INTRODUCTION

The economic development and technical progress have transformed the product demands from the mass and quantitative consumption to customised and perceptual consumption. Despite the importance of technology capability, technology is not

DOI: 10.4018/978-1-5225-5023-5.ch005

only what it takes to captivate customers (Boatwright and Cagan 2010); they want products that are functional at a physical level, usable at a psychological level and attractive at a subjective and emotional level. Hence products are meant to satisfy some functional requirements such as aspiration, cultural, social, and emotional needs.

Research by Boatwright and Cagan (2010) revealed that people pay for products that address their cultural and emotional needs in all types of businesses. This is because; emotion in its various forms is a strong driving factor for a consumer want. Thus, a sustainable user-centred design approach can help to establish an emotive connection and resilience of relationship between product and user (Chapman 2009; Overbeeke and Hekkert 1999). Although collaborations between manufacturers and supplies are becoming increasingly more usual, it is still far from common for the customer to be considered as a fundamental participant in the collaborative design chain (Camarinha-Matos and Afsarmanesh 1999). The different types of customers that will filter the product throughout its life cycle are stakeholders in the process and should be given preferences from the very earliest design stages.

Against this backdrop, this chapter investigates the knowledge regarding how user-centred design approach can be employed to develop emotive products. Furthermore, the chapter elaborates on the three-design level and the resulting appraisals that elicit emotions. Based on this understanding, an experimental study was conducted where two products were designed using computer-aided design (Solidworks software). The 3D CAD models were evaluated to know the emotive response towards the design and the result shows some interesting findings by demonstrating the theories in practice and also reveals that 3D computer-aided design can help to engage potential customers as fundamental participants in the collaborative design chain.

Computer-Aided Design as Creative Tool in Product Development

Improving productivity has been one of the major economic growth that has developed the world; that is, getting more output per worker. The advent of twenty first century has brought alongside the digital revolution where the process of making things has gone digital. The use of computer aided tools has impacted significantly in the execution of technological innovation in almost all spheres of industrial design products development. With the use of computer-aided design (CAD), product ideas can now be expressed all through the whole spectrum of the production process; from design initiation and decision making through to technical design and subsequent link to the machinery ("Computer Aided Design" 2008).

According to Fredriksson (1994), up to 80 per cent (%) of the total cost is often committed in the concept development phase; this is because decisions made at the development stage play a significant role in the final product performance. Atoms (physical prototypes) are weighty, and so are the consequences of their failure (Anderson 2012); this makes the development process technically sophisticated, costly and time-consuming (Kai and Basem 2009).

Therefore, instead of building multiple physical prototypes and then testing them to see if they'll work, companies can conduct testing digitally throughout the process by using Digital Prototyping (CAD) to catch design problems up front. With this, manufacturers can reduce the number of physical prototypes they need to create before a product can be manufactured, which will reduce the cost and time needed for physical prototyping (Aberdeen 2006). As illustrated in Figure 1, CAD can be applied in product development process along three aspects (i.e. implementation, form and quality).

Digital prototyping helps product developers to design, iterate, optimise, validate and visualise their products digitally throughtout the product development process (Autodesk 2013). Companies often adopt digital prototyping with the goal of improving communication between product development stakeholders, getting products to market faster and facilitating product innovation. A digital prototyping workflow involves using a single digital model throughout the design process to bridge the gap that typically exists between workgroups such as industrial design,

Figure 1. Application of CAD in Product development process (Folasayo Enoch Olalere and Aziz 2012)

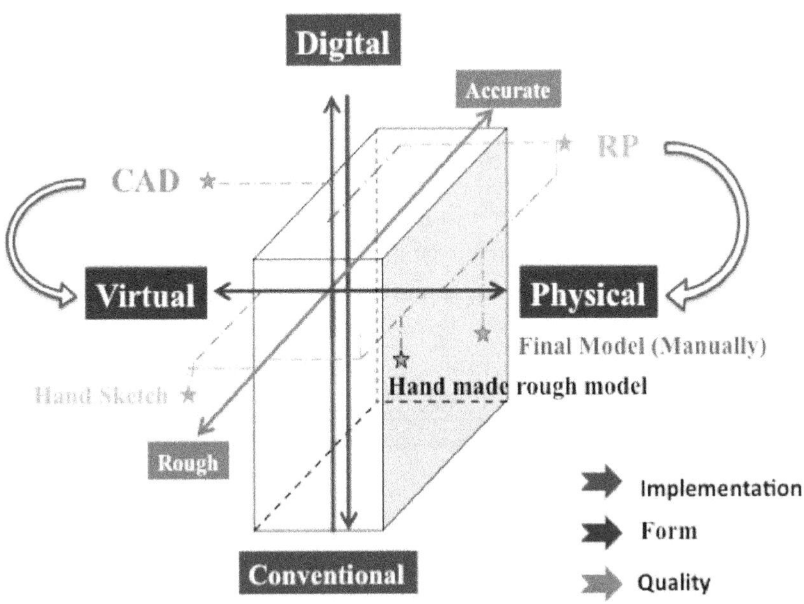

engineering, manufacturing, sales and marketing. Product development can be broken into the following general phases at most manufacturing companies:

- **Conceptual Design:** This phase involves taking customer input or market requirements and data to create a product design. In a digital prototyping workflow, designers work digitally from the very first sketch, throughout the conceptual design phase.
- **Engineering:** In this phase, engineers create the product's 3D model (the digital prototype), integrating design data developed during the conceptual design phase. At this phase, the digital prototype is a fully realistic digital model of the complete product. Engineers test and validate the digital prototype throughout their design process to make the best possible design decision and avoid costly mistakes. Using the digital prototype, engineers can:
    - Perform integrated calculations and stress, deflection and motion simulations to validate designs.
    - Test how moving parts will work and interact.
    - Evaluate different solutions to motion problems.
    - Test how the design functions under real-world constraints.
    - Conducts stress analysis to analyse material selection and displacement.
    - Verify the strength of a part.
- **Manufacturing:** This input helps engineers and manufacturing experts' work together on the digital prototype throughout the design process to ensure that the product can be produced cost effectively. Manufacturing teams can see the product exactly as its intended, and provide input on manufacturability.
- **Customer Involvement:** Rather than waiting for a physical prototype to be completed, companies that use digital prototyping bring customers into the product development process early. They show customers realistic renderings and animations of the product's digital prototype, so they'll know what the product looks like and how it will function. This early customer involvement helps companies get sign-off up front, so they don't waste time designing, engineering and manufacturing a product that doesn't fulfil the customer's expectations.
- **Marketing:** 3D CAD (digital prototype) can help companies to create realistic visualisations, renderings, and animations to market products in prints, on the web, in catalogues, or in television commercials.

Levels of Design

Norman (2004) breaks down emotional responses into three levels of design, which are visceral design, behavioral design and reflective design. According to Luke (2009), the three levels come from within our instincts, visceral, from use, behavioral, reflective and from outside influence & aspiration. Each of the three levels of design is as important as the other; however, they require a different approach by the designer (Norman 2004).

- **Visceral Design:** Visceral design evokes our inner instincts and human drives. It's a foundation level of product emotions. The principles underlying visceral design are wired in, consistent across people and cultures (Norman 2004). According to Luke (2009), when something triggers an emotion at a visceral level, it has an immediate unknowing impact to a person. At a visceral level, physical features (look, feel, etc.) dominate (Norman 2004), and this result in an appealing appraisal that elicits fascinating emotion. According to Desmet & Hekkert (2002), the fascination is a sense of desire emotion called "*desire of presence,*" which is brought about with a focus on the object itself and its visual aesthetic attributes. Therefore, attractiveness is a visceral level phenomenon where the response is entire to the surface look of an object (Norman 2004). For example, when a consumer takes one look at something and says 'I want it', before asking 'what does it do' and 'how much it cost' (Luke 2009). Thus, it can be deduced that the attractiveness/ fascination is as a result of the appealing appraisal, which originates from product concern known as attitude.
- **Behavioral Design:** Behavioral level is entirely about the use and performance of the product; appearance is less relevant (Lin, 2007; Olalere & Ab. Aziz, 2014). However, the appearance in the context of the use is a contributing factor. The behavioral design addresses four main components; function, understandability; usability and physical feel. The function comes first while understandability focuses on establishing a proper conceptual model. However, a product that does what is required and is understandable may still not be usable. Therefore, usability needs to be put into consideration; it is the critical test of a product to know how the product will perform and how comfortable it feels to use. Physical also feels matters because we are biological creatures that interact with our sensory systems and the environment; therefore, they are critical to our behavioral assessment of a product. These four components are the key to a product's usefulness (Lin 2007) that leads to products being appraised as motive compliant, which elicits satisfactory emotion (Desmet & Hekkert, 2002; Desmet, 2012). This is because it will fulfill a goal (product

concern) and will, therefore, be appraised as motive compliant. According to Desmet & Hekkert (2002), the satisfactory emotion elicited is a sense of desire brought about through the *desire of consequence* (that is, an anticipated event of using a product that fulfils a goal).

- **Reflective Design:** Reflective design points to our culture, the meaning of a product or its use (Norman 2004). It's how we see the products reflecting our self-image and aspirations to others (Luke 2009). This is due to products playing an important role in the statements they make to others. According to Luke (2009), the reflective level is similar to Jordan (2002) Ideo-pleasure, which is where one appreciates the aesthetics, or the quality or perhaps the extent to which a product enhance life and respects the environment. However, the reflective level product is about long-term customer experience and interaction. Reflective design elicits admiration emotion; this is because the product matches self-ideal standard and is therefore appraised to be legitimate. Thus, the reflective level product will bring about emotional desire known has 'desire of identity'.

Study

This study designed products (mugs) using CAD (Solidworks) and evaluated the emotive response towards the products. A mug was selected for the study because it is not only a highly ritualistic object used for drinking but also an excellent indicator of our personal and meaningful relationships with objects (Lacey 2009). Besides, the study by Lacey (2009) also shows that mug is a universal and emotive object that penetrates personal experience and transcends gender, background, and circumstance.

The basic concept of Post-modernism also affirms that: "knowledge claims must be set within the conditions of the world today and in the multiple perspectives of class, race, gender and other group affiliations (Creswell 2007; Bloland 1995). Therefore, ceramic mugs were used for the study, since it met the condition of the world today (contemporary needs) and used across class, age, gender, culture and religion. The study includes two stages: concept development and Pre-Production Evaluation (PPE).

- **Concept Development:** At this stage, product ideas were generated using the Don Norman's level of cognitive processing (visceral, behavioural and reflection design). As previously explained, this was applied because it elicits desire emotion (Desmet & Hekkert, 2002), which is the fundamental motivation of all human action ([1]Hobbes, 1588–1679). Ideas were generated into 2D and 3D design (digital prototyping) using Solidworks software (version 2011). Ideas were developed based on two concepts; first, to create functional ceramic mugs that would nudge the user into exploring the object consciously.

To achieve this, the design was conceptualised with features that deviate from existing mugs. According to Desmet & Hekkert (2002), a product that deviates from users' knowledge or expectation is usually appraised as novel/innovative; and this result in an emotion such as pleasant surprise or fascination. Based on this concept, Product 1 was developed, which is a *handle-free* mug with a double-layered wall that acts as a cooling feature when used (Figure 2). The double-layered wall prevents heat transfer by conduction, and this feature is aimed to achieve a behavioural design level that focussed specifically on the function rather than the appearance. This product has been registered with the Intellectual Property Corporation of Malaysia (IP registration number: MY13-00342-0101)

Figure 2. 2D & 3D CAD of the handle free mug

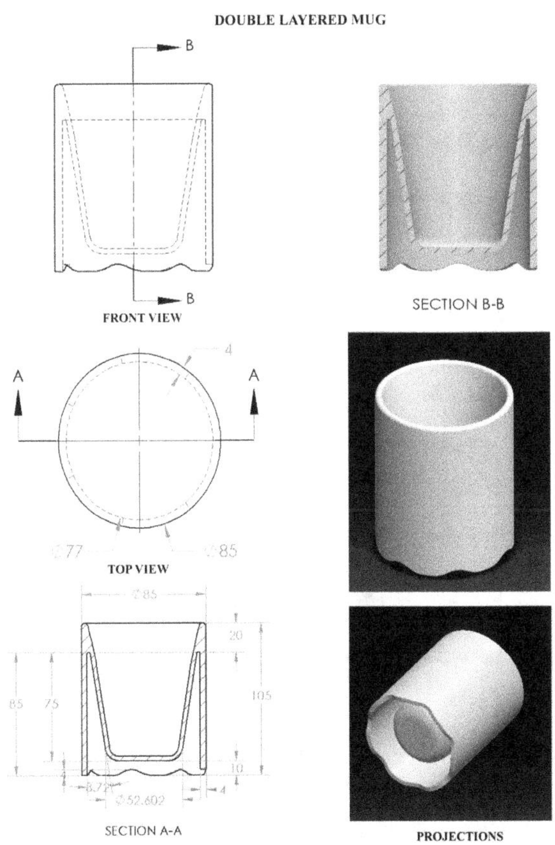

The second concept was to redesign mug as a tool for social conditioning; thus, a *training mug* (Product 2) was conceptualised with novelty feature that makes it only usable when held with the right hand (Figure 3). The mug is designed with a barrier that covers 300° of the mug tip. This design concept is in-line with the sociocultural norms and dining etiquette across Asian, Africa and Middle East (Prakash 2012). The right-hand rule in these communities establishes a sense of hygiene and respect, where people are expected to eat (including holding cup & utensil), shake hands, give and receive with the right hand. Although it is a simple and obvious rule, however, it seems to be very easy to forget and may take a bit of practice. Hence, the training mug is designed to achieve a reflective design level that will build the system overtime to become a way of life. This product has also been registered with the Intellectual Property Corporation of Malaysia (IP registration number: MY13-00340-0101) (Appendix D).

Pre-Production Evaluation (PPE)

The stage evaluates the designs (CAD Models). The products were evaluated based on the emotions from the three level of design features by Norman (2004), which are fascination, satisfaction and admiration (Desmet & Hekkert, 2002). These were

Figure 3. 2D & 3D CAD of the training mug

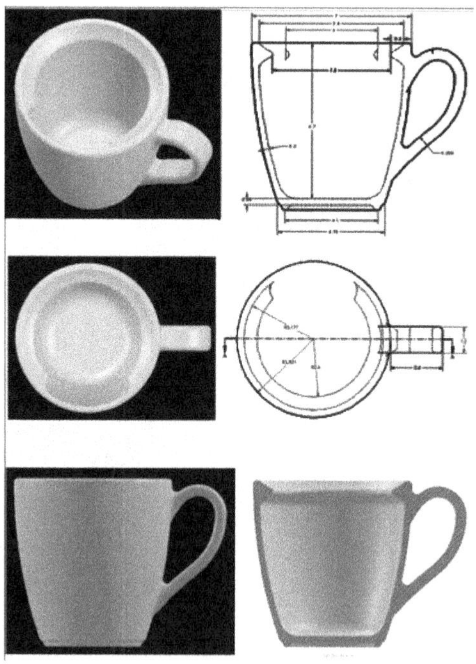

adopted for the evaluation because the product ideas were conceptualised based on these three design features.

The evaluation employed *Semantic Differential* scale (-3 to 3). Semantic differential (originated by Charles Osgood) measure the psychological meanings of objects to an individual (Osgood, Suci, and Tannenbaum 1957). According to Gable & Wolf (1993), it scales people on their responses to pairs of bipolar adjectives in relation to objects or concepts. Semantics differential has been a successful evaluation method used for measuring affective responses in a wide variety of projects (Heise 1970). Thus, this research adopts the method to evaluate the emotive (affective) responses towards the products.

Random sampling technique was used as suggested by Azlan (2013) and respondents were asked to rate their feelings (fascination, satisfaction, admiration) towards each design. The ratings refer to the following; -3, -2, -1 are levels of negative feelings, 0 means neutral while 1, 2, 3 are levels of positive feelings.

The survey questionnaire was designed, developed and administered through Qualtrics online survey software. This was used in order to reach a wider range of respondents (Luke 2009). The survey link was sent through emails to a total of 350 respondents. This was determined using the FPC factor as recommended by S. Smith (2013); where F is the confidence level, P is the percentage of picking a choice (also known as the standard of deviation) and C is the confidence interval.

For the purpose of this survey, the confidence level (F) was estimated to be 90% (Z-score = 1.645); standard of deviation (P) is 0.5 while confidence interval (C) is 5%, as recommended by S. Smith (2013). Thus, the *Necessary Sample Size* (SS) is calculated as follows:

$$SS = [(\text{Z-score})^2 \times P(1-P)] \div (C)^2$$

$$= [(1.645)^2 \times 0.5\,(1 - 0.5) \div (0.05)^2 = 270.6025$$

Sample Size (SS) = 271

Hence, since 271 respondents were needed, the researcher sent the survey to 350 people. Finally, 284 responded; 18.3% of these respondents were people from the arts and design professions, 86.5% of the total respondents are Malaysians while 60.6% of the respondents are age between 25 to 35, 22.5% are between 36 to 55 and 8.5% are above 55 or below 25. IBM SPSS Statistics 20 Software for Windows was used to store, analyse and test the reliability of the data collected. The analysis was presented in tables, bar charts and graphs while descriptive statistical analysis was used to analyse the data as suggested by Lovelace & Brickman (2013) and Marcus-

Roberts & Roberts (1987) that it is appropriate for analysing ordinal responses or rank data.

Result

The result of the evaluation shows that Product 2 elicited the highest emotive response with a mean value of 2.32 for admiration (Figure 4). This can be associated with the reflective design feature of the product, which points to culture and meaning of product usage. That is; the product trained to obey cultural norms and dining etiquette. Hence, since the product matches the self-ideal standard, it is appraised as legitimate and therefore elicits admiration emotion (Desmet and Hekkert 2002). This shows that cultural norms and beliefs have a holistic influence on product perception, as opines by [2]Michael Eyquem De Montaingn (1533-1592) that, "*there are nothing men readily give themselves to than pushing their own beliefs.*' This means that belief drives human's behaviour.

The chart (Figure4) also shows that Product 1 elicited the highest emotive for both fascination and satisfaction with mean value of 2.12 and 1.9 respectively. The visceral level of emotion was because the product deviates from users' knowledge or expectation (i.e., double-wall) and it was appraised as novel/innovative, which result to an emotion such as pleasant surprise or fascination.

Figure 4. The average mean values of visceral, behavioural and reflective responses

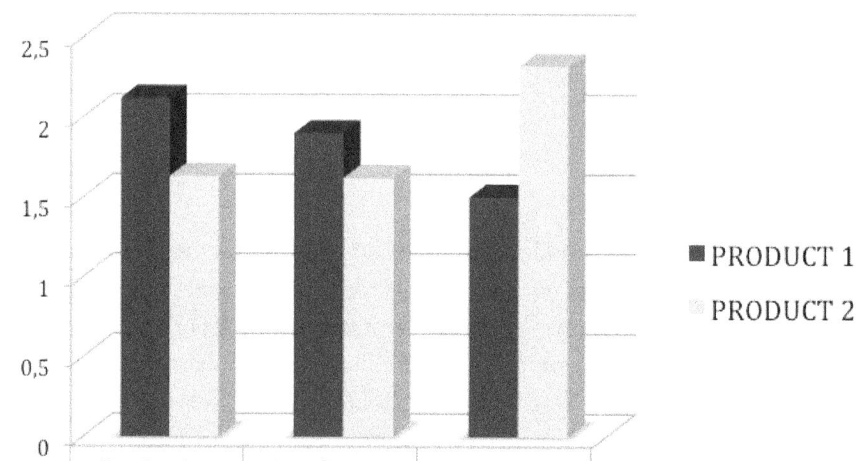

	Fascination	Satisfactory	Admiration
PRODUCT 1	2,12	1,9	1,5
PRODUCT 2	1,63	1,62	2,32

Furthermore, the behavioural level of Product 1 is apparent due to the conditions of satisfactory emotion as explained by Desmet & Hekkert (2002) & Lin (2007) that a product with features that enhance performance is a behavioural design, which is appraised as motive compliant. Therefore, the doubled-wall feature of the mug was appraised as motive compliant, as it enhanced the performance of the mug. Consequently, the participants anticipated the event of using the mug, and this elicits satisfactory emotion.

Based on the thirty-three (33) items in the evaluation, Cronbach's Alpha is used to calculate the reliability of the measures, which recorded a value of 0.767 (Table 1). According to Hair et al., (1998), a Cronbach's Alpha of more than 0.7 indicates that items are homogenous and measuring the same constant. Thus, the Cronbach's Alpha of 0.767 recorded, which is higher than the recommended value as suggested by Hair et al. (1998) indicates that the measures in this study are reliable.

Table 1. Reliability Statistics

Cronbach's Alpha	Cronbach's Alpha Based on Standardised Items	N of Items
.767	.745	6

Figure 5. Rapid prototypes of product 1& 2

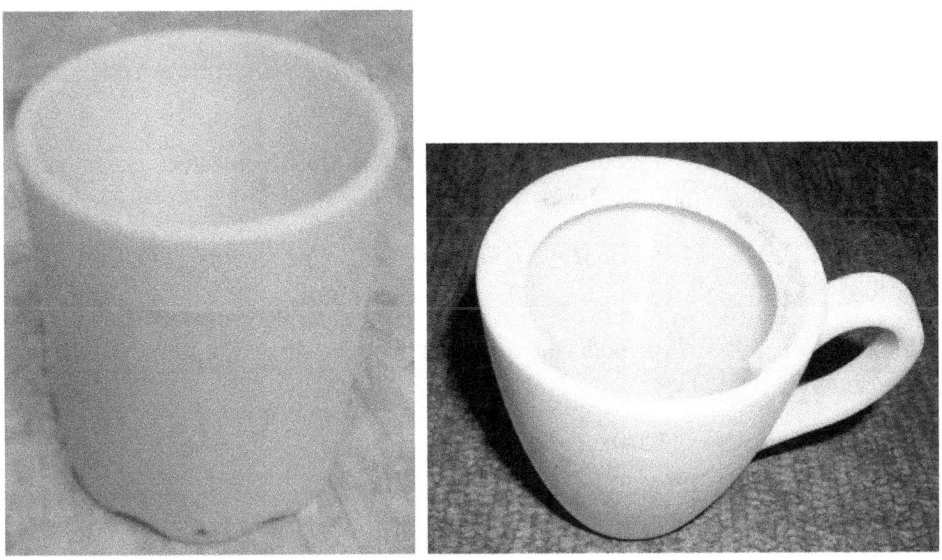

Prototype Development

The physical prototypes were developed using rapid prototyping technology (uPrint 3D Printer). The 3D models created with Solidworks software were saved to STL format and send to the 3D printer to build the prototypes.

CONCLUSION

The process of developing physical products has started to look similar to the process of making digital designs. With the advent of computer-aided design (CAD) tools, 3-D printer, and 3-D scanner; designers can now design 3-D objects digitally and produce them at any scale. Besides, designers can also apply computer aided design models (digital prototypes) to gauge the emotive response of intended user even before producing the physical prototype; explore design alternatives; verify suitability of design and also evaluate design to identify possible faults. The early customer involvement will help to advance the connection of the product with the customers while extending the brand and reputation of the company within current and potential customers. Furthermore, it will help manufacturer avoid production based on misguided vision.

REFERENCES

Aberdeen. (2006, September). The Transition from 2D Drafting to 3D Modelling Benchmark Report. *Design Bulletin*.

Anderson, C. (2012). *Makers: The New Industrial Revolution*. New York: Crown Publishing Group.

Autodesk. (2013). Digital Prototyping. *Autodesk*. Retrieved from http://usa.autodesk.com/digital-prototyping/

Azlan, N. I. (2013). *Research Methodology in Landscape Architecture*. Singapore: Trafford.

Bloland, H. G. (1995). Postmodernism and Higher Education. *The Journal of Higher Education, 66*(5), 521–559. doi:10.2307/2943935

Boatwright & Cagan. (2010). *Product Emotion: The Way to Captive Customers*. Academic Press.

Camarinha-Matos, L. M., & Afsarmanesh, H. (1999). The Virtual Enterprise Concept. *Proceedings of the IFIP TC5WG5.3/PRODNET Working Conference on Infrastructures for Virtual Enterprises: Networking Industrial Enterprises*, 3–4. doi:10.1007/978-0-387-35577-1_1

Chapman, J. (2009). Design for Emotionally Durability. *Design Issues*, *25*(4), 29–35. doi:10.1162/desi.2009.25.4.29

Computer Aided Design. (2008). In *The Columbia Encyclopedia*. Columbia.

Creswell, J. W. (2007). *Qualitative Inquiry and Research Design*. Thousand Oaks, CA: SAGE Publications.

Desmet & Hekkert. (2002). *The Basis of Product Emotions, no. 1988*. Academic Press.

Desmet, P. (2012). *Faces of Product Pleasure: 25 Positive Emotions in Human-Product Interactions*. Academic Press.

Fredriksson, B. (1994, November). Holistic Systems Engineering in Product Development. *The Saab-Scania Griffin*.

Gable, R. K., & Wolf, M. E. (1993). *Instrument Development in the Affective Domain* (2nd ed.). Boston: Kluwer Academic. doi:10.1007/978-94-011-1400-4

Hair, J. F., Anderson, R. E., Tatham, R. L., & Black, W. C. (1998). *Multivariate Data Analysis* (5th ed.). Prentice-Hall International.

Heise, D. R. (1970). The Semantic Differential and Attitude Research. In G. F. Summers (Ed.), Attitude Measurement (pp. 235–253). Chicago: Rand McNally. Retrieved from http://www.indiana.edu/~socpsy/papers/AttMeasure/attitude.htm

Jordan, P. (2002). *Designing Pleasurable Products: An Introduction to the New Human Factors*. London: Taylor & Francis.

Kai, Y., & El-Haik Basem, S. (2009). Design for Six Sigma: A Roadmap for Product Development (2nd ed.). McGraw-Hill.

Lacey, E. (2009). *Contemporary Ceramic Design for Meaningful Interaction*. Academic Press.

Lin, R. T. (2007). Transforming Taiwan Aboriginal Cultural Features into Modern Product Design: A Case Study of a Cross- Cultural Product Design Model. *International Journal of Design*, *1*(2), 45–53.

Lovelace, M., & Brickman, P. (2013). Best Practices for Measuring Students' Attitudes toward Learning Science. *CBE Life Sciences Education, 12*(4), 606–617. doi:10.1187/cbe.12-11-0197 PMID:24297288

Luke, W. (2009). *The Structure, Influence and Application of Emotion in Product Design*. Leicester, UK: De Montfort University.

Marcus-Roberts, H. M., & Roberts, F. S. (1987). Meaningless Statistics. *Journal of Educational Statistics, 12*(4), 383–394. doi:10.2307/1165056

Norman, D. (2004). *Emotion Design: Why We Love (or Hate) Everyday Things*. New York, NY: Basic Books.

Olalere, F. E., & Ab, S. A. (2014). Computer-Aided Ceramic Design: Its Viability for Building User-Centered Design. In Contemporary Advancements in Information Technology Development in Dynamic Environments. Hershey, PA: IGI Global. doi:10.4018/978-1-4666-6252-0

Olalere, F. E., & Ab, S. B. A. (2012). The Use of Computer Aided Tools for Prototyping in Traditional Ceramics Production in Malaysia. *Proceedings of the 1st International Conference on Design and Innovation, ICDI2012*.

Osgood, C. E., Suci, G. J., & Tannenbaum, P. H. (1957). *The Measurement of Meaning*. University of Illinois Press.

Overbeeke, C. J., & Hekkert, P. (1999). Editorial. In *Proceedings of the First International Conference on Design and Emotion*. Delft: Department of Industrial Design.

Prakash, A. (2012). The Rules for Eating with Your Hands in India, Africa and the Middle East. *Food Republic*. Retrieved from http://www.foodrepublic.com/2012/11/19/the-rules-for-eating-with-your-hands-in-india-africa-and-the-middle-east/

Smith, S. (2013). *Determining Sample Size: How to Ensure You Get the Correct Sample Size*. Retrieved from https://www.qualtrics.com/blog/determining-sample-size/

ENDNOTES

1 Thomas Hobbes is an English Philosopher and the founding father of modern political philosophy. He is best known for his political thought and vision of the world, which is original and still relevant to contemporary politics.

2 Michael Eyquem De Montaingn (1533-1592) is a French Philosopher, also known as the great French Renaissance thinker.

Chapter 6
Playing With Text in Space

Elif Ayiter
Sabanci University, Turkey

ABSTRACT

This chapter delves into the creation of text and its visualization through typographic design inside three-dimensional virtual worlds known as the metaverse, with a particular focus upon the way in which these environments may place the usage of text within a context that stands in contradiction to its traditional one – namely legibility. Readability can be displaced through the usage of text and typography as playful devices that may manifest as visual structures, the contents of which are meant to be understood through means other than reading – as it is understood in the traditional sense of the word. Instead, such spaces are aimed to bring about states of heightened engagement, wonder, and "play" by being immersed within the typographic conglomerations that constitute their essences. This subject is explored through the discussion of three such typographic playgrounds that the authors created in the metaverse.

INTRODUCTION

Metaverse are three dimensional, collective, online virtual worlds in which (unlike their gaming counterparts) all content is user-created. A further name by which these worlds can therefore be called is the term 'builders' worlds', since an important attribute of these spaces is that not only is their content user-defined and created, but also the purpose of residing in these worlds is entirely up to and decided upon by their users, since the developers of these platforms provide no narrative that is to be followed, and there are no system defined goals or quests. Instead, metaverse residents are expected to formulate the raison d'être for their virtual sojourn out of

DOI: 10.4018/978-1-5225-5023-5.ch006

a self-motivated inquiry. (Boellstorff 2008) From this it may follow that one of the most widely embraced reasons to stay in the metaverse is to be creatively active: These worlds possess attributes that are grounded in creative endeavor to the point where the best developed of them to date, namely Second Life®, has been defined as "a wildly provocative experiment in user generated content." (Cervieri 2007)

Creative activity stands center stage from the onset, from the moment that metaverse land has been appropriated; initiating itself through the very shaping of the virtual geography, upon which virtual architecture is then built, to be traversed with many different kinds of vehicles, ranging from space craft to realistic replicas of Real Life carriers, such as cars, helicopters and sea craft. An important note is that all of these artifacts as well as the geography on which they are placed are essential constituent elements of an elaborate system of 'play,' that has been remarked upon and discussed by Brown and Bell, who researched play and sociability in online virtual worlds through an in-depth examination of 'There,' a collaborative online virtual world that can be seen as the partial precursor of the concept of the metaverse. Their findings point at strong ties between sociability, objects and play states; saying that play and sociability rely upon a number of interactional 'building blocks' for their satisfactory fruition. While talk, topic and identity all work together to make 'play' possible, an equally essential element is the interaction that comes about through objects. (Brown and Bell 2006) Such objects can manifest in many forms and address many different needs and typographic objects should certainly be counted amongst them.

In my 7 years in the metaverse I have engaged in many of the creative activities listed above. In many of my projects typography has held a center-stage position – which is of no surprise, since as a graphic designer I have had a life-long love relationship with the display of semantic material. It was therefore a foregone conclusion that I would carry my fascination with typography into my virtual three dimensional explorations as well – albeit, embedded into an altogether different format than the intrinsic one that holds typography as an informational device through which the content of the written word is to be eloquently expressed. What I have chosen to do instead is to place type and text as frivolous, oftentimes nonsensical game objects into my constructions. The following are some of the deliberations and inspirations through which I contextualize text and typography as playful artifacts in the metaverse, leading into a brief discussion of a recent work that was built under these precepts.

VIRTUAL 3D AND TEXT

In her book 'The Visible Word: Experimental Typography and Modern Art, 1909–1923' Johanna Drucker offers a model for the materiality of visual-linguistic signs that looks forward into the future through what she calls the "nearly proto-electronic and cybernetic sensibility of F. T. Marinetti," who seems to have heralded what Richard Lanham has called "the complete renegotiation of the alphabet/icon ratio inherent in desktop publishing." (Lanham 1993) This extends itself to the same questions of subjectivity that are problematized by virtual reality. Insofar as it brings "visual presence" to meet "literary absence," this notion of materiality is supported by a theoretical model which contains certain internal and irresolvable contradictions: Johanna Drucker notes that the typographically rendered page is an image while at the same time it is also language. Furthermore the reader is also a voyeur, a viewer, or a 'screener.' These simultaneities operate equally for the production of visual pattern and semantics – both are integral to signification, and both inform Drucker's "materiality of interpretation," which is a potent model for a digital visual poetics, whose object is not merely 'text' – and especially never so when the text comes about as a conglomeration of three-dimensional typographical objects in the quadri-dimensional hermeneutic space of an electronic visual simulation. (Drucker 1994)

When it comes to the screen based viewing in which a 3D environment is displayed on a flat surface, working with text in 3D has noteworthy issues that are related to legibility, which according to Kevin Lynch can be defined as the ease with which parts of a whole can be recognized and be organized into a coherent pattern. "Just as this printed page, if it is legible, it can be visually grasped as a related pattern of recognizable symbols." (Lynch 1960)

The Virtual Shakespeare Project held the entire text of 'A Midsummer Night's Dream,' to be traversed and read inside a 3D environment that was displayed on a 2D screen. As such, the project is possibly one of the most noteworthy experiments conducted in this area, especially when it comes to working with large amounts of text. The project was developed during the mid-1990s, jointly between IBM and the MIT Media Lab; and has been deliberated upon extensively by one of its collaborators, David Small, who starts his discussion of the project by saying that "by escaping the confines of the flat sheet of paper, we can arrange information into meaningful landscapes that exhibit qualities of mystery, continuity, and visual delight," (Small 1996) after quoting Stephen and Rachel Kaplan wrote about the experience of mystery in landscapes in their book Cognition and Environment: Functioning in an Uncertain World:

In the case of mystery, the new information is not present; it is only suggested or implied. Rather than being sudden, there is a strong element of continuity: the bend in the road, the brightly lighted field seen through a screen of foliage—these settings imply that new information will be continuous with, and related to, that which has gone before. Given this continuity one can usually think of several alternative hypotheses as to what one might discover. (Kaplan and Kaplan 1982)

Small has drawn attention to a number of design issues, including the use of perspective, scale, space, and interaction to create simple and flexible visualizations of information. One of the foremost challenges when it comes to 3D virtual text is the fact that traditionally letters have been designed to be viewed directly on a flat 2D surface. When it comes to screen-based 3D however, there are an uncounted number of uncontrollable viewpoints, through which a continuous perspective distortion that is less than ideal for reading is created. Thus, it is not always possible to guarantee the angle of the view relative to the angle of the text and one cannot be certain of maintaining the integrity of the letterform in screen based 3D environments: Each new angle will result in a differently shaped letter and at extreme angles the image can even be reduced to a line. Furthermore, when the camera moves behind the text, what is to be read looks reversed as though seen in a mirror. While certain word shapes can still be recognized in less than ideal circumstances, in general there are few viewpoints from which text holds its legibility.

A further issue in terms of legibility also addresses size:

A graphic designer can use size differences to visually distinguish certain elements in a text. In a 3D space however, you cannot always resolve the relative size of two objects. If one object appears smaller in the picture plane, it could actually be smaller, or it could be the same size and farther away, or it could even be much larger and very far away. So, in the design of an information space, one must be careful about using size as a differentiating variable. (Small 1996)

Small's findings from the Virtual Shakespeare Project lead to the conclusion that attempting to engage in creative activity involving text as informational content in a screen-based three-dimensional environment is not an easy option. Equally compelling however is an understanding of the metaverse as a playful environment that resides upon tenets which are closely related to make-belief, indeed to absurdity, to the solving of puzzles. All of these considerations have impelled a need for a textual strategy that may fulfill the demands of 'play,' make room for personal readings and interpretations: Text that is meant to be 'felt' as 'play'/artwork, rather than to be 'read' as informational content. This has led to an examination of concrete poetry, especially as it relates to digital poetry.

SHAPE POETRY

Concrete poetry, also sometimes referred to as visual poetry or shape poetry, is a literary artifact in which visual attributes are as important as the semantic content delivered in the plain text itself. This prominence is achieved through giving typographical arrangements a major role in conveying the intended effect, often bypassing the meaning of words, of rhythm, and rhyme and so on. Concerns with space and surface are deemed to be major features of concrete poetry – it is through these that the relationships between linguistic elements and their spatial interactions gain structural and meaningful significance.

According to Rebecca Sharp, concrete poetry by necessity involves a reduction of language which is dismantled in terms of form, design and function. Significant also is the new relation to space; no longer an impotent backdrop against which text must rigidly stand, the new writing breaks out onto the entire surface of the page, allowing the reader/viewer to form personal patterns and make new, independent inferences. "In so undermining the security between word and page, concrete poetry also dislocated word from meaning and furthermore, destabilized the traditional relationship between the text and the reader." (Sharp et al. 2007)

Thus, one of the most significant features of concrete poetry is that in order for meaning to be communicated, the reader must actively engage in the text by deciphering its layout and typography, by interpreting its shapes and spaces, viewing the work as an object and thereby playing an active role in its creation – to acknowledge what Gomringer called the presence of a visio-linguistic "Eigenwelt" in which concrete poems are no longer "about something, but instead are autonomous realities in themselves." (Gomringer 1968)

While Gomringer's visio-linguistic 'Eigenwelt' was at the core of the work and the musings of pre-digital era concrete poets, the advent of computation added unforeseen opportunities which were integrated as almost a natural continuation of a tradition that had its origins in neo avant-garde experimentation. Friedrich W. Block sees contemporary digital poetry as an extension of avant-garde and concrete poetry, pointing out that almost all vital concerns of digital poetry – such as the spatio-visual arrangement of text, multilinearity, and indeed even movement and interactivity – can be traced back to its historical predecessors in which these were as much key concerns as they are in digital poetry. (Block 2009)

Lori Emerson says that digital poetry attempts to exploit the medium of the word to represent our desire of a full experience through language as a form of life – in the case of digital poetics this also comes about through movement, generation, interactivity. What is fundamental in the perceptual shift brought on by the digital is that the digital realm offers us the opportunity to conceive of space in different or expanded terms than that of paper-based writing; and, further, this sense of space requires that we come up with a different set of terms for the interpretation of certain digital texts. (Emerson 2006)

Especially compelling, when considered in relationship to digital poetry that has been generated in a virtual world where such displays are lived and experienced through avatars, Emerson also quotes Steven Wallace who has said – at a time much before the onset of virtually embodied lives – that the subject matter of poetry is not that

Collection of solid, static objects extended in space but the life that is lived in the scene that it composes. Reality is not that external scene but the life that is lived in it, is things as they are. The word proliferates its special senses within such lived-in-realities. It is a jungle in itself. (Stevens 1965)

The question to ask then, is how we are to anticipate meaning when a poem is based upon a conception of space as a "multiple, variable, and vibrant," plane; one where the literal ground is ever shifting and heterogeneous? Or, put in another way, what if the milieu upon which the poem is built is not, as Brian Rotman puts it, "an ideal realm, untouched by change, independent of energy and matter, beyond the confines and necessities of space and time," (Rotman 2000) but instead materializes as a mathematically determined, artificially fluid space – as an online virtual builder's world inherently is – in persistent flux, both through the very nature of the frenetic building activity therein; and as importantly, through the further artistic appropriations of consumer-avatars who joyously engage in manipulations of existent artworks, creating self-perpetuating chains of transformative creative activity.

CONSTRAINED PLAY: 5555555 55555 555

The project that provides the first example to this discussion came about when artist/curator avatar Giovanna Cerise invited me to contribute to her collaborative Second Life project revolving around numbers. The framework set by Cerise was that participants were asked to pick a number between 0 and 9 and everything that was to be built had to be made out of that single digit. Given my preoccupation with typography in three dimensional spaces that is fed by my interest in concrete poetry, I gave a most enthusiastic 'yes' as a response to Cerise's invite and built an architecture made only out of 5s which was exhibited as part of her show in Second Life in the Fall of 2013.

What was especially appealing about Cerise's brief was the constraint that it placed upon the builder since an easy correspondence for it can be found in constrained writing, a technique which binds the writer by some condition that forbids certain things or imposes a pattern. Constrained writing can be said to contain a sense of play as is evidenced in works such as 'Gadsby,' an English-language novel consisting of 50,100 words, none of which contain the letter 'e'; the 2004 French novel 'Le Train de Nulle Part' by Michel Thaler which was written entirely without verbs; or Christian Bök's 'Eunoia' – a univocalic poetic treatise that uses only one vowel in

Figure 1. "5555555 55555 555." Typographic Architecture, Elif Ayiter, Second Life, 2013. Photograph by Elif Ayiter.

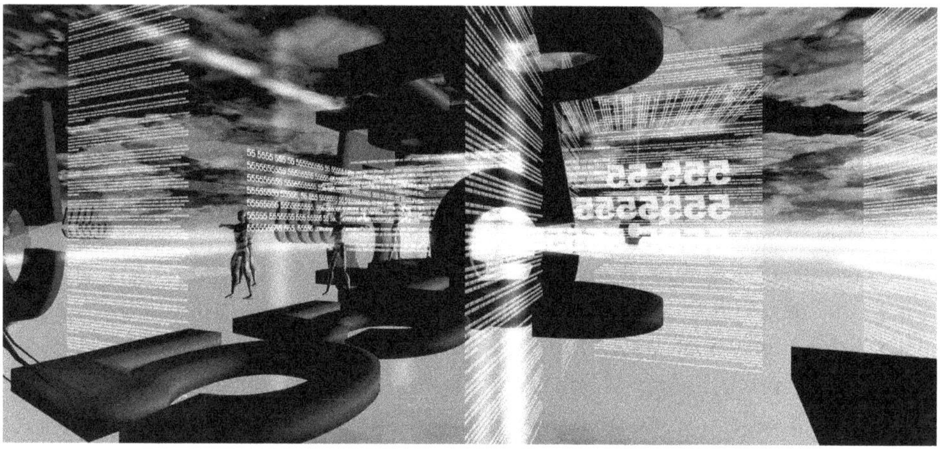

each of its five chapters. Oulipo – a French literary group dedicated to the practice – defines the term as "the seeking of new structures and patterns which may be used by writers in any way they enjoy." (Bogost 2012) Constraints, within this context, are used as a means for triggering ideas and inspiration, most notable amongst them being Georges Perec's 'story-making machines/devices,' as well as techniques such as lipograms, palindromes, and restrictions based upon mathematical formulae.

My professional specialization as a graphic designer is layout. Accordingly, the space that I built might be called a three-dimensional magazine spread of sorts, complete with body text, subheadings and headlines. I wondered if text composed of a single glyph would still give a sense of 'content' when placed in such a way, in a logic that followed the typographic hierarchies of a standard page layout. In order to make this conglomeration of 5s resemble as closely as it could to a text that is comprised of real words, I replaced all the letters of a lengthy piece of writing with the glyph 5, using the 'find and replace' feature of a standard word processing software. Headlines and subheads were generated in similar fashion, and these, together with the body text, became the components of an architecture that came about through a constraint. The typographic hierarchy was achieved by placing the body text on planes and by building the heading and the subheadings as three dimensional 5s that were clustered together to form words, around which the body text walls of the architecture were grouped – very much in the manner that a layout designer arranges elements of text on a page by using techniques such as negative space, clustering, proximity and directional axes. These 'pages' were then set at right angles to each other to form the rooms of a building that could be and inhabited and played in by avatars.

However, the project took on a life of its own after I completed the building, when I started to think about how avatars would behave if they could only talk through a text which was strung out of 5s. What sort of a world would that be? Would a restraint like this provoke a far more visual type of communication since equal-valued strings (equal size, weight, placement, etc) of a single glyph would not suffice to express what one wished to bring across? Would, in other words, concrete poetry emerge out of such a limitation? Would text inevitably become a deconstructed artifact whose visual attributes compensated for its semantic absences if our utterances were to be placed under such a constraint?

To investigate I sent three members of my coterie of avatars into the architecture and situated them in an encounter where their exchanges had to be decipherable (even if what they actually said was not) in order for a tangible storyline to develop. The tale is very simple: A lonesome avatar appropriates a strange world where one can only talk in 5s. But then she gets company – which, at first, she rejects since she does not want to share what she holds to be solely her own. But then she is talked into playing with the incomers. Has a good time. However, just at the height of the fun and games “they” turn around and go. Leaving her bereft.

Although 5555555 55555 555 was initiated as an architecture that could be immersively experienced in a virtual world, it evolved into a project of many parts for which ‘frozen moments in time’ of the original three dimensional manifestation were taken back to the two dimensional desktop, where they were further manipulated through image editing software as well as other applications: I took many virtual photographs of this storyline to which I added two dimensional speech bubbles (made out of strings of 5 to resemble words) as a further layer. These textual snippets were deconstructed in such a way that they would express some sort of ‘emotion’ – to make up for the circumstance that conveying actual ‘meaning’ was beyond their capabilities. Out of these typographically enhanced screenshots were made several displays of the tale since an important part of the playful experiment was too see whether I could narrate the same events through different means: A storyboard made with an online infinite grid, a virtual flipbook and a 2.5D application that lets viewers follow the tale through a zooming user interface in which some scant clues regarding the progression of events are also provided.

Thus, it has to be acknowledged that my interest in 5555555 55555 555 has transitioned from three dimensionality to the flat surface of the screen, materializing as an ongoing online project – what can in some ways be called a simulation of a simulation.

URANOMETRIA

Uranometria was an installation that I was commissioned to do as part of a major group exhibit that took place in the Second Life region Aire Ville Spatiale, sponsored by the French National Costume Museum in Moulins, France and curated by Marc Blieux. The exhibition took place in 2011 and was developed around the concept of synchronicity, which also gave the show its name: SynchroniCity.

The term synchronicity brings us back into the jurisdiction of the aleatoric, since the concept resides upon the experience of two or more events, that are apparently causally unrelated, and are yet observed to occur together in a meaningful manner whereby just as events may be grouped by cause, they may also be grouped by meanings that are structured in their own logical way and thus give rise to relationships that are not causal in nature. These relationships can manifest themselves as simultaneous occurrences that are related to one another through their meanings.

In honor of Carl Jung and his concept of synchronicity, Uranometria brought together two unrelated subjects: Excerpts from Jung's introduction to Richard Wilhelm's translation of the I Ching (Jung 1951), and the astronomical maps from Johan Bayer's Uranometria, which is the first Western celestial atlas, created in 1603. Although at the time of the atlas's creation the Age of the Enlightenment was still

Figure 2. "Uranometria." Typographic Architecture, Elif Ayiter, Second Life, 2011. Photograph by Elif Ayiter.

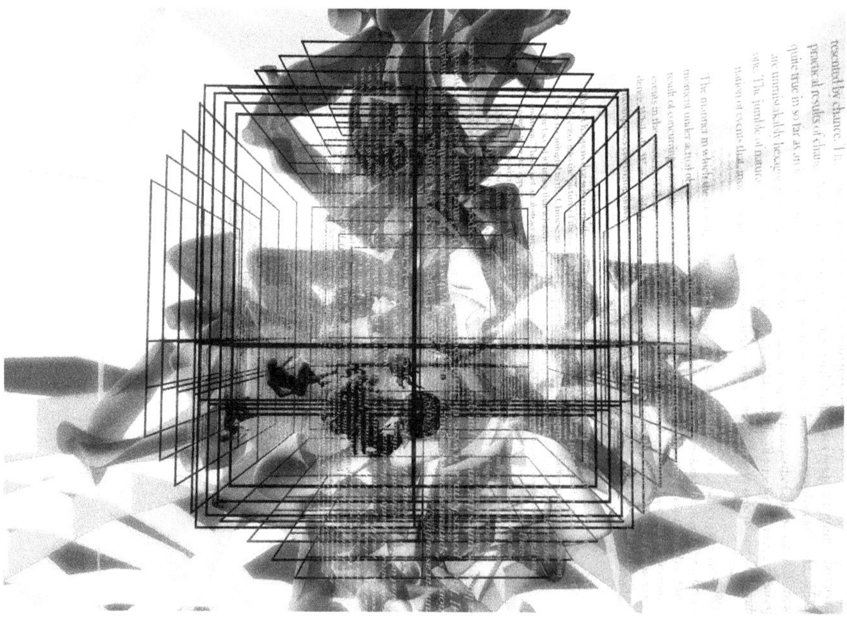

centuries away, it would be safe to say that Uranometria was an early harbinger of it, presaging a mindset that was grounded in causality and reason. This also led me to thoughts on the entire nature of western astrology with its mechanistic, clockwork universe wherein the rigidly structured and predetermined sojourn of the planets chart out seemingly immutable patterns of influence upon human life – a world view that is so very different from the fluid, ever changing universe of the I Ching which instigated Jung to coin the term 'synchronicity.'

Since I cannot read Chinese Calligraphy, and since my knowledge of Chinese Cultural artifacts is very limited in general, it did not seem appropriate to use visual elements that would refer to the I Ching by setting up associations through visual signifiers that would imply affiliations to the 'Orient.' What was important was to provide the notion of fluidity, of change, of acausal connections that lies at the essence of Jung's foreword. I decided to use this text itself – however in such a way that it's meaning would be revealed, would be 'felt rather than read' through typographic deconstruction.

A typographic work can be called 'deconstructed' when it exposes and transforms the established rules of writing, interrupting the sacred 'inside' of content with the profane 'outside' of form. Ellen Lupton sees deconstruction in graphic design as a process - an act of questioning typographic practice itself. In Derrida's original theory deconstruction asks several questions that are crucial to typographic design as well: How does representation inhabit reality? How does the external appearance of a thing get inside its internal essence? How does the surface get under the skin? (Lupton 1994)

Consequently, deconstructionist typography exhibits a fascination with appropriation, juxtaposition, détournement, montage and collage, as well as repetition, all of which are facilitated by or reflect upon the capabilities of digital technologies in these regards. It is thus of no surprise that the outcome oftentimes resonates upon a world of diffused and distributed communication mediated through networks of powerful information technologies. Even when the artifact itself is presented as a static printed page the reference to a cyberspace driven by hypertext is very often implicit, underscoring that "communication for the deconstructivist is no longer linear, but involves instead the provision of many entry and exit points for the increasingly over-stimulated reader." (Cahalan 1994) Thus the page is no longer to be just 'read' but also to be 'perceived', beyond the pure textual content, into all of its associative conjunctions: We are meant to 'feel' rather than just to 'read' a page.

In Uranometria deconstruction was achieved through two means: Layering and motion. The text was placed upon huge transparent sheets that were tethered to a cage and were in perpetual motion due to the wind factor of the metaverse. Long excerpts from Jung's introduction were typeset with the font Garamond and as such no deconstruction would have been in evidence since the typesetting itself was undertaken in a straightforward manner – same point size, same line spacing, left alignment throughout, set in solid white, albeit upon a transparent background. It was this element of transparency, together with motion, that deconstructed the text since this brought together and superimposed different parts of Jung's words as the sheets swung to and fro. The entire typographic system remained in a state of constant flux since the wind saw to it that novel alliances and configurations between the text would be coming about constantly – nothing was fixed, the flow remained in continuous change, effectively deconstructing meaning and causality through wind: 'Deconstruction through wind' was seen as a befitting symbology of the entire notion of synchronicity and the I Ching, i.e., the Book of Changes.

In contrast to this was the cage itself to which the text flow was bound. This cage and its inhabitants represented the flip side of the coin – causality, reason and determinism – the fundamental tenets of the Western Age of the Enlightenment. To further accentuate this contrast the construct also held a number of petrified avatars who were dressed up as Uranometric zodiac signs, held captive by their star bound fate which was represented by the similarly attired and caged humanoid objects that revolved around them. In their vicinity however Jung's words floated freely, tossed hither and thither by virtual winds.

THE ASEMIC AVATAR

A standard practice in graphic design, and one that most designers cherish due to the freedom that it gives the design process, is the usage of so-called meaningless text during preliminary phases of page layouts that stands in lieu for the actual text that it is meant to be replaced with at later stages. The text most commonly used for such purposes is the famed 'Lorem Ipsum' compilation that has been the typesetting industry's standard dummy text ever since the 1500s, when an unknown printer took a galley of type and scrambled it to make a type specimen book. Although used as a placeholder Lorem Ipsum is not random text. Instead, it has its roots in classical Latin literature, making it over 2000 years old. Richard McClintock, a Latin professor at Hampden-Sydney College in Virginia, looked up one of the more obscure Latin words, 'consectetur,' from a Lorem Ipsum passage, and going through the cites of the word in classical literature, discovered the source: Lorem Ipsum comes from 'de Finibus Bonorum et Malorum' (The Extremes of Good and Evil) by Cicero, written in 45 BC.

The usage of the 'Lorem Ipsum' aggregate as a placeholder for 'real' text resides in the notion that since the dummy text is embedded into an overall design system in which it is complemented through images, shapes and color, the viewer will be able to assign meaning through the associations which the surrounding visual

Figure 3. "Asemia." Typographic Environment. Elif Ayiter, Second Life, 2012. Photograph by Elif Ayiter.

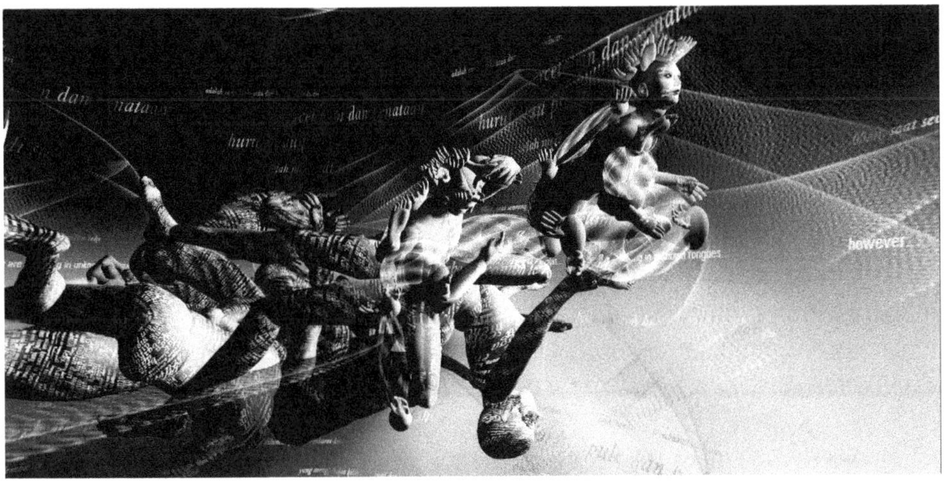

material provides to what essentially makes no sense when read as a standalone piece of writing.

I wished to take this notion into 3D, however with text that did not even have the association to 'Lorem Ipsum,' in other words by using text that was truly meaningless. To this end I compiled by own version of Lorem Ipsum by generating text in several of the less widely read languages which I then combined and further garbled up and in a word processing software. The result led me to the notion of asemic writing which is a wordless form of writing, deriving from the Greek word 'asemic;' that can be translated as 'having no specific semantic content'. Through this nonspecificity comes a vacuum of meaning which is left for the reader to fill in and to interpret since asemic writing has no inherent verbal sense. Nonetheless, asemic writing artifacts may have a clear textual sense which is brought about through formatting and structure, and as such they may visually suggest 'real' documents. Drawing meaning from these depend upon the reader's overall sense and knowledge of writing systems, or alternatively such meaning can also be understood or felt through sole aesthetic intuition.

A wide spread manifestation of asemic writing seems to materialize not through typographic means, but instead as a hybrid form of output that resides somewhere between drawing, doodles, signs or icons which are arranged and aligned in such a way that they resemble text, such as seen in the work of poets Henri Michaux and Christian Dotremont which is discussed by Nina Parish (Parish 2008). A very good visual collection of this genre can also be found in Michael Jacobson blog 'thenewpostliterate', which explores asemic writing in relation to post-literate culture.

Although these artistic explorations which are engaged in creating literature out of visual form are highly compelling; my personal preoccupations, which ultimately reside in my background in graphic design, lead me in a direction in which I do not wish to forsake the aesthetic refinement which I find in typographic forms as well as in the overall structural beauty of textual output. I wished take my quest in a different direction by bringing together the concept of asemic writing and that of aleatoric poetry, which resides upon the chance encounters of words. Within this context Christian Bök, refers to Gilles Deleuze by saying that

Writing by means of an aleatory protocol almost fulfills the dream of Deleuze, who imagines an ideal game of chance, one whose rules are themselves subject repeatedly to chance, resulting in an aimless outcome so futile that we have no choice but to dismiss the game as a nonsensical dissipation of time itself. (Bök 2006)

However, Bök takes the notion of chance altogether beyond nonsense when he further quotes Deleuze who tells us that "if one tries to play this game other than in thought, nothing happens, and if one tries to produce a result other than the work of art, nothing is produced" (ibid).

The work came about as a huge sphere and a landscape complete with inhabitants that resided in this sphere. The work was exhibited in Second Life in the Spring of 2012, as part of a collaborative art project sponsored by the Linden Endowment for the Arts.

Almost all of the elements of the habitat, as well as its dwellers were carriers of 'text,' however text that was devoid of meaning, that did not resemble any known language and that repeated itself throughout, mapping itself equally on 'living' beings as well as the many solid rocks that were scattered throughout the space and that these beings were intertwined with and/or embedded into. The only exception to this asemic assemblage was a single line of text that wound its way around the center of the sphere and that could be read as 'alas, they are speaking in unknown tongues.'

The environment was grouped around a central tree, whose foliage was comprised of the same asemic text. Deconstruction, in the case of this installation, was not achieved through motion, transparencies, layering or any other such visual means, but instead the confusion was achieved through the very semantic absence itself. It was hoped that this lack of semantic content would enable the visitor to draw their own meaning by taking clues from what surrounded them, and how the text was used in conjunction with other visual elements. Yet another addition was a custom designed soundscape, consisting of human sounds such as screams, grunts, sighs, and cries – all of which were devoid of words and meaning, but instead conveyed emotions such as being bound, being helpless, being petrified within a fate.

Although the Asemic Avatar is quite different from its precursor Uranometria, both visually as well as through the presence of this soundscape; nevertheless there are strong conceptual overlaps; to the point where it could be said that they are sister projects: In both locations what I aimed at was to bring about a sense of immutable fate and claustrophobia. In the case of Uranometria this was represented through the petrified avatar sculptures that could not join the free floating words of Jung but were confined to living inside the cage, in fixed positions and poses. These petrified sculptures made a re-appearance inside the Asemic sphere, however this time as the main protagonists of the piece, heaped in masses all over the floor. In both installations visitors could join the fate of these solidified avatar sculptures by donning the attire that was provided at both locations, thus increasing the sense of presence, immersion, identification and empathy that both domains called for.

CONCLUSION

In the end, is 'play with text' a common occurrence in the metaverse? Sadly, it isn't. As previously said, most of the text that one encounters in virtual worlds will be informational, i.e. will either be signage, packaging/display or advertising – just as it is in Real Life. Mapped onto clean 2D surfaces no less – just as it is in Real Life.

My guess is that the reason for such information based usage is much the same as the one in Real Life, which, at the end of the day is nothing but yet another 3D world in which playing with text objects that obey the laws of gravity, that float with the wind, that are layered onto one another as transparencies is just as difficult – if relaying information is the primary objective. And, information seems to be a mindset as pervasive in virtual worlds as it is in RL, when it comes to text.

What this compels me to ask is whether this isn't rather unexpected? I believe that it is unexpected – indeed startlingly so: No matter what the state of affairs may be in a Real Life that is abundant with 'seriousness,' when it comes to the metaverse one would expect that in a 'world of play' textual objects would have to succumb to 'pure play' as pervasively as all other artifacts seem to do so most enthusiastically. So, why do they not?

Could it be because of all the associations to 'seriousness' that text may carry intrinsically? After all, don't we have to actually sit down and 'learn' text when we first go to school – at that sad little moment when 'pure play' must slowly give way to 'serious play' as we embark upon the long road to adulthood.

Could it be that these connotations are overwhelming enough to override 'play' when it comes to text, regardless of where we are?

Yet another query when thinking about typography in the virtual realm may reside in Nelson Goodman's taxonomy of creative output whereby he distinguishes between what he calls 'autographic' works (such as painting, drawing and calligraphy), in the case of which there exists only one original artifact that is based upon intrinsic hand strokes and as such is hard to replicate; and 'allographic' artifacts in which a symbol system carries the work and multiple instances of the original work are therefore possible. The second category involves abstract media (such as sound, speech, performance etc) and the route to capturing such states of abstraction is to incorporate formal notation. (Goodman 1976) Taking Goodman's definitions into the digital realm, McCullough notes that computing is an abstract medium that is based upon symbol notation, and the modus operandi of the computer applies to all output generated through computational means, regardless of whether this output appears to be gestural or is evidently based upon a symbol schema (such as musical notes or letters). Thus, the digital environment renders (previously) autographic media through allographic means, in effect transforming 'things' into abstractions. (McCullough 1996)

McCullough's application of Goodman's definitions are particularly compelling when it comes to text and typography since by its very nature and function text essentially visualizes the spoken word through means which almost appear to me to be transfer stations upon a complex sojourn: This is initiated by the sounds that our human voice brings forth in order to create words and meanings, which are translated into an allographic medium (i.e., text which is based upon notation), and then from there further transform themselves into a quasi-autographic medium through their display as typographic artworks in the physical realm.

Although typography does of course base itself in the allographic properties of the text that it manipulates, in many cases this basal medium (particularly in the case of deconstructive typography) is re-worked and re-interpreted in such a way that the boundaries between what is notational and what is gestural can become blurred in an attempt to overcome the boundaries between the 'alive' nature of speech and the 'dead' nature of text that were noted upon by Derrida, as mentioned earlier in this text. While they may be, and often are, copied through printing, nevertheless it seems to me that the original artifacts that are duplicated through printing or similar means do possess autographic as well as allographic properties.

Thus, certain types of typographic output (such as deconstructive typography), would appear to be fields of endeavor that carry inherent dichotomies in their manifestations. In regards to these, computational creativity may well be bringing further ingredients into what is already a rich conceptual mix between allographic and autographic characteristics - to which virtual 3D may be contributing additional properties that may be worth further investigation from a creative as well as a design theoretical perspective.

REFERENCES

Bense, M. (1969). *Einführung in die informationstheoretische Ästhetik: Grundlegung und Anwendung in der Textheorie* (Vol. 320). Hamburg, Germany: Rowohlt.

Block, F. W. (2009). *How to Construct the Genre of Digital Poetry. A User Manual.* Retrieved from http://www.netzliteratur.net/block/digital_poetry.pdf

Boellstorff, T. (2008). Coming of Age in Second Life. Princeton University Press.

Bogost, I., Bruns, G. L., Campbell, T., Cixous, H., Conan Doyle, A., Becker, E. J., & Levin, D. (2012). Many Subtle Channels. In *Praise of Potential Literature*. Cambridge, MA: Harvard University Press.

Bök, C. (2006). Aleatory Writing: Notes Toward a Poetics of Chance. Public: Art | Culture | Ideas, 33, 24 - 33.

Brown, B., & Bell, M. (2006). *Avatars at Work and Play.* In R. Shroeder & A. Axelsson (Eds.), *Play and Sociability in There: Some lessons from online games for collaborative virtual environments* (pp. 227–245). Springer.

Cahalan, J. M. (1994). The guilty forgiving the innocent: Stanislaus, Shaun, and Shem in Finnegans Wake. *Notes on Modern Irish Literature*, *6*, 5–11.

Cervieri, M. (2007). *User Generated Content in Second Life with Cory Ondrejka*, . Retrieved from http://www.scribemedia.org/2007/03/20/2nd-life/

Drucker, J. (1994). The Visible Word: Experimental Typography and Modern Art, 1909–1923. University of Chicago Press.

Gomringer, E. (1968). *The book of hours, and Constellations: Being poems of Eugen Gomringer*. Something Else Press.

Goodman, N. (1976). *The Language of Art*. Indianapolis, IN: Hackett.

Jung, C. G. (1951). *Foreword to Richard Wilhelm's translation of the I Ching*. Routledge & Kegan Paul.

Kaplan, S., & Kaplan, R. (1982). *Cognition and Environment: Functioning in an Uncertain World*. Westport, CT: Praeger.

Lanham, R. A. (1993). *The Electronic Word: Democracy*. University of Chicago Press. doi:10.7208/chicago/9780226469126.001.0001

Lupton, E. (1994). A Post-Mortem on Deconstruction? *AIGA Journal of Graphic Design*, *12*(2), 45–47.

Lynch, K. (1960). *The Image of the City*. Cambridge, MA: MIT Press.

McCullough, M. (1996). *Abstracting Craft: The Practiced Digital Hand*. Boston: MIT Press.

Parish, N. (2008). Between Text and Image, East and West: Henri Michaux's Signs and Christian Dotremont's 'Logogrammes'. *RiLUnE*, *8*, 67–80.

Rotman, B. (2000). *Mathematics as Sign: Writing, Imagining, Counting*. Stanford, CA: Stanford University Press.

Sharp, R., McKinney, P., & Ross, S. (2007). *Visual text: Concrete poetry, hyperfiction and the future of the narrative form*. Glasgow, UK: Humanities Advanced Technology and Information Institute. Retrieved from http://www.ed.ac.uk/iash/vkpublication/Sharp_McKinney_Ross.pdf

Small, D. (1996). Navigating large bodies of text. *IBM Systems Journal*, *35*(3/4), 514–525. doi:10.1147/sj.353.0514

Stevens, W. (1965). *The Necessary Angel: Essays on Reality and the Imagination*. Vintage Books.

Chapter 7
Imagination and the Phenomenology of Virtual Practice

Denise Doyle
University of Wolverhampton, UK

ABSTRACT

The experience of creating in technology-mediated spaces through an avatar form brings the phenomenological experience of the body into the act of creating itself. Ways of explaining this creative process and its relationship to imaginative experience highlights a phenomenology of virtual practice of artists working in this realm. This chapter explores the phenomenology of the imagination in relation to embodied experience and considers the implications of phenomenology as a methodological approach and presents the case for investigating artistic and imaginative experience through adapting phenomenological research methods.

INTRODUCTION

We insert ourselves into a philosophy and it transforms us from the inside out (Kozel, 2007, p.4)

According to Edward Casey 'imagination (in Western thought) is not securely situated in regard to such decidedly different acts as sensory perception and conceptual thinking' (Casey, 2000, p.19). Perhaps this is because 'imagining is easy enough to enact as experience, but it is extremely difficult to capture in midair for the purposes of scrutiny and examination' (Casey, 2000, p.4). This chapter considers

DOI: 10.4018/978-1-5225-5023-5.ch007

the importance of investigating imaginative experience in relation to the process of artistic creativity in virtual worlds. Through adapting existing phenomenological research methods it discusses the ways in which the 'virtuality' of the space and the art created have impacted on the analysis in what is termed here as a *phenomenology of virtual practice* of artists working in virtual space.

Since it's inception artists have been exploring the virtual space of *Second Life* where the act of creating is already mediated through technology. In a review of artworks created in *Second Life* in the last fifteen years the sheer array of work created across a broad range of artistic disciplines such as art, architecture, performance, film, and media arts is salient. From early *Second Life* music performances by Susanne Vega (2006) and Duran Duran (2006), to the Chinese artist Cao Fei's *Second Life* projects including *iMirror* (2007), an installation and three-part documentary about Fei's avatar identity, *China Tracy*, and *RMB City* (2008-ongoing), to Paul Sermon's public interactive installation *Liberate your Avatar* (2007) and his more recent collaborations with Charlotte Gould in *Mirror on the Screen* (2012): these all point towards a retesting of artistic principles in virtual world spaces. Artist and filmmaker Lynn Hershman worked with Stanford University to develop *Life to the Second Power* (2007-ongoing), documenting and archiving two of her projects, *The Dante Hotel*, and *Roberta Breitmore*. A number of performance-based works relied on a re-enactment of historical performance art, such as *Seven Easy Pieces* (2005) which Marina Abramovic performed at the Guggenheim, re-enacting seven significant historical and noted Performance Art works, and Eva and Franco Mattes *Synthetic Performances* (2009-10), exploring what they term an 'imponderable medium' (Lichty 2008, p.6). More recently works such as

But what of the artist's experience of creating mediated through the presence of an avatar form?

BACKGROUND TO THE RESEARCH

On the Virtual and Imaginary

Casey commented in his book *Imagining: A Phenomenological Study* (Casey 2000) that, at the time of its first publication in 1976, a 'concerted phenomenological study of the imagination had yet to be done' (Casey 2000, p.xi). In this research it has been found that any substantial study of the virtual and the imaginary together has yet to occur. There have been some passing associations, such as philosopher Brian Massumi's link to the imagination as a 'mode of thought' that is most suitable to the virtual (Massumi 2002, p.134), or that of Pierre Levy, describing the imagination as one of the three vectors of the virtual (Levy 1998, p.28). As a starting point for the dialogue on the virtual and the imaginary they are placed in direct relation to each other (with the two terms on an equal footing). This pairing throws up some initial thoughts on what pulls or pushes them together, what attracts them, and conversely what pulls them apart. When paired together, as relatives, or at least as an associated grouping, three shared characteristics or impulses are revealed. Firstly, both terms are often associated with an 'elsewhere' or to a place or space not immediately associated with the real. Secondly, each appears to be multi-faceted, whose meaning changes quite dramatically when seen as an act or as a description, when a verb (imagine), a noun (virtuality, imaginary), or equally as an adjective (virtual, imaginary). Finally, each term can be as elusive and fleeting as the other. When identifying what would differentiate them, one can make the distinction in what may be the very impulse that stirs or moves them. In fact, other than the imagination being more closely aligned with creativity, or a creative impulse, (which does not necessarily incorporate or include notions of virtuality), one may quickly flounder in attempts to keep them separated, rather falling back on the transitional or liminal qualities that they both share. Yet each term cannot necessarily be exchanged as freely as an initial analysis may suggest.

On the Imaginary and Space

Massumi writes very little of the imagination compared to his discussions on the virtual. However, he suggests that 'imagination is the mode of thought most precisely suited to the differentiating vagueness of the virtual':

Imagination can also be called intuition: a thinking feeling. Not feeling something. Feeling thought – as such, in its movement, as process, on arrival, as yet unthought-out and unenacted [...] Suspended. Looped out. Imagination is felt thought, thought only-felt [...] Outside any given thing, outside any given sense, outside actuality. Outside coming in. The mutual envelopment of thought and sensation, as they arrive together. (Massumi 2002, p.134)

But, is imagination something more than a 'thinking feeling', more than an equivalent of intuition? This chapter argues that intuition is an unexplained knowledge that is, as Massumi places it, a knowledge that is outside coming in. Yet the 'imaginary' is not a static knowledge, if it is even knowledge at all. Is there an imaginary experienced as sensation, as well as image? Massumi's consideration of the imagination as mutual envelopment of thought and sensation is helpful, and points to why virtual world spaces lend themselves so strongly to the study of the imaginary and the imagination. In particular, it is the dual experience of being there, and being somewhere else simultaneously that echoes and replicates the process of the development of the imaginary in lived experience.

How do we understand our experience of the imaginary in the context of our own experience of a lived reality? Grosz suggests that things that are possible now would have literally been 'unimaginable' in Bergson's time (Grosz 2001, p.117). The relevance of the term the imaginary in this chapter relates to our fundamental lived experience of reality, and has a trace of the unseen, the movement in the shadows, the sense out of the corners of our eyes, the impulse of another space, somewhere. Through the emergence of mediated spaces such as *Second Life* and other virtual worlds it is apparent that this could be a space where the imaginary could be investigated, or as Grosz describes it, the tabula rasa, The Thing (Grosz 2001, pp.169-170).

Bodies and Imagination

Author and psychoanalysist Robert Bosnak, in writing of the dreaming brain, agrees with a current set of 'stories' of neuro-science as expounded by both Mark Solms and J. Allan Hobson, that dreaming is related to an experience of space (Bosnak 2007, p.36). In fact, a key element of what Bosnak terms the 'embodied imagination' is that there is 'an *inversion* of the notions of inside and outside [that] *changes the very nature of the space* [original emphasis]' in which we find ourselves (Bosnak 2007, p.20). Giving an explanation of how to visualize the experience of multiple embodied emotions, Bosnak explains that if dreaming was considered to be a narrative of simultaneous embodied states that unfold along a timeline, then equally, if the timeline 'tilts [...] into a vertical *axis* [original emphasis], a cross-section of multiple embodied stories [...] making its many story times visible at a glance' (Bosnak 2007, p.39). In suggesting that emotions are 'fully embodied states existing throughout the physical body' he further explains that dreaming could be considered to be 'a simultaneous spatial experience of multiple embodied emotions' (Bosnak 2007, p.38). He uses the term 'image-presence' to explain the environment, or space, we might find ourselves in when experiencing the embodied imagination. For Bosnak the process of embodiment precedes any mental or emotional knowing. For him 'embodiment is the fundamental archaic way of knowing' (Bosnak 2007, p.71).

How does this approach to, or view of, embodiment through Bosnak's embodied imagination relate to the avatar-mediated experience? In contrast to the space experienced or even created with the dreaming brain, through a phenomenological analysis of our experience of virtual space, it is Don Ihde, a philosopher of technology, that concludes that there is:

> *[...] a variation between what would be called full or multidimensional experience and a visual objectification of presumed body experience. Where does one feel the wind? Or the vertigo in the stomach? Can it be felt "out there" in the disembodied perspective? The answers quickly show partial primacy to the embodied experience. (Ihde 2002, p.4)*

In exploring these ambiguities in virtual space, Ihde suggests that this duality of experience lies between what he terms the here-body and the image-body. In particular, when our presence is identified through a third person avatar perspective, Ihde suggests that we move between the full multidimensional experience of the here-body and the disembodied experience of the image-body perspective, as represented by the image of the avatar (Ihde 2002, p.6). The here-body is where we can have a full, multidimensional experience, 'whereas the visual objectification out there is spectacle like' (Ihde 2002, p.4). The image-body is where the body of the avatar lies. As Ihde explores the ambiguities experienced in virtual space, particularly when our presence is identified through a third person avatar perspective, he suggests that this is 'the opening to a sliding perspective from the multidimensional experience of my here-body toward the image-body perspectives [that] lie within these ambiguities' (Ihde 2002, p.6).

PHENOMENOLOGY OF VIRTUAL PRACTICE

From 2007 until 2014 Kriti Island[1] acted as a laboratory space for ongoing artistic and imaginative practice research in *Second Life*. Central to the development of the space was the community of experimental artist practitioners who relied on a particular form of collaborative co-operation. The research process began with a grass roots development from connections made within the space itself. As an artist-researcher the main strategy was to create a presence on the platform and invite others to participate. Kriti acts as a space for the presentation of the process of thinking in relationship to the key themes of the imaginative potential of the space and the exploration of the relationship between physical and the virtual spaces.

A variety of methods were developed for the research, one of which was the development of the virtual exhibitions (*Kritical Works in SL* and *Kritical Works in SL II*) and their subsequent analysis (see Figure 1). Following the realisation of the exhibitions, interviews were conducted with six of the participating artists and analysed for the artist's insight in the context of creating immersive and interactive experiences in the *Second Life* space. The accepted format of the artist's interview traditionally explores the tacit knowledge base of the artist's experience and draws out any impact on the artistic process. However, a further method was used for the analysis that exposed a previously unacknowledged layer of experience for evaluation. This adapted the phenomenological method of Imaginative Variation to analyse the interview transcripts to uncover the essence and meaning of the experience of being a creative practitioner in the *Second Life* space. This focused on the artists account:

- Of particular experiences of time, or of space, or time and space;
- Of bodily concerns as exemplified by an experience of embodiment, or a particular relationship or experience towards the artist's avatar;
- Of the challenges or understanding of the issue of materiality and/or our understanding of the physical/virtual world relationship;
- Of the relationship to others experienced in the space;
- Of the relationship to the self as exemplified by an exploration of an avatar identity.

In addition to this the experience of 'vividness' and 'imaginative experience' was particularly noted. Further, the relationship to and experience of avatar mediation as a participator and maker/creator in the *Second Life* community was considered. The subsequent articulation of a *phenomenology of practice* of the artists working in the space is the main focus of the research presented here.

Figure 1. Map of the Kritical Works in SL Project and Art Exhibition on Kriti Island, Second Life (2009) © Denise

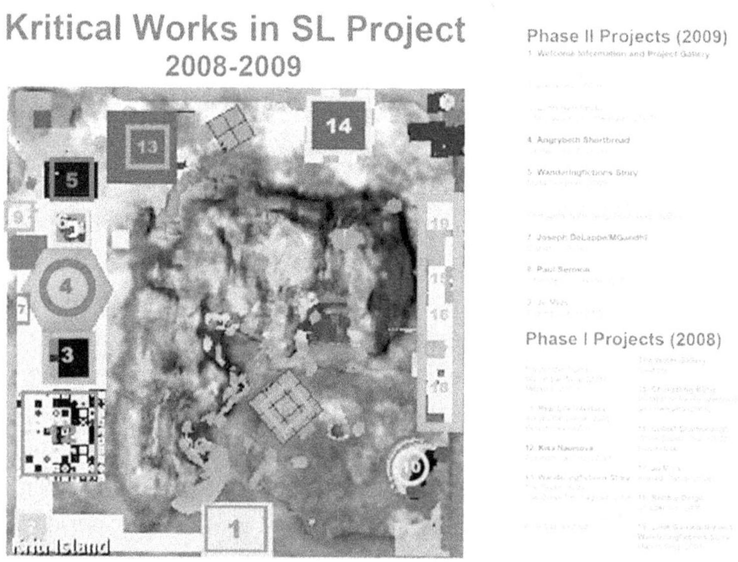

ART AND HEURISTIC PHENOMENOLOGY

In the study of the Philosophy of Technology (Ihde, 1993, 2002) it may be no surprise that new experiences created in the field of Art and Technology would be researched as the study of the phenomenon itself. Casey remarks that 'the phenomenologist's basic attitude is: no matter how something came to be in the first place, what is of crucial concern is the detailed description of the phenomenon *as it now appears*' (Casey, 2000, p.9). Philosopher and dancer Susan Kozel terms this a pre-reflective stage and considers that this should be built into any process of phenomenology in relation to technology:

As we assess the artistic, social, and corporeal implications of developments in computational devices and ambient systems, we need to have a method and an intent to receive information from our own bodies and from the bodies of others. In giving a voice to diverse corporeal states we can overcome naive distinctions between matter and spirit, between body and mind. (Kozel, 2007, p.23)

A key concept of heuristic phenomenology is approaching the subject of the research in the act of wonder. At the base of all heuristic discovery Moustakas suggests is 'the power of revelation in tacit knowing' (Moustakas, 1990, p.20). He observes that:

Through phenomenology a significant methodology is developed for investigating human experience and for deriving knowledge from a state of pure consciousness. One learns to see naively and freshly again, to value conscious experience, to respect the evidence of one's senses, to move toward an intersubjective knowing of things, people, and everyday experiences. (Moustakas, 1994, p.101)

In his article, *Phenomenology of Practice* (2007), van Manen proposes that the phenomenologist is akin to, or not unlike the poet, as she 'directs the gaze toward the regions where meaning originates, wells up, percolates through the porous membranes of past sedimentations' (van Manen, 2007, p.11). A phenomenology of practice may also be considered in, and of, artistic practice, although what this relationship means in the field of phenomenology and in the field of artistic practice may vary. In presenting a phenomenological interpretation of someone else's experience Kozel suggests a 'phenomenological philosophy provides [both] the conceptual framework and methodological sketch for interpreting the experiences of others' (Kozel, 2007, p.58). In fact, a second realization becomes apparent in that 'sensibility and interpretive power comes from the physical experience of the phenomenologist, in this case the second-person phenomenologist, or heterophenomenologist' (Kozel, 2007, p.58). In the question of how phenomenological research is conducted, and how first, second and third person methodologies necessarily position the researcher in relation to the experience being researched, Kozel remarks that Merleau-Ponty does not say or indeed discuss how he conducted his own phenomenological research. It is of note that Casey conducted his research on the imagination as a phenomenological first-person study (Casey, 2000).

The Method of Imaginative Variation

Casey points out that a phenomenological method takes 'objects, events, or acts – whether real or imagined – as *exemplifying* an essence or essential structure', in actual fact, 'insight into essences is a mediated affair [original emphasis]' (Casey, 2000, p.24). Moustakas observes that 'the uncovering of the essences, the focusing on pure possibilities, is central in the Imaginative Variation process. In this [...] process the structures of the experience are revealed; these are the conditions that must exist for something to appear' (Moustakas, 1994, p.98). The principle key structural themes within this method draw on what 'we imagine [as] possible structures of time, space, materiality, causality, and relationship to self and to others' and that through this method 'countless possibilities emerge that are intimately connected with the essences and meanings of an experience' (Moustakas, 1994, p.99). Imaginative Variation appears to be a most notable method in which to understand experiences such as time, space, and materiality. The Van Kaam method of Imaginative Variation, as presented by Moustakas, was followed through from (1) the listing of every expression relevant to the experience; (2) reduction, to reveal the *horizon of experience*; (3) the clustering of the core themes of the experience; and (4) final validation, that the core themes are expressed explicitly or are compatible with the interview itself [my emphasis] (Moustakas, 1994, pp. 120-121).

THE INTERVIEWS

Following the interviews a textural-structural description was created for each artist and includes their horizon of experience of being a creative practitioner in the *Second life* space (three of which are presented below)[2]. The format of the interviews presented includes a short description by the artist of the work presented on Kriti Island. All artist interviews were conducted in 2010.

Interview One: Joseph DeLappe

The artist Joseph DeLappe was invited to participate in the second phase of the Kritical Works project in the spring of 2009 and a request was made to include *MGandhi 1* as part of the Kritical Works project.

Artist's Description of the Artwork

MGandhi 1 is an 8" rapid prototyped 3D print of my Mahatma Gandhi avatar from Second Life. The figure is the first of a series of three physical, sculptural reproductions of my avatar, MGandhi Chakrabarti (the other two include a 15" MGandhi 2, finished in genuine gold leaf, and a 17' tall monumental sculpture entirely constructed from cardboard and hot glue). The Gandhi sculptures were created after a mixed-reality performative reenactment of Gandhi's famed 1930's Salt March which took place in Second Life [3] *(DeLappe, 2009, p.28).*

Figure 2. Joseph DeLappe's MGandhi 1 with Wanderingfictions Story as the Meta-Dreamer in the background (2010) Photograph © Denise

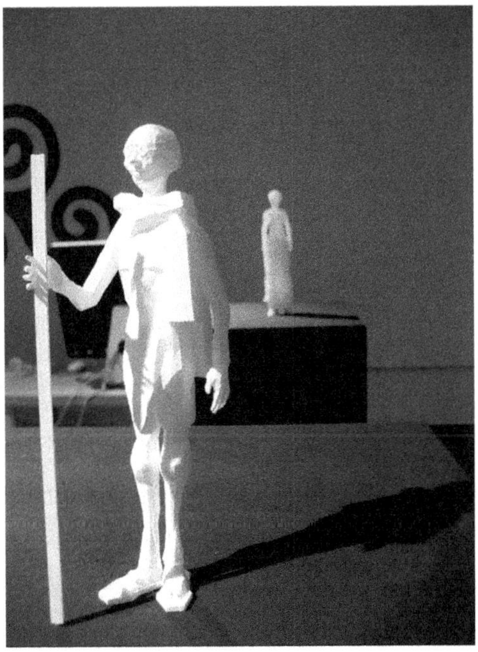

Textural-Structural Description

For the artist Joseph DeLappe, the main theme that was indicated as prominent in his 'horizon of experience' of the *Second Life* space was the overriding impact of embodying the *MGandhi Chakrabarti* avatar. Another major theme was DeLappe's experience of time and duration in undertaking the *Salt March to Dandi* (2008) project and related works. Spending such a prolonged period of time in *Second Life* during the re-enactment eventually impacted on DeLappe's physical world experience. He explains:

After about a week and half of doing this every day [the re-enactment], I would be walking back to my apartment in New York City, and thinking I could click on people.[4] It's what I did everyday. It got into my psyche, my consciousness, on such an intense level (DeLappe, 2010, p.247).

The motivation to create artworks in the space was initially rather mixed, as DeLappe had 'heard about *Second Life* early on in, though it didn't really engage me until I did this march. There was something more about this experience' (DeLappe in, 2010, p.247). However, he commented that once the *Salt March to Dandi* walk had been completed, 'I was really sad when it was over and I really had a great sense of accomplishment that I had actually done something' (DeLappe in, 2010, p.246). Of the experience of embodying the Mahatma Gandhi avatar, DeLappe notes that he had never spent so much time with one avatar before. He commented at the outset of the interview that:

The idea of playing Gandhi was ridiculous and absurd and really audacious, and the notion of me, a middle aged white professor playing Gandhi, was really, really scary and dangerous in a way, and that was interesting […] There is this rhetoric in Second Life that you can be anything, your wildest imaginings: if that's true then why not push this as far as it can go, [I thought] lets see what this would be like [to be Gandhi] (DeLappe, 2010, p.243).

During the interview DeLappe described a number of times that he was 'walking with Gandhi', and in the process of the re-enactment, a real connection had been made through the physical experience of walking on the treadmill at the Eyebeam Gallery. He was surprised at how connected he became to *MGandhi Chakrabarti*:

There was something about the physical control, and this connection with walking with him. Sometimes he would be walking off a mountain and I found myself almost falling off the treadmill - several times that happened, it was such a visceral connection (DeLappe, 2010, p.244).

DeLappe commented that after Gandhi finished the Salt March and he was gone from *Second Life*, people would ask him when they would see Gandhi again. He commented 'I brought him back a couple times for some special events and I felt I was violating the principle of the concept. I didn't feel right just walking around with him without the treadmill' (DeLappe, 2010, p.248). This suggests a deeper sense of the presence of Gandhi in the avatar representation. For DeLappe, 'with Gandhi it felt like I was a different person. There is something about an avatar that is a bit like putting on a clown's suit, you become this something else in a way' (DeLappe, 2010, p.247).

The experience of time in the Gandhi project was a second major aspect that was prominent for DeLappe and the experience of duration gave him a sense of accomplishment through a number of aspects of the project. When building the cardboard *MGandhi 3* for the third time for the ISEA2009 exhibition in Belfast he noted that 'each of these building processes has been this durational event, like really physical, and again reflecting the labour and the results being something that reflects the entirety of that experience' (DeLappe, 2010, p.245). DeLappe already explained that the durational time spent walking in *Second Life* gave him a different relationship to the space, and yet 'the whole teleporting thing and all that is really kind of disjointed, but in a way, with the Gandhi thing [it] was a very analogue experience of that space' (DeLappe, 2010, p.251).

Interview Two: Annabeth Robinson

Annabeth Robinson was first approached in 2008 to participate in the first phase of the *Kritical Works in SL* project after becoming known through the *Second Life* Educators Forum[5] (SLED) list as a practitioner and educator based in Leeds in the UK. In 2009 she developed *Gestalt Cloud* specifically for the exhibition and for presentation on Kriti Island.

Artist's Description of the Artwork

Gestalt Cloud is a Second Life™ multi-user installation that invites online social play. When an avatar is within the Cloud's generating 10m³ space, it reacts by creating cubes engulfing the avatar producing a low resolution 3D cloud that follows the user about. When a group of four or more avatars are in close proximity, their combined cloud will be rewarded with precipitation, beginning a sequence of change to the space around them (Robinson, 2009, p.10)[6].

Figure 3.The Gestalt Cloud rain forming (2010) © Annabeth Robinson

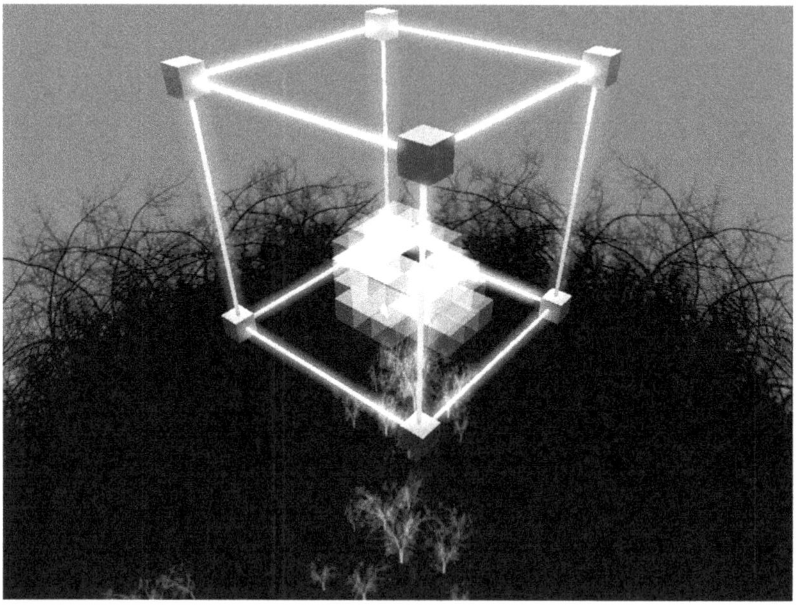

Textural-Structural Description

Themes that were indicated as prominent in Robinson's horizon of experience of the *Second Life* space were her relationship to her avatar *Angrybeth Shortbread*, the potency of the experience of interacting with other avatars in virtual space, and her experience of playfulness within the *Second Life* space. She explains her experience when in *Second Life* as 'actually [being] in a big toy box or a theme park: a hyper-real place' (Robinson, 2010, p. 260). Other themes were her sensitivity to the time and space logic of the physical world and the implications of this to the space itself and the desire to make projects that depend on human relationships. She observes that, 'I always love that serendipity of how people react to things' (Robinson, 2010, p.256). Of her experience of *Second Life* and the memory of the space as distinct from the experience itself, she explains that 'if I'm not looking at it [*Second Life*] straight away, I am left with the impression which I think I don't get when I am first hand experiencing it' (Robinson, 2010, pp.260-261). In terms of the space itself Robinson comments that 'if you don't have a project it's a very strange place to be in' (Robinson, 2010, p.257). For Robinson it is the engagement on a creative level with the space that plays a central role in her experience.

Robinson acknowledges that she has a 'kind of weird disjointed relationship' with her avatar. Of her relationship to *Angrybeth* she notes that:

When I am using Second Life my head is aware that I am in two spaces at once. I am a reader viewing the viewer and when I look at Angrybeth, it's a puppet [...] I certainly don't have any kind of wish fulfilment that that is me in there. It is more in tune with the word avatar itself; it's an aspect of me that's in concurrent time (Robinson, 2010, p.258).

She commented that 'I can get quite easily sucked into the empathy of it having some kind of emotion', but notes that 'when I am not real-time inhabiting Angrybeth I think of it as a mouse-pointer [and] when I'm reminiscing and I'm not online, I think of it as the experience I had' (Robinson, 2010, p.259). When the projects she has undertaken have been intensive or she has spent a long time working on a project in *Second Life*, Robinson remembers them as though they were real events. In fact, she notes that:

There was a period [of time] a couple of years ago when I would actually have Second Life dreams. Not that it was about flying or things like that, it was more about the people I met but I knew I had never met in real life, they were real enough relationships [to me] (Robinson, 2010, p.262).

Observing that 'this is a world that is constantly living, even though it isn't tangible in our real world sense', she comments that 'what's nice about it is, it is certainly timeless' (Robinson, 2010, p.260). She remarks, that 'even though it attempts to replicate a physical world there are certain levels that, things don't reflect. Everything feels close but when you walk to it [it] actually takes longer, because there is no depth of field'. In her experience of the playfulness of the space, Robinson explains the role her imagination plays:

When I'm building it's like when you are a child, a cardboard box can be any story and I think when I am doing my own stuff, my imagination fills in the gaps accordingly [...] When I'm not in that playful state of mind that I think I need to be in to actually allow myself to mentally be set free in the space and react to it like I'm in a real space (Robinson, 2010, p.261).

Interview Three: Jacquelyn Ford Morie

The artist Jacquelyn Ford Morie's participation in both exhibitions contributed two major projects, *Traceroutes* (2009) and the *Remembrance and Remains* (2008), to the Kritical Works project and to Kriti Island itself. An early adopter to *Second Life*, Morie has participated in the art community as an artist, artist-curator and as a researcher.

Artist's Description of the Artwork

In 2007 I exhibited a virtual environment entitled The Memory Stairs. This artwork was a fully implemented virtual environment that could be experienced in stereo-vision via a head-mounted display, and navigated through with a game controller. It featured full surround sound and scents delivered at appropriate places in the experience by means of a bespoke scent collar. Traceroutes is work is based on that piece, but expands its visual and interactive scope within the online world space (Morie, 2009, p.27).

Textural-Structural Description

For the artist Jacquelyn Ford Morie, themes that were indicated as prominent in her horizon of experience of the *Second Life* space were the playfulness and exploration around her avatars in relation to her identity, her associated sense of being an artist in the *Second Life* space, and her experience of embodiment in the space itself. In fact, Morie notes that she was surprised by how immersive *Second Life* could be compared to her experience in Virtual Reality spaces, and curiously, her experience of immersion was present without the isolation from the real world that would occur in work such as Char Davies' *Osmose*, or *Ephémère* (Morie, 2010, p.282). Giving an example of her experience in *Second Life* she explains that, 'when I was in *Ping*

Figure 4. The entrance to Traceroutes (2010) © Jacquelyn Ford Morie

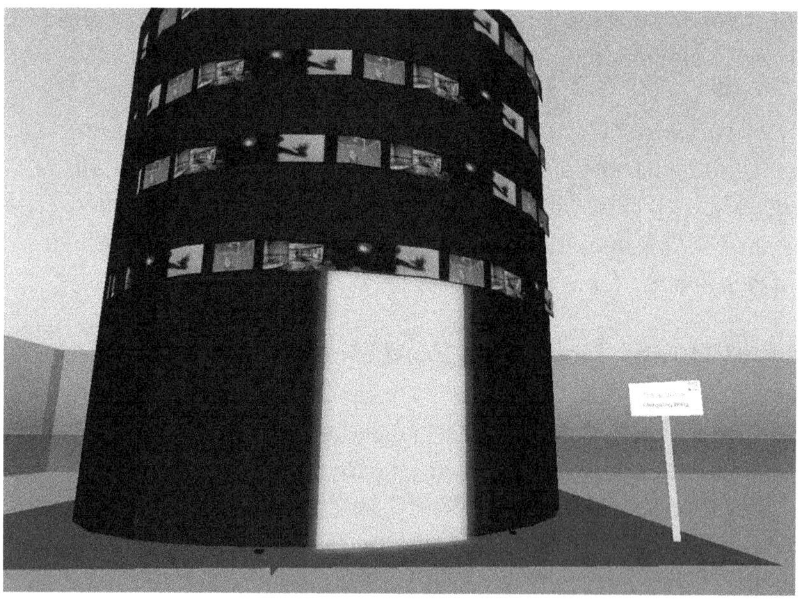

Space[7], I was not aware of the rest of the room I was in: *I was there*' (Morie, 2010, p.281). Morie's connection between the *Second Life* space and her experience of embodiment remains central to her virtual world experience. Drawing comparisons to her experience in the physical world she explains that, 'to wander through a [Dr Seuss] forest […] in the real world would probably [have to] be an illustration. It wouldn't be something that I could take my embodied self into' (Morie, 2010, p.281). Morie describes her experience:

[I was] amazed how immersive things can be in Second Life, even on a small screen and the fact that I can experience something that I couldn't experience in the real world, because it just can't be built in the real world. It is like a tangible imagination world (Morie, 2010, p.282).

The experience of being an artist in the *Second Life* space relies on Morie's imaginative responses to the space and to a familiarity she experiences with the tools inherent in the platform itself. She comments that, 'I love the modelling system in *Second Life* - something I knew from a long time ago was based on the same kind of modelling system. I just took to it right away' (Morie, 2010, p.279). Morie explains her imagination in the artistic process, 'if I imagine something it's more I am imagining the effect it could have. [If] I want to create this windy lonely plane that makes me feel like there's rain on my face and I am on the edge of a cliff that's not what I am going to build, that's just what I want to evoke' (Morie, 2010, p.282). Of the artists interviewed it was Morie who had the greatest sense of the playfulness and exploration of her identity through her avatars. She explains that:

The first thing I did (when I first went into Second Life) was to really mess with my avatar and really try and evoke something within myself, some part of my identity that's not usually allowed to come out […] the third avatar I made was as a child; [although] I didn't do that for very long. I really liked being a child because it's definitely about my own persona (Morie, 2010, p.276).

A PHENOMENOLOGY OF VIRTUAL SPACE

In the final step of the Imaginative Variation method the research brings together the experience of all of the participants (in this case the six artists working in the *Second Life* space[8]) and is presented in what is described as a Textural-Structural description of the variations and essence of the experience. In this phenomenological account of the experience a poetics of writing is fashioned to enable the possibility of what Bachelard terms phenomenological reverberation (Bachelard, 1994, p.xxiii).

Textural-Structural Description of the Artist Group

The artists' experience of time and space in *Second Life* was one of an increased and heightened awareness, whether that was in simply reflecting the experience of the physical world or conversely being set free of the constraints of the laws of the physical world. Time and space are both reinforced and inverted in the artists' experience of virtual space. Here there were implications both creatively and personally. There was a freedom associated with the 'timelessness' of the space, of 'not being handcuffed to a situation', rather 'being unlocked from time'; of having the opportunity of 'being in a different space and time' and 'to be set free in the space'. There was an imaginative freedom described with the chance to play with time and space for the audience, or to have the experience of being embodied in someone else's imagination. Time and duration are closely linked in some of the artists' experiences. It appears that in the early stages of engaging in *Second Life*, when the space is new, it takes time to normalise the experience - and the imagination can be seen at its most active. This links imagination and cognition in a most interesting way.

Space is invariably experienced as one that does follow the logic of the physical world and this is only reinforced by the 3D nature of the grid itself. There is a sense of the known and there is something comfortable about the logic of the space. An artist comments that '*Second Life* does have that kind of traditional perspective sensibility to it. And the way that things represent in the space is very much according to traditional perspective visualisation and it feels very comfortable in that sense'. Conversely, the space is also experienced as a lived space and evokes other experiences such as the experience of a mirrored or other world, or a world that both evokes and reflects the experience of dream states, one that is 'half clear and true and half a story that is made up'. The playfulness of the space was experienced by a number of the artists, and for some it felt like 'a big toy sand box' or like being in 'a theme park'. The space evoked both a sense that 'it can be a strange place to be' and at the same time there 'was something in there [*Second Life*] that clicked' with a number of the artists. Each of the artists has maintained an interest in the space. Some artists immediately engaged in the space and one artist began creating 'within the first hour' of being in *Second Life*. Another very quickly began experimenting with their avatar to try and evoke 'something within myself'. For others *Second Life* didn't really engage them until a creative project enabled there to be 'something more' about the experience.

Closely linked to the experience of space is the sense of embodiment in the space and the ways in which the artists developed and engaged with their avatar(s). The intense sense of immersion through the embodiment of the avatar was a common experience amongst the artists. For some, the avatar is a way of being 'in two spaces at once', or to experience an aspect of themselves 'in concurrent time'. For another the avatar is an extension 'of myself that allows me to be in a different space and time'. Some artists experienced a playfulness and desire to explore their own identity through the avatar, and for others embodying the persona of another person (whether real or imagined) made them feel like they were 'a different person'. Experiences of embodiment through the avatar are also linked to the relationship to others, and for some, this encouraged new relationships or new ways of experiencing relationships. For one artist the nature of the space actually magnified the desire to make projects that depend on human relationships. A number of the artists identified the network element of the space as its most important aspect. For others it was the ability to experience the immersive qualities of the space as a solitary experience. Overall, the experiences described by the artists tended to be solitary in nature in *Second Life*, although all artists commented on how the dynamic of the space changes through the social elements of the space.

Sometimes a sense of immateriality in the space reinforced the materiality of the physical world, and inversely, at times a sense of materiality of virtual space itself was suggested. This tension is expressed in the 'lack of physicality', or a 'stepping away from reality', and 'an almost physical way of creating the imaginary and the impossible'. This 'tangible imagination world' all challenged the artists' sense of what it means to experience the material. At the centre of the responses to *Second Life*, of this 'hyper-real place', and what is expressed above in the vividness of the imaginative experience, is a 'freer and more expressive space' in order to be mentally 'set free in the space'. Finally, in what is described as 'a real but not real' space, one artist suggests that 'when you could tap into a virtual world and in the future onto an augmented reality, the fact that you could walk around a real space and go back in time, that's going to be a whole new situation to lose your imagination in'.

FUTURE RESEARCH DIRECTIONS

Adapting existing phenomenological methods, a particular approach taken in this research, was a strategy that uncovered the true richness of the interview material. Using this second method of analysis on the data revealed further findings and this brought into focus another layer of the artist's experience of practice in the space. In writing a phenomenological account of the artists' experience of creative practice in *Second Life*, a richness and variety of experience was able to be included in the final textural-structural description. For example there was a variation in the response to the sense of embodiment of the avatar. The avatar experience was a critical and essential part of the phenomenological experience of the space but the reasons for this were varied. For some it was like embodying another character or persona entirely, for others the avatar was experienced as a mouse pointer or an aspect of themselves in concurrent time. Conversely, the avatar was described as an unexplored aspect of their own identity. The opportunity to play with time and space were seen as an important aspect of the potential of the space for creative investigation and there was a freedom associated with this loosening of time-space relationships. The heightened experience of the imagination was consistently seen as something difficult to describe. This almost physical, almost tangible, was a particularly interesting expression used by some of the artists - the almost tangible is presented as if the artist could almost make real their imagination or experience the imagination of others in embodied form.

Approaching mediated space as a phenomenon in its own right (with its own potential *essences* of experience) raised some epistemological assumptions that should be acknowledged when using reductive phenomenological methods. However, using the method of Imaginative Variation as a phenomenological uncovering of the essences of the experience was appropriate to the study of the imagination (and creativity) in this context. The relationship of the artists to the space that they are simultaneously, inhabiting and creating in, suggests that the space is experienced somewhat differently. This points to the phenomenological experience of the imagination and the opportunities that arise in technology-mediated space for the experience and understanding of creative practice and the extent to which the body, and in turn the avatar-mediated body, informs those creative and imaginative processes. Consideration for research examining 'mediated technology space' as a phenomenon in its own right is worthy of future study and with particular reference to the growing use of mixed realities in contemporary creative practices.

REFERENCES

Bachelard, G. (1994). *The Poetics of Space* (M. Jolas, Trans.). Boston: Beacon Press. (Original work published 1958)

Bosnak, R. (2007). *Embodiment: Creative Imagination in Medicine, Art and Travel*. East Sussex, UK: Routledge.

Casey, E. (2000). *Imagining: A Phenomenological Study*. Bloomington, IN: Indiana University Press.

DeLappe, J. (2009). Kritical Works in SL II. Wolverhampton, UK: CADRE (Centre for Art, Design, Research and Experimentation).

DeLappe, J. (2010). *Art and the Emergent Imagination in Avatar-Mediated Online Space* (PhD Thesis). University of East London.

Fei, C. (2008). *RMB City. Second Life* Sim. Available at: http://rmbcity.com/

Grosz, E. (2001). *Architecture From the Outside: Essays on Virtual and Real Space*. Cambridge, MA: MIT Press.

Hershman, L. (2007). *Life to the Second Power (L2)*. Available at: https://exhibits.stanford.edu/women-art-revolution/about/life-squared

Ihde, D. (1993). *Philosophy of Technology*. Paragon House.

Ihde, D. (2002). *Bodies in Technology*. Minneapolis, MN: University of Minnesota Press.

Kozel, S. (2007). *Closer: Performance, Technologies: Performance, Technologies, Phenomenology*. Cambridge, MA: MIT Press.

Levy, P. (1998). *Becoming Virtual: Reality in the Digital Age*. New York: Plenum Trade.

Lichty, P. (2008). *Why Art in Virtual Worlds? E-Happenings, Relational Milieux & "Second Sculpture." CIAC Electronic Magazine*.

Massumi, B. (2002). *Parables for the Virtual: Movement, Affect, Sensation*. Durham, NC: Duke University Press. doi:10.1215/9780822383574

Mattes, E. (2009). *Synthetic Performances*. Available at: http://0100101110101101.org/synthetic-performances/

Mohanty, J. N. (1991). Method of Imaginative Variation in Phenomenology. In T. M. Horowitz (Ed.), Thought Experiments in Science and Philosophy. Rowman & Littlefield Publishers.

Morie, J. F. (2009). Kritical Works in SL II. Wolverhampton, UK: CADRE (Centre for Art, Design, Research and Experimentation).

Morie, J. F. (2010). *Art and the Emergent Imagination in Avatar-Mediated Online Space* (PhD Thesis). University of East London.

Moustakas, C. (1990). *Heuristic Research: Design, Methodology, and Applications*. Newbury Park, CA: SAGE Publications. doi:10.4135/9781412995641

Moustakas, C. (1994). *Phenomenological Research Methods*. Thousand Oaks, CA: SAGE Publications. doi:10.4135/9781412995658

Robinson, A. (2009). Kritical Works in SL II. Wolverhampton, UK: CADRE (Centre for Art, Design, Research and Experimentation).

Robinson, A. (2010). *Art and the Emergent Imagination in Avatar-Mediated Online Space* (PhD Thesis). University of East London.

Sermon, P. (2007). *Liberate your Avatar. Public Performance Installation*. Available at: http://www.paulsermon.org/liberate/

Sermon, P., & Gould, C. (2012). *Mirror on the Screen*. Interactive *Second Life* Installation. Available at: http://www.paulsermon.org/mirror/

Van Manen, M. (2007). Phenomenology of Practice. *Phenomenology & Practice*, *1*, 10.

ENDNOTES

1 Kriti Island was funded by the University of Wolverhampton until 2014.

2 Typically, a full study such as this would be with twelve to fifteen co-participants and a final textual-structural description of the whole group would be created (this being the overall outcome of the research). However, in this study, the method was used to inform a broader study of the artist and the artist's work in avatar-mediated space.

3 Delappe spent 26 days re-enacting Gandhi's Salt March to Dandi by walking 240 miles on a treadmill at the EyeBeam Gallery, New York that controlled the Gandhi avatar in *Second Life*.

4 Right-clicking on an avatar in *Second Life* brings up extended details of the avatar such as their friendship status and profile.

5 This active forum was set up by Linden Labs to help support educators working in *Second Life*. https://lists.secondlife.com/cgi-bin/mailman/listinfo/educators

6 A temporal virtual forest grows below the newly formed cloud.

7 Annabeth Robinson's *Ping Space* (2008) project was her first contribution to the Kritical Works in SL exhibition.

8 For the purpose of this chapter only three interviews have been presented here. The full interview transcripts of the six interviews can be found in the appendix of (2010).

Chapter 8
A Vitruvian-Inspired Theoretical Framework for Architecture in Virtual Worlds

Viviana Barneche-Naya
Universidade da Coruña, Spain

Luis A Hernández-Ibáñez
Universidade da Coruña, Spain

ABSTRACT

Virtual worlds are an environment for the realization of multiple social activities ranging from entertainment to education, economics, or even medicine. Such activities require virtual spaces and buildings for their development and are inhabited by avatars that stand in place of the user. The architecture of these virtual buildings is, for the most part, a replica of architecture in the real world, and yet the realm of the virtual is replete with opportunities to develop novel and imaginative forms of buildings. It is possible to better meet the requirements and needs of the users if they use formulas adapted to the potentialities of this world of information. In this emerging field of architectural design, there is still no theoretical framework that can guide the creators of virtual buildings. For this reason, the authors propose a possible starting point as it was done in classical times with the real-world architecture, following the example of Vitruvian triad, firmitas, utilitas venustas, which is complemented by the virtual and exclusive domain of virtual architecture.

DOI: 10.4018/978-1-5225-5023-5.ch008

INTRODUCTION

Technological advances are present today in all fields of our daily lifes. Nowadadys, the world is permanently connected by means of WiFi networks, smartphones, GPS and social networks. Connected minds, direcly linked to the non-material cyberspace world once foretold by William Gibson, are present everywhere. (Gibson, 1984)

In this technological ecosystem, metaverses are one of the technologies that have evolved more in the last decade. Metaverses are three-dimesional, persistent and immersive virtual worlds which allow real-time interaction, telepresence and synchronic communication among users and in many cases, they also permit content creation by users.

These virtual environments, also known as MUVE (Multi-User Virtual Environments), follow a format derived from MMORPG (Massive Multiplayer Online Role-Playing Games) but with an important difference; virtual worlds do not rely on a predifined set of goals or roles such as videogames. Rather, they are paidia spaces, emphasizing creativity and participation. This distinction is related to the game classification stated by Caillois (1961) which distinguish clearly between *paidia* and *ludus*. *Ludus* is related to a game with a clear set of rules and goals, while *paidia* refers to free play, spontaneous amusement, free of conventionalisms.

Virtual worlds are navigable spaces that facilitate experiences of a spatial nature. Users inside those virtual environments have the ability to share life experiences with other people connected from different points of the planet, sharing a common "virtual physical place" and stablishing social ties.

Users are represented an avatar. An avatar's movement is essential to the user experience of the virtual space. The environment is perceived in relation to the avatar size, position and speed of displacement and is viewed using first, or preferably third, person view. Space is felt with the same qualities as with real world space.

Undoubtly, multiuser virtual environments constitute a field of research and development and is one of the great contributions of digital technologies applied to disciplines including education and learning evaluation (Bower, Lee & Dalgarno, 2016; Gregory et al, 2015; Barneche Naya & Hernández Ibáñez, 2014; Duncan, Miller & Jiang, 2012; Dalgarno & Lee, 2010; Hew & Cheung, 2010); medical simulation (Cook & Winkler, 2016; Hack, 2016; Ghanbarzadeh, Ghapanchi & Blumenstein, 2015; Creutzfeldt et al, 2010) social behavior and communication studies (Huvila, Ek & Widen, 2014; Stichter et al, 2014; Kandalaft et al, 2013; Schmidt et al, 2012; Williams, 2010), economy and business (Castronova, 2005), narrative (Ryan, 2003), game studies (Bartle, 2004), education in collaborative design (Alahuhta et al 2014; Merrick, Gu & Wang, 2011; Ehsani & Chase, 2009) among others. Some authors believe that this massive exodus to virtual spaces is due to the fact that they are, increasingly and to a greater extent, satisfying real human needs. (McGonigal, 2011).

Virtual worlds open up a wide range of possibilities since they permit development of all kinds of shared activities. There is a new need for meeting spaces and virtual buildings to host such activities in a new form that includes interaction, media, social networking and other concepts related to cyberspace. That is the point where Architecture comes into play.

However, in spite of their innovative features, certain types of constructions of the real world Architecture have been used as clichés by some of the users of these environments as a way of representing a "physical" universe recognizable by the presence of elements of the real world that they are familiar with.

The Architecture of Cyberspace, or Cyberarchitecture, is understood today as something more than just a form of representation; it is a new way of designing spaces in which to carry out existing or new activities in a virtual space. It is not a scenography nor does it have illusory or fantastic forms. Rather, it is a substantial part in the operation, narrative, navigation and interaction of the user with the virtual world that has certain spatial qualities. In this new universe, it should be the mission of the architects before any other profession to work on the analysis and to form a theoretical corpus that supports the architectural ideation of the virtual buildings.

ARCHITECTURE IN CYBERSPACE

The great paradigm shift in the techniques of architectural representation which took place in the late eighties was enabled by the development of Computer Aided Design (CAD). The development of these new systems allowed an evolution from the classical representation modes based in plans, elevations, cross-sections, perspectives and physical models to a new, more complete, three-dimensional form existing in virtual space which can provide almost unlimited views of a design.

As a matter of fact, the use of modern visualization tools constitutes an essential part of today's architectural practice. There is little doubt of the excellent visual quality of renderings and animations which can be achieved from a three-dimensional model of a building. Rendering algorithms allow the modeler to attain a level of hyperrealism so that the digital image can be indistinguishable from the real.

Computer architectural visualization constitutes one of the two main branches of the emerging field of Digital Architecture. The other main branch is related to the design of spatial forms not necessarily related to a future materialization in the real world but created only to exist in cyberspace hosting human activities. (Bertol & Foell, 1997).

In the first case, architectural visualization is a tool, a representation technique at the service of the real world Architecture, which is the ultimate aim of architectural design. The second case implies a very different approach; the architectural representation is the utter architectural fact. If we consider Architecture as the art and science of designing spaces to host human activities, is not that the case when we create virtual environments for social interaction, education or collaborative work?

Architectural research in cyberspace is not new; it has been around for more than twenty years. The concept itself has been evolving from the digital abstract space metaphor, flowing and dynamic, (Spiller, 1998; Novak, 1996; Lynn, 1996; Ostler, 1994; Anders, 1994; Novak, 1991, Benedikt, 1991) towards the influence of computational paradigms in the architectural discourse and to the definition of architectural space in the digital age (considering the same terms that define real world Architecture) (Spiller, 2009; Bullivant, 2005; McCullough, 2005; Mitchell, 2003; Mitchell 1996). On the other side, many architects are attracted by digital technologies and hybrid experiences that combine human interaction in virtual and real spaces (Biloria & Chang, 2013; Bier & Knight, 2010; Dollens, 2009; Oosterhuis & Jaskiewicz, 2007; Fischnaller & Hill, 2005; Serafijn & Spuybroek, 2004; Guynup, 2003; Asymptote, 1999).

Currently, examples of virtual architectural objects designed to accomplish a program and designed with an intended use can be found only in virtual worlds.

There are several experiences in building design inside virtual worlds, such as Second Life or OpenSim, that make the most of the distinctive features of cyberspace to break free from real world rules and limitations such as gravity, climate, or the laws of Euclidean geometry. (e.g. Brouchoud, 2013; Hernández-Ibáñez. & Barneche-Naya, 2013; Tea Cup Communication, 2008; Lubell, 2008; Schultz, 2007)

Nevertheless, novel proposals in virtual Architecture are not easily adopted by common users, mainly due to difficulties in comprehending their spatial keys. Abstract forms and unfamiliar shapes bewilder unknowledgeable users. In this respect, it seems necessary to investigate better and adequate representation guidelines for virtual Architecture that supply the visual clues and metaphors that the user needs to intuitively understand the spaces, their meaning, purpose and the way they function.

Virtual worlds offer innovative aspects that compose a novel field of research in architectural design from a formal and spatial point of view. User interaction capabilities, immersion, sense of presence and the ability to move through the different spaces by means of an alter ego (i.e. avatar) are important considerations to be taken into account in the design of this kind of Cyberarchitecture, but they are not the only ones. New uses and requirements appear, together with new spatial configurations, to respond to an avatars' needs as they play in a medium where, as in the real world, there is a socio-cultural context to be considered in the design.

The use of the term Cyberarchitecture to refer to architectural design in virtual worlds may sound bizarre and may even raise a smile in certain architectural circles. The complex relationship between materiality and immateriality in Architecture has contributed to the resistance from part of the architectural research community towards the inclusion of virtual and media spaces as objects of study.

Although materiality has been part of the base of the architectural discourse, the immaterial is as important for Architecture as the material. Johnathan Hill pointed out in Architecture: The subject is the Matter that the actual discourse of Architecture is too biased to materiality and suggests that architectural matter may not be always physical or constructive. But other forms of material may exist such as media, information networks and software. Architecture may be built today using words, bricks, sounds or pixels. (Hill 2001)

It seems clear that mediated spaces and presence design should be adopted by architects. As technology advances, there is every indication that in a few decades computer simulation systems will be able to provide the user with sensations of virtual presence and immersion in an extremely vivid way that is not available today. There is little doubt the mission of Architecture and architects is to contribute their theoretical work, talent and knowledge to design the spatial experiences of future Internet users in virtual buildings.

The ability to interact with the architectural object, the freedom to control the point-of-view and move into space receiving all sensorial information through sight and hearing brings the sensation of "being your own being" inside the virtual world. The link that is generated between user's actions in real space and their effects on the virtual space is – in Wilhelmsson terms- the game-ego. A concept based on experiential cognitive theory that is defined as an extension of the body's sensorial motor system, which allows the user to exercise his control, interact and even modify the represented space and manipulate the point-of-view by means of touch and movement (tactile motor/kinesthetic link) (Wilhelmsson 2008)

We are still missing a complete frame of reference for the analysis and study of Cyberarchitecture that can encompass each and every aspect which defines the field, not only the formal and aesthetic aspects. At a time when architects are beginning to understand Virtual Architecture more as a discipline within formal Architecture than a mere experimentation with shapes and spaces, and as the debate about the question of Cyberarchitecture being Architecture at all is arising, it will be interesting to compare the qualities of these virtual constructions with the qualities that Vitruvio (Vitruvius Pollio, 23-27adC) defined as necessary to consider a constructed object to be Architecture.

This look to the classics for methodology and theoretical basis in times of innovative change in architectural design is not new. Indeed, in his book "Towards a New Architecture" (1923), Le Corbusier weaves together a theoretical logic from classical historical examples such as: The Pantheon, Hadrian's Villa, and The Parthenon; analyzing and selecting those aspects that should be incorporated in

Modern Architecture. Utilizing, for example, the use of axes, proportions, rhythm, articulation of the parts, functional organization, and principles of construction without losing sight of innovative technologies of that historic moment: the car or the packet boat, for their practical and functional design.

A VITRUVIAN APPROACH TO THE ANALYSIS OF CYBERSPACE

From the foregoing, it is apparent that many concepts associated with Virtual Architecture do not have parallels in real world construction and others that have a parallel are modified when applied to the metaverse. Without a doubt, the presence of these new concepts affects the formulation and design of the virtual building, limiting in some cases but expanding the creative possibilities of Cyberarchitecture in many others. But how many and what new concepts which must be taken into account?

Based on the three fundamental conditions that should complete l'aedificato as defined by Vitruvio in De Architecturae, we look here to establish how the parameters that define Firmitas, Utilitas and Venustas in the real world are affected by the introduction of the Virtualitas, and the manner in which these translate, adapt, and compare to the new conditions, necessities and uses in cyberspace.

This comparative analysis of Real Architecture and Virtual Architecture is presented here in the form of a table (Table 1), considering the principles enunciated by Vitruvio and applying to each case and enhancing a previous work developed by the authors. (Hernández-Ibáñez. & Barneche-Naya, 2013)

The detailed description of each one of the points carries a scope much greater than this article, as each deserves detailed study and debate. With that in mind, the table can serve as a guide for the analysis of Cyberarchitecture and its multiple facets.

Each of the three Vitruvian aspects are discussed briefly below as related to being adapted to the new meaning they acquire when exposed to the nature of the virtual world: the Virtualitas.

Firmitas

The firmness: Grouping the concepts derived from stable and durable construction. Involving the derived aspects of the construction processes and systems, the selection of materials and terrain, strength and durability, as well as protection against the elements and aggression and a rational use of necessary economic and energy resources in construction and maintenance. Extrapolating these concepts of real world Architecture to the virtual world should be considered along the following lines:

1. **Stability:** Stable virtual Architecture would be sized within the limits of the structural system that supports it. In a computing environment, this involves the geometric and textural complexity and the number and characteristics of the processes that are carried out to run the virtual building which could, in case of exceeding the capacity of the server, cause system failure.
2. **Systems and Process of Construction:** The topological relationship of constructed space in the virtual world is established in a manner similar to the

Table 1. Adapting concepts related to the Vitruvian triad to Virtualitas

AEDIFICATIO	Related concepts	Real World Architecture	VIRTUALITAS
FIRMITAS	Stability	Application of physical laws.	Physical laws are optional.
		Selection of inner structure and basement.	Building stability is more related to computer stability; avoiding overloading machine and network resources.
		Design of constructive details to stand against external actions.	Protection of design against unauthorized modification.
	Constructive systems and processes	Architect: design. Builder: construction.	Architect: design and construction. Architect and/or Programmer: interaction.
		Complex constructive processes.	Modeling > Texturing> Lighting > Programming interactivity
	Materials selection	Materials selected for visual appeal, protection and durability.	Material selection based on appearance only. All visual aspects of RW materials can be emulated. New material properties: Mutability, multimedia and procedurality.
	Terrain	Terrain influences the building design.	Terrain can be designed to fit building needs.
	Durability	Aging, decay and wearing.	Aging by means of obsolescence, incompatibilities through versions, broken links.
		Temporary. Archaeology.	Timeless. Reconfigurable obsolescence according to needs.
	Protection	Shelter for atmospheric agents.	Unnecessary.
		Security against external aggressions.	Access control. Closure against malicious activities
	Economy	Construction units (number of bricks, cubic meters of concrete, etc)	Economy of data. Amount of vertices, polygons, size of textures, etc.
		Energy (electricity, fuel, manpower).	Bandwidth, CPU and GPU use, cost of modeling and programming. Information storage and retrieval
		Maintenance of materials and systems.	Maintenance of server and network.

continued on following page

Table 1. Continued

<table>
<tr><th>AEDIFICATIO</th><th>Related concepts</th><th>Real World Architecture</th><th>VIRTUALITAS</th></tr>
<tr><td rowspan="22">UTILITAS</td><td rowspan="7">Program</td><td>Living.</td><td>Virtual residence. Personal space.</td></tr>
<tr><td>Leisure.</td><td>Games. Social casual activities and relationships.</td></tr>
<tr><td>Business and work.</td><td>e-Business, telecommuting.</td></tr>
<tr><td>Exhibition.</td><td>Virtual exhibits and museums.</td></tr>
<tr><td>Transport.</td><td>Teletransport integrated into virtual buildings.</td></tr>
<tr><td>Public space – Private space.</td><td>Public–private–personal spaces.</td></tr>
<tr><td>Fora.</td><td>Meeting places, game lobbies.</td></tr>
<tr><td rowspan="6">Spatial Organization</td><td>Euclidean Space</td><td>Logical Space</td></tr>
<tr><td>Static. Fixed.</td><td>Dynamic. Motion. Reconfigurable.</td></tr>
<tr><td>Topological qualities of building space: open, closed, centric, directional, articulated, etc.</td><td>Same plus new capabilities derived from new geometries, mutability, weightlessness, practically limitless dimensions and teleportation.</td></tr>
<tr><td>Function – Form relationship.</td><td>Empowered by adaptability of shape of building elements to functions, both static and dynamically.</td></tr>
<tr><td>Wayfinding and navigation.</td><td>Navigation: cognitive maps and aids to navigation (Visual anchors, marks within the virtual environment, maps, text, etc)</td></tr>
<tr><td>Antropometry and Ergonomics.</td><td>Avatar sizes not necessarily in the human limits or not necessarily human-like at all.
Design to support new actions, i.e. flying.</td></tr>
<tr><td rowspan="2">Conditioning</td><td>Domotics.</td><td>Every aspect of the building is intrinsically programmable.</td></tr>
<tr><td>Installations and systems.</td><td>Lighting control. Multimedia control. Data flow control.</td></tr>
<tr><td>Accessibility</td><td>Local/ Territory</td><td>Global/ Network</td></tr>
<tr><td rowspan="2">Environment</td><td>Relation with landscape.</td><td>Construction of landscape.</td></tr>
<tr><td>Orientation to Sun and winds.</td><td>Control of Sun and wind.</td></tr>
<tr><td rowspan="2">Ecology</td><td>Environmental and visual impact.</td><td>Control of contents exposed. Fitting to in-world local building trends and customs.</td></tr>
<tr><td>Waste treatment. Recycling.</td><td>Control of objects and data left behind by users.</td></tr>
<tr><td rowspan="3">VENUSTAS</td><td>Composition rules</td><td>Order, proportion, symmetry, modulation, rhythm, etc.</td><td>Research of new composition rules and relations.</td></tr>
<tr><td>Space qualities</td><td>Expressivity of materials, light, color, distance, size, direction.</td><td>Spaces that produce sensations and feelings through seeing and hearing. Different perception of variables such as close and far, up and down.</td></tr>
<tr><td>Formal languages. Styles</td><td>Expressionism, Rationalism, Modernism, Postmodernism, Minimalims Deconstructivism,</td><td>New emerging concepts: Liquid Architecture, Transvergence, Cybrid. Responsives Architectures. Styles still to consolidate.</td></tr>
</table>

real world but not its geometry as the virtual medium allows for greater freedom of form. The construction processes of the virtual world are much simpler than in the real world: modeling, texturing, illuminating, programming. The architect not only designs but often constructs as well. The classic role of the builder disappears in the virtual world and in its place appears the programmer responsible for coding interactivity as well as the computer system.

3. **Selection of Materials:** The visual aspect of the building is a result of the selection and combination of materials, which is true in both worlds. All the construction materials of the real world can be emulated in the virtual world. Here, the virtual adds three new characteristics to the building materials: mutability, multimedia and purely computational entities that can be used as materials such as volumetric lights, particle systems, etc.
4. **Selection of the Terrain:** In real world Architecture, the terrain is a determining factor for design of the building. In virtual Architecture, the terrain and even the landscape may be a design decision at the level of the building itself.
5. **Durability and Sustainability:** The virtual building should be protected so far as possible from aging, that is: the obsolescence of technology, the disappearance of online references, data loss and backward incompatibilities.
6. **Security and Protection:** The building should protect itself and its occupants from external agents. The virtual world is not so different in this respect. But where weather does not threaten, the security of the data contained in the virtual world and the integrity of the building should be ensured through an adequate system of access permissions and barriers of protection against unauthorized access and malicious activities.
7. **Economy:** Good design in virtual construction, as in the real world, should adapt to economic and energy criteria. A well-constructed virtual building has a good balance of built spaces and the quantity of data necessary for elaboration, such as the number of polygons or the quantity of textures, resulting in an efficient rendering, an adequate consumption of bandwidth and adaptation to a wide range of graphic hardware. Additionally, virtual Architecture requires a specific space for information storage and retrieval.

Utilitas

1. **Utility:** Brings together the concepts related to the capacity of the building to satisfy a program and the method by which this is carried out. Fulfilling the Utilitas is the requirement that more deeply influences the way in which space in the construction is organized. Different uses give a place a different spatial organization and, in that sense, virtuality brings uses unknown in real world Architecture. Other aspects relative as this compliance is carried out

are in this category, such as: adaptation to the environment, the conditioning of the building and the environmental impact of the construction.

2. **Program:** The program of a building describes the use requirements that must be satisfied. It is linked with the needs of the user, lifestyles and modes of inhabitation, desires and habits. There are multiple real world uses that can be satisfied through virtual buildings, such as training, business and labor, museum, and recreation. Other uses, such as housing, can exist in symbolic form as can be seen in some Metaverses where the user-inhabitants construct their personal space; an extension of their real world private space with characteristics very similar to this. As for urban spaces, virtual environments are based on the encounter and communication, giving a new essence to the classical concept of the forum, places of interlinguistic and intercultural encounter and interchange, both socially and economically.
3. **Spatial Organization:** The spatial organization of the building and cities of the virtual world follow the same guides that organize real world space in terms of opening, closing, centralization, directionality, articulation, etc, but with new potentials that emerge. On the one hand, the capacity of generating spaces of any size and shape, on the other hand the mutability of the geometry that permits the form to adapt to the function in each moment, such that the same space can completely transform to adapt to various uses. Domotics is not an option here but is an intrinsic part of the essential, programmed space.

When experiencing complex spaces, it is important to include some aids to the navigation in the virtual environment by means of the use of "visual hooks" which help identify the different regions of the virtual space. Maps and mini-maps are also a good way to help the user in the process of creating his cognitive maps and to provide him with a quick general view of the spatial structure and the existing connections. Since this information can be overlaid in the screen to the tree-dimensional space, the user can navigate the virtual world and the map at the same time.

The anthropometric relationship between the user that can be represented in first and third person and the architectural space in which he is displayed are also very different because the avatar does not have necessarily to follow the metrical patterns of human means, so that ergonomic issues must be studied for each distinct case in the virtual world.

4. **Accessibility:** In real world Architecture, the building is located in a precise geographical location, while in virtual Architecture the building is not linked to a point in the territory. The virtual building is accessible from any place in the planet if the user has the right permission.

5. **Adaptation to the Environment, Ecology:** The relationship of the virtual architectural program with its environment sees this as more an element of design.

Environmental aspects such as sunlight, are configurable and can be adapted as part of the creation of the constructed element. However, when the virtual space is shared by various buildings, it is vulnerable to questions of the visual impact and other problems that emerge in the making of the virtual building which is essentially a container of information and can be a transmitter of unacceptable contents in some cases and therefore require some form of control. Neither is the building free of problems of waste since the actions of the users can leave permanent marks and scrap elements that necessarily should be filtered and eliminated.

Venustas

Beauty: Vitruvius refers here partly to the capacity of Architecture to produce aesthetically pleasing sensations that in contemporary context would be generalize to sensations of any kind, provided they are infused with intent and meaning.

The classical rules of architectural composition, based on synthetic relationships between the spatial elements that form the building such as order, symmetry, rhythm and modulation. Relating the parts and the whole are still valid but should be reinterpreted in the composition space where the variables that define space, distance, size, and time have been relaxed to contain values unthinkable in the physical world.

It this sense, there exists an important, common criticism that can point out the limitation of the ways of the real world. Although it could be important to provide the user of Cyberarchitecture with orienting marks that facilitate comprehension of the space in which the user is immersed, this can result in a burden that slows the appearance of new languages of architectural composition exclusive to the virtual world.

New forms of Virtual Architecture inside the cyberworld must respond to a diversity of requirements that extend from the definition of the spatial limits to the establishment of the building behavior, the pursuit of computational efficiency and the control of contents. Although many of these requirements are different from those of a real building, a parallel exists for most of them to concepts associated with a building that is constructed in the virtual world, being equally possible to group them completing the different aspects which, following the Vitruvian triad, allow us to define an object as being Architecture.

This yields support to the idea that design of Cyberarchitecture can be considered another form of architectural design and not just a simple formal game. Hence, it can find its place in the field of Theory of Architecture, beginning with the fulfillment of the classical requirements. The classification previously described enables one to use it as a theoretical framework for analysis and study in aspects that go beyond aesthetic and formal issues, dealing also with constructive and commodity aspects. The table is nevertheless open to the inclusion of new concepts.

CONCLUSION

Although there are parallels with many aspects present in the architectural design of the real world, there are others that are typical of cyberspace that do not have a parallel in real world Architecture.

Based on this approach, the first sections analyzed the forms of Cyberarchitecture created within the virtual worlds and how they respond to the different requirements, ranging from the definition of their spatial limits and the behavior of the building through its Computational efficiency and the control of its contents.

From this first analysis, the intention was to reflect on the characteristics of these new types of Architectures in comparison with their equivalents in the real world, through the lens of the three classic classical conditions of Architecture such as Firmitas, Utilitas and Venustas defined by Vitruvio and how these are affected, modulated or modified in the virtual world with the introduction of the Virtualitas concept.

Without a doubt, the design of virtual architectural spaces, despite its intangibility, poses its own challenges in the formal responses to the needs of the avatars, who as virtual representations of the users, interact in cyberspace requiring environments in which to meet, learn or have fun. They are spaces perceived in the third person, but understood with the same qualities with which the real space is experienced. In this sense, research in the design of innovative proposals for Architecture in cyberspace allows exploring new opportunities and experimenting with unconventional ideas for the various possible activities.

It is evident that we are in an initial but promising moment for the creation and the architectural design within the metaverse. For cyber-architects, these virtual worlds represent a challenge filled with new design formulas to be explored, and waiting to be exploited.

REFERENCES

Alahuhta, P., Nordbäck, E., Sivunen, A., & Surakka, T. (2014). Fostering team creativity in virtual worlds. *Journal of Virtual Worlds Research*, *7*(3).

Anders, P. (1994). The architecture of cyberspace. *Progressive Architecture*, *75*(10), 78.

Asymptote Arch. (1999). *Virtual Guggenheim Museum*. Retrieved June, 2017, from http://openbuildings.com/buildings/guggenheim-virtual-museum-profile-2437

Barneche-Naya, V., & Hernández-Ibáñez, L. (2014). Evaluating user experience in joint activities between schools and museums in virtual worlds. *Universal Access in the Information Society*, *14*(3), 389. doi:10.1007/s10209-014-0367-y

Bartle, R. A. (2004). *Designing Virtual World*. Berkeley, CA: New Riders.

Benedickt, M. (1991). *Cyberspace: first steps*. Cambridge, MA: MIT Press.

Bertol, B., & Foell, D. (1997). *Designing Digital Space: An Architect's Guide to Virtual Reality*. John Wiley & Sons.

Bier, H. H., & Knight, T. (2010). Digitally-driven architecture. *Footprint: Delft School of Design Journal*, (6).

Biloria, N., & Chang, J. R. (2013). Hyper-Morphology: Experimentations with bio-inspired design processes for adaptive spatial re-use. *eCAADe 2013: Computation and Performance Proceedings of the 31st International Conference on Education and Research in Computer Aided Architectural Design in Europe*.

Bower, M., Lee, M. J., & Dalgarno, B. (2016). Collaborative learning across physical and virtual worlds: Factors supporting and constraining learners in a blended reality environment. *British Journal of Educational Technology*.

Brouchoud, J. (2013). *New Virtual Learning and Workplace Campuses for Avaya with Gronstedt Group*. Retrieved June 2017, from http://archvirtual.com/2013/01/29/new-virtual-learning-and-workplace-campuses-for-avaya-with-gronstedt-group/

Bullivant, L. (2005). *4dspace: interactive architecture (Architectural Design)*. John Wiley & Sons.

Caillois, R. (1961). *Man, play, and games*. University of Illinois Press.

Castronova, E. (2005). *Synthetic Worlds: The Business and Culture of Online Games*. Chicago: University of Chicago Press.

Cook, N., & Winkler, S. L. (2016). Acceptance, Usability and Health Applications of Virtual Worlds by Older Adults: A Feasibility Study. *JMIR Research Protocols*, *5*(2), e81. doi:10.2196/resprot.5423 PMID:27256457

Creutzfeldt, J., Hedman, L., Medin, C., Heinrichs, W. L., & Felländer-Tsai, L. (2010). Exploring virtual worlds for scenario-based repeated team training of cardiopulmonary resuscitation in medical students. *Journal of Medical Internet Research*, *12*(3), e38. doi:10.2196/jmir.1426 PMID:20813717

Dalgarno, B., & Lee, M. (2010). What are the learning affordances of 3-D virtual environments? *British Journal of Educational Technology*, *41*(1), 10–32. doi:10.1111/j.1467-8535.2009.01038.x

Duncan, I., Miller, A., & Jiang, S. (2012). A taxonomy of virtual worlds usage in education. *British Journal of Educational Technology*, *1*, 1–16.

Ehsani, E., & Chase, S. (2009). Using virtual worlds as collaborative environments for innovation and design: lessons learned and observations from case studies in architectural projects. In *Computation: The New Realm of Architectural Design* (pp. 523-531). Estambul: eCAADe.

Fischnaller, F., & Hill, A. (2005). CITYCLUSTER. From the Renaissance to the Megabyte Networking Age: A Virtual Reality and High-Speed Networking Project. *Presence (Cambridge, Mass.)*, *14*(1), 1–19. doi:10.1162/1054746053890251

Ghanbarzadeh, R., Ghapanchi, A. H., & Blumenstein, M. (2015). Characteristics of Research on the Application of Three-Dimensional Immersive Virtual Worlds in Health. In *International Conference on Health Information Science* (pp. 213-224). Springer International Publishing. doi:10.1007/978-3-319-19156-0_22

Gibson, W. (1984). *Neuromancer*. New York: Ace Edition, Berkley Publishing Group.

Gregory, S., Scutter, S., Jacka, L., McDonald, M., Farley, H., & Newman, C. (2015). Barriers and enablers to the use of virtual worlds in higher education: An exploration of educator perceptions, attitudes and experiences. *Journal of Educational Technology & Society*, *18*(1), 3–12.

Guynup, S. L. (2003). *Bespace, GaTech*. Retrieved June, 2017, from http://bespace.lmc.gatech.edu/single/

Hack, C. J. (2016). The benefits and barriers of using virtual worlds to engage healthcare professionals on distance learning programmes. *Interactive Learning Environments*, *24*(8), 1836–1849. doi:10.1080/10494820.2015.1057743

Hernández-Ibáñez, L., & Barneche-Naya, V. (2013). The Building as the Interface: Architectural Design for Education in Virtual Worlds. In *International Conference on Virtual, Augmented and Mixed Reality* (pp. 152-161). Springer Berlin Heidelberg. doi:10.1007/978-3-642-39420-1_17

Hernández-Ibáñez, L., & Barneche-Naya, V. (2013). Vitruvius in Cyberspace. *International Journal of Art, Culture and Design Technologies*, *3*(1), 36–47. doi:10.4018/ijacdt.2013010104

Hew, K. F., & Cheung, W. S. (2010). Use of three-dimensional (3-D) immersive virtual worlds in K-12 and higher education settings: A review of the research. *British Journal of Educational Technology*, *41*(1), 33–55. doi:10.1111/j.1467-8535.2008.00900.x

Hill, J. (2001). *Architecture: the subject is matter*. Psychology Press.

Huvila, I., Ek, S., & Widen, G. (2014). Information sharing and the dimensions of social life in Second Life. *Journal of Information Science*, *40*, 237-248. DOI: 0165551513516711

Kandalaft, M. R., Didehbani, N., Krawczyk, D. C., Allen, T. T., & Chapman, S. B. (2013). Virtual reality social cognition training for young adults with high-functioning autism. *Journal of Autism and Developmental Disorders*, *43*(1), 34–44. doi:10.1007/s10803-012-1544-6 PMID:22570145

Le Corbusier. (1923). *Vers une architecture*. G. Crès et Cie.

Lubell, S. (Dec. 4, 2008). *Original Sim*. Retrieved June, 2017, from http://www.nytimes.com/2008/12/07/style/tmagazine/07secondlife.html

Lynn, G. (1996). Blobs, or why tectonics is square and topology is groovy. ANY: Architecture New York, (14), 58-61.

McCullough, M. (2005). *Digital ground: Architecture, pervasive computing, and environmental knowing*. The MIT Press.

McGonigal, J. (2011). *Reality is broken: Why games make us better and how they can change the world*. Penguin Group.

Merrick, K., Gu, N., & Wang, X. (2011). Case studies using multiuser virtual worlds as an innovative platform for collaborative design. *Journal of Information Technology in Construction*, *16*, 165–188.

Mitchell, W. J. (1996). *City of bits: space, place, and the infobahn*. MIT Press.

Mitchell, W. J. (2003). ME, the cyborg self and the networked city. MIT Press.

Novak, M. (1991). Liquid architectures in cyberspace. In M. Benedikt (Ed.), *Cyberspace: First Steps* (pp. 225–254). Cambridge, MA: MIT Press.

Novak, M. (1996, December 9). *TransArchitecture. Building the edge of thought*. Retrieved June, 2017 from https://www.heise.de/tp/features/transArchitecture-3445869.html

Oosterhuis, K. (1998). *Spacestation. Hyperbody*. Retrieved June, 2017, from http://www.onl.eu/projects/space-station

Oosterhuis, K., & Jaskiewicz, T. (2007). 798 Multiplayer Design Game. A New Tool for Parametric Design. In Space Time Play. Computer Games, Architecture and Urbanism: The Next Level (pp. 358-361). Basel: BirkhäuserPublishing.

Ostler, T. (1994). Architecture in cyberspace. *Architects'. Journal, 199*(21), 33–35.

Ryan, M. L. (2003). *Narrative as Virtual Reality. Immersion and Interactivity in Literature and Electronic Media*. Baltimore, MD: Johns Hopkins University Press.

Schmidt, M., Laffey, J. M., Schmidt, C. T., Wang, X., & Stichter, J. (2012). Developing methods for understanding social behavior in a 3D virtual learning environment. *Computers in Human Behavior, 28*(2), 405–413. doi:10.1016/j.chb.2011.10.011

Schultz, R. (2007). *Designing the Architecture of Architectural Design Collaboration: Studio Wikitecture 2.0*. Retrieved June, 2017, from http://studiowikitecture.wordpress.com/2007/04/15/wikitecture-20-designing-the-architecture-of-architectural-design-collaboration/

Serafijn, Q.S., & Spuybroek, L. (2004). Emotional landscapes in the City. D-Tower in Doetinchem. *(In) Security*, (6), 142-150.

Spiller, N. (1998). *Architecture in Cyberspace II*. Academy Editions.

Spiller, N. (2009). Plectic architecture: Towards a theory of the post-digital in architecture. *Technoetic Arts: a Journal of Speculative Research, 7*(2), 95–104. doi:10.1386/tear.7.2.95/1

Stichter, J. P., Laffey, J., Galyen, K., & Herzog, M. (2014). iSocial: Delivering the social competence intervention for adolescents (SCI-A) in a 3D virtual learning environment for youth with high functioning autism. *Journal of Autism and Developmental Disorders, 44*(2), 417–430. doi:10.1007/s10803-013-1881-0 PMID:23812663

Tea Cup Communication. (2008). *Tea Cup Communication Enters Second Life Business Partnership with Atelier Ten.* Retrieved June, 2017, from https://www.gmo.jp/en/news/article/?id=261

Vitruvius Pollio, M. (23-27 adC). *De Architecturae, Liber primus.* Retrieved June, 2017, from http://www.thelatinlibrary.com/vitruvius1.html

Wilhelmsson, U. (2008). Game Ego presence in video and computer games. In Extending Experiences. Structure, analysis and design of computer game player experience (pp. 58-72). Lapland University Press.

Williams, D. (2010). The mapping principle, and a research framework for virtual worlds. *Communication Theory, 20*(4), 451–470. doi:10.1111/j.1468-2885.2010.01371.x

Chapter 9
Soundscaping

Michael Johansson
Kristianstad University, Sweden

ABSTRACT

In this chapter, the author presents how he developed different processes for collectively producing a series explorative soundscapes through interface creation and mechanical artifacts using specific constraints influenced by theories of art, design, and architecture. He shows how he worked with a design methodology that brought together an editor and the final expression of the artwork into one surface of interaction and execution using a virtual cityscape as an iterative ground for sound and music explorations, and gives some examples of different prototypes and iterations. The author also discusses how he tweaked/iterated with the parameters of the framework, the sounds, and the final visual expression to match his artistic intention, and finally to bring some noise.

BACKGROUND

In 1999 we ended an art project called "from an indefinite point in the Cartesian space" that had generated 2000 low-res and 550 high-resolution models of buildings, interiors, objects and exteriors split up in over 50 scenes. Here was a unique possibility to do just that. Therefore, we extracted all of the models from the separated scenes and placed them on top of a superimposed infrastructure of sixteen different formula one tracks. We show how we used personas, role-playing (GURPS) and conceptual mathematical formulas to be able to explore and furniture the world. We named the virtual world "the city of Abadyl", and made it to the initial venue for the project.

DOI: 10.4018/978-1-5225-5023-5.ch009

Against the self-evident – a thorough indefiniteness, a defined obscurity A ”wild thinking” aiming to undermine the present and prevalent must nevertheless have a starting point, and a location in which to perform its laboratory work. Such a location was placed unintentionally on the map of the possible in the mid-seventies when Swedish Public Broadcasting, educating their listeners how to manage the new stereo technique, we're establishing that: my voice will now be coming from the right my voice will now be coming from the left my voice will now be coming in between the loudspeakers my voice will now be coming from an indefinite location in the room.

This indefinite location in the room is something completely different than the outside location of the natural sciences, the point from which reality is measured and translated into objectivity. [This point too has proven itself absurd (even if strikingly efficient). Gödel, Heisenberg, Bohr etc] Then instead of an indefiniteness within the room, and a voice imperatively calling forth its own elusive presence. Within the room but not clear where, in many ways resembles the location of the potential in the prevalent, given. A floating possibility hidden in the persistently present.

From the exhibition catalog Filedasy text Galley Skanes konst

THE CITY OF ABADYL

Abadyl is a proposed city, a fantasy, a set of codes and models, a library of artifacts and prototypes and foremost it is it's co-creators. From the beginning attempts to understand and redefine our world in a situation where information is lacking. This lack of information is used as a resource. By providing ambiguous fragments as a starting point, the scenarios put no constraints on imagination. It has become a large database of material that is interlinked through the architecture of a city, regardless of its incompatibilities. It creates an open-ended way of working with art and design based on scenarios extracted from the framework and rule-set of the city. It has deliberately become a detailed environment for our worldmaking experiments, where we have played around with both real and fictitious assumptions. Using as a narrative structure for this design fictions, we created scenarios, plotting a series of human dilemmas that are applicable in relation to what it means to be human, how to plan and live our lives and to find some purpose in our journey across different contexts and situations, both future and present. We have also at the same time concluded a series of creative methods that can govern shared and inter disciplinary explorations and sometimes the creation of prototypes and artifacts. See Figure 1

Creating such a world and communicating it with others is not easy. The task both complex and demanding and the risk of being disappointed on what you will achieve is evident. The basic idea was to establish a space where to practice a critique of art, culture, and society, through an investigation of philosophy and criticism

Figure 1. Images from the first exhibition Gallery Skanes Konst 2003

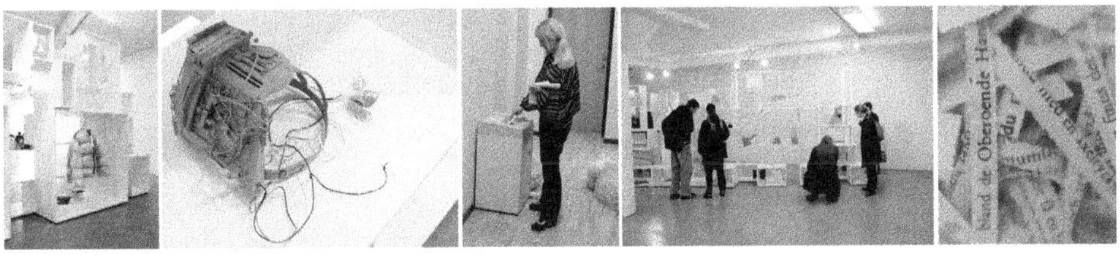

in a dynamic material in a mixed reality space. So to be able to establish such an infrastructure we came up with the idea to set up a closed but yet complex space, an area with blurred borders to the surrounded unknown. So to support this we looked for metaphors that could host this delimited space. The analogy to racing tracks was obvious; with roads that just looped themselves through their environment; Together with five architect students, I had to find, write and draw a set of characteristics for every part of the city. See Figure 2 A sort of simplified framework or pattern language *Alexander, C., S. Ishikawa (1977)* had to be developed. It was set up as follows:

A seven scale system that we introduced to handle the event space of the city it described the different levels on which objects and events can occur.

1. Environment
2. Building
3. Room

Figure 2. The city of Abadyl framework

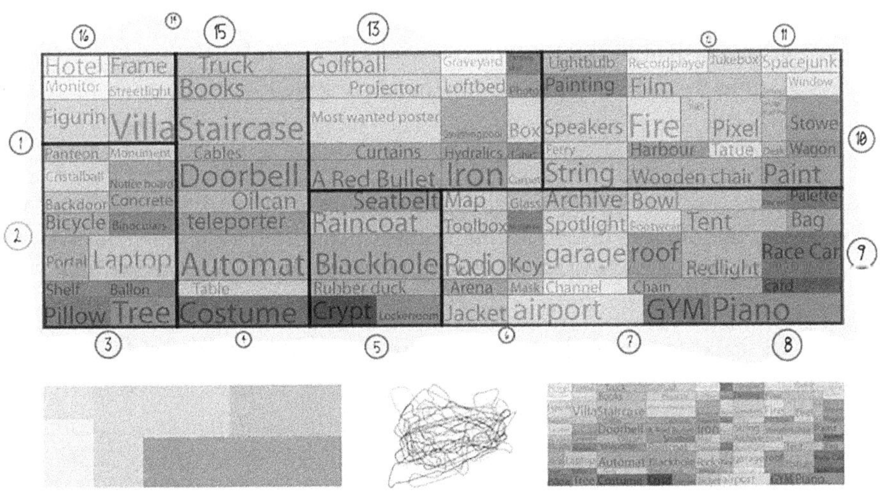

4. Furniture
5. Tool
6. Interface
7. Idea

Sixteen locations from different parts of the world constituting the starting point and were initially held apart. But later they were merged into a connected city. As joining infrastructure the series of sixteen Formula One tracks was chosen, piled on top of each other to a joint figure. Hence, aiming to expose architecture and cityscape to extreme strain (= life, unpredictability) The formula one track became our point of departure, shaping an interesting ornamentation of roads just waiting for be driven on. The city then divides into sixteen parts with the idea to host their own separate ideology, architecture, fashion, sports etc.

A Hundred objects to represent the city or the world was inspired by Peter Greenaways (1995) idea of an encyclopedic approach. In the various quarters of the city hundred objects would be found, each devised in a way that constituted an elegant description and was manifesting the aspects of their very own district. These objects is helping us shape the differences on a broader range of levels in every part of the city and serves as a series of obstacles for the explorations. The purpose is to interfere with both the already built objects and the activities that are going to occur here later on. In this way we happened to examine our original objects once again, when we were depicting them in every detail. To be very thorough in investigating its form, visualizes other layers in the object and creates conditions for both deconstruction and new narratives. To force out its meaning by depicting/modeling – to sort of philosophize from the street level and up. Over the years working with different projects I have learned about and experimented with how to transfer methods from creative writing and worldmaking (Goodman, N. 1978) into my own work practice in the fields of art and design. In Postscript to the Name of the Rose, Umberto Eco (1984) writes on the generative logic he has adopted, a logic which both limits and expands creativity. The fundamental parameters of this logic guide what can and cannot be included in a fictional but plausible universe. In my own work and together with others I have chosen to focus on the generative itself of this logic; that is, it is not about parameters resulting in a consistent, watertight universe, but rather what can be generated from numerous predetermined real and fictitious parameters. These parameters resonate through a mix of creative writing, physical artifacts, virtual objects, and modified game engines in a constant loop that results in novels, film, interaction designs, and art installations. By letting them evolve in these different settings they will separately host detailed and comprehensive perspectives, which incorporate surprising visual and technical proposals. “At the heart of creativity lie constraints: the very opposite of unpredictability. Constraints and unpredictability, familiarity and surprise, are somehow combined in original thinking." Boden (1997). So the idea of the framework is to create and support a design process that resulting in knowledge acquired during the process, and an artifact storing that knowledge (Armour 2000). The design process can thus be viewed as a process of knowledge acquisition or learning from the previously unknown. The knowledge acquired pertains not only to the particular domain of the artifact but to the design process itself. That is, we acquire knowledge on how to evolve both artifact and design process (Imai 1997).

Bring the Noise

Since 2009 started a series of collaborations called hit strip music, Timescapes and Abadyl of tunes with musicians and sound artists where we tried to explore the soundscape of Abadyl and bring some noise to the city, and also to provide some ideas of how the locations in Abadyl sounds like. And at the same time investigate how interface, interaction and visual expressions in different ways are tied together into an operational logic (Wardrip- Fruin, N. 2009) "When a work of digital media operates, this can be seen as an interplay between the elements of the model: data, process, interaction, author, and audience. Is the system actually doing what it is described as doing? What outspoken assumptions are built into the ways in which operations proceed."

Hit Strip Music

In Hit strip music we follow a musical composer on the day when he decides to stop play and write music. How he turns to his paper shredder to get rid of all the scores and notes, but suddenly rediscover them in a way he never imagines. In this scenario, the co-creators explore how short musical fragments connected to physical paper strips (that are tagged with rf-id tags) can be used to explore music making by play and combine musical fragments in real-time in different ways, and use them for composing new music.

Timescape

In this project we use a physical object based on three independently rotating discs with different graphical elements that picks up external information on the internet and local network connected to different activities, statistics, news etc that will influence and change the expression of the object, the disc's relation to each other and at the same time be able to trig different media in an unpredictable way

– mystical machine that both collect, express and actual sounds. See figure 3 that shows a 3d printed prototype.

3D Soundscapes

In 2017, I started a collaboration with the musician Kevin Kerney where we stage different sounds and musical elements in a 3d world that constantly evolves and the sound listener (what the listener hears) also constantly moves in a loop that changes based on outside world conditions and therefore the soundscape created here will be ever-changing. The soundscapes here consists of environmental sounds recorded in many locations in the world, composed music by Kevin Kerney and a series of monologues and dialogues written and recorded by me. We here try to build a fictive site specific experience of a place that never actually existed.

Abadyl of Tunes

The quest in this project and that is going to be presented more thoroughly in this article as follows, is to look at how to create an awareness for the invited co-creators of the underlying programming structure that makes the interaction and execution of the music possible and at the same time make it visible interesting to play, interact, watch and listening to - where interface and the final expression is the same.

Co-Creation

So the question is not only how you go about exploring a complex digital space but also how that work is set up and how the actual production is conducted. We have for over ten years used the framework of Abadyl and the fieldasy method (Johansson & Linde, 2004) to stage different events and spaces in the form of written scenarios. The scenarios are handed over to the invited temporary citizens and co-creators. They can then act in relation to the scenario, in and by themselves chosen environment.

Figure 3. Time scapes 3D printed prototype

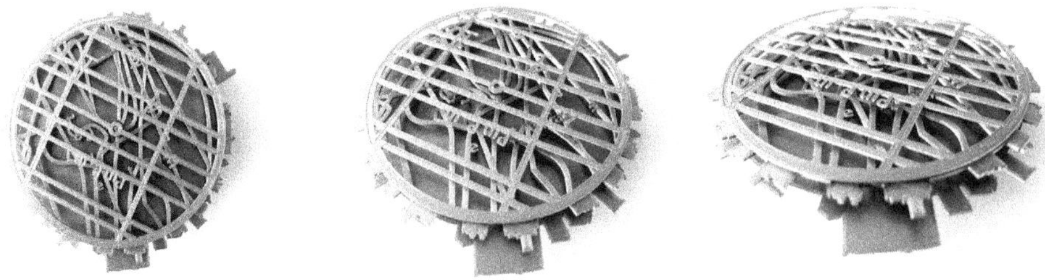

That, in the end, helped them produce the artifacts. Our scenarios try to slowly create a discreet dynamic tension and/or displacement between a person, objects, time, places, work-process and events that are not usually — if ever — associated into new and surprising conjunctions. By using scenarios we are able to provide detailed and specific data, which the co-creator can use as background material for their action. Scenarios relation to the overall project is loosely defined as to allow the creation of artworks, that though enriching the database, still are autonomous from the mother project in the sense that they can be exhibited by themselves. They also act as generators while they generate new and unforeseen processes which extend into new and likewise unforeseen contexts. As an Artist/Designer one should always be aware that one never design the actual co-creation experience, only the framework wherein that experience can take place.

So in 2011, I started a collaboration with the media designer Jim Hall.I was introduced to his work Isle of tune by one of my former students who saw his work and said that it had some similarities with The city of Abadyl and how we over the years used racing tracks to compose voice and sounds for different investigations and exhibitions. The idea was that we would make an Abadyl modification of the original Isle of tunes and produce sixteen soundscapes, one for each city part and a soundtrack for the city of Abadyl and release it as multiple art pieces over two years, and at the same time explore Abadyl sound and music wise. The proposed modification included different crossroad configurations such as random, clockwise and anti clockwise, to make variations of how a specific tune could be played and how the music developed over time.

Loops

When we are trying to follow the artistic intention or concept, we want and try to build our different prototypes to produce qualities not known beforehand. But to achieve this the prototype has to reach a certain state of a complexity, a complexity max (Hoberg, 2006) so we can begin to explore it thoroughly. Therefore, a production environment such as the Abadyl framework, with its, constraints, obstacles, limitations, stories, is rewarding for conducting practice based research in the area of art and new media. To many art projects are just illustrations of technology, and to many design projects, unreflected use convention and qualities from art. We do art well, and computing more dubiously; we do computing well but the art is questionable. (Bardzell, 2009)

The underlying shapes of racing tracks or loops in the city have been used by us and other in different projects in time-based media. In our loop-based scenarios, we try to employ parallel and repetitive elements, we can establish and create an open narrative field where creative input from the user can add new elements to the experience. The loops are static and dynamic and can be varied in length and tempo and be interlinked and/or crossed by other loops. The loop also, in its basic version can be seen as a tool for reflection, a moment to reverse the constant streams of attractions and desires created and recreated in every digitally performed action and story. Since new media itself has matured, the process is no longer depended on the predecessor's more traditional and linear methods of authoring, instead, every part of the process is constantly changing the way we author, script and express a multi threaded open work.

When working with 3d graphics and sound we choose from a partly new, partly re-appropriated, a palette of narrative tools as the loop. Were the time-based linearity can be substituted with other means of dramatic tension, in a mix of controlled, generative and random execution functions, accessed through the interface and at the same time creating the visual expression of the artwork itself. By closely integrating the interface, the visual expression and the programmable structure we wanted to facilitate an open tool for our collaboration. Where the constraints initially can be challenging since it breaks with the user's conceptual space (Boden, M, 1990) but after a couple of days of use it reveals a dense array of possibilities hidden in its obvious limitations both sound and image vise.

Compositions

For Abadyl of tunes, we used two basic concepts for the design of our editor on a more philosophical level. Firstly we used something that already been explored in the Abadyl project in earlier works. Coming from Scandinavia with a firm tradition of participative design, we wanted to contrast our design work in the relation to that by using some of Sol Lewitts (1969) Sentences on conceptual art. Sentence no: 29

The process is mechanical and should not be tampered with. It should run its course.

Based on Sol Lewitt's sentence we designed a fixed and nearly mechanical setup where only two options were available for the co-creators — adding sounds to the database and ordering of the interactive graphical elements, houses, trees, objects, roads, and cars. Working with explicit and visual recognizable constraints triggers creativity. The "mapping" of a conceptual space involves the representation, whether at conscious or unconscious levels, of its structural features. Much as a real map helps a traveler to find — and to modify — his route, so mental maps enable us to explore and transform our conceptual spaces in imaginative ways. Boden (1997) At first it looked simple and trivial to the co-creators but they soon learned how complex a constraints' system can be its simple programming principles were revealed for the user. The paradox is that constraint enables creativity. By constraining the generative system into an appropriate conceptual space, a writer gains a conceptual structure that can be systematically explored and transformed. A conceptual space can be explored by testing the sounds by the existing constraints. Sharples (1996) In this way of constructing a city space for musical execution, the co-creators soon learned how in different ways make use of the space for exploring and designing their sound scape. See figure 4 that shows some of our design iterations of the story engine.

Secondly, we went back to one of the basic ideas we worked with when Abadyls architectural space was designed. Through Boden (1994) Creativity cannot only and simply be reduced to novel combinations of old ideas. For example, In Architecture and Disjunction, Bernhard Tschumi outlines a framework for multiple combinations and substitutions that exist simultaneously, including the following three operations:

- **Cross-Programming:** The use of a space not as intended.
- **Transprogramming:** The combination of disparate programs.
- **Dis-Programming:** Disparate programs that contaminate each other.

These operations possibilities of combinations and permutation of existing programs as well new programs. In the same book, he also stated that there are four ways of working with architecture.

1. Design a masterly construction, an inspired architectural gesture (a composition)
2. Take what exists, fill in the gaps, complete the text, scribble in the margins (a compliment)
3. Deconstruct what exists by critically analyzing the historical layers that preceded it, even adding other layers derived from elsewhere, from other cities, other parks (a palimpsest)
4. Search for an intermediary — an abstract system to mediate between the site (as well as all given constraints) and some other concept, beyond city or program (a mediation)

Figure 4. Formula one story engines in collaboration with Jon Ovander

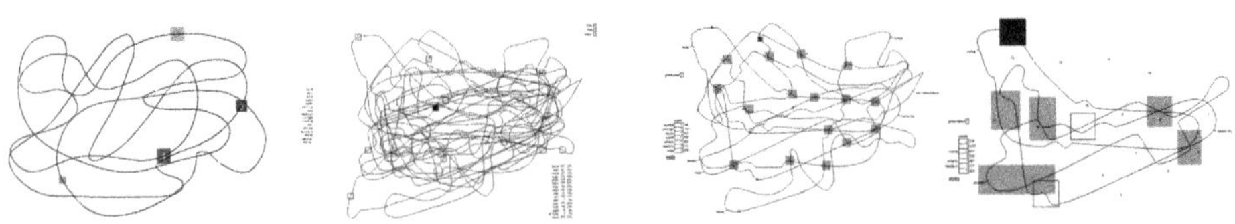

For this project, we choose to work with the first one since the sounds and graphical elements in its final form have all the characteristics of a composition, a score, a sound player and a visual expression capture in one single unit — it plays what you see, an example is shown in Figure 5

In Recombinant urbanism (David Graham Shanes, 2005) the Seven "-ages" of postmodernism maps well how this project operates in the 7th one of them called Rhizomatic assemblage. "Rhizomic assemblage differs from another assemblage in that it reintroduces the narrative path as an important element of the design process." In our case the scenarios written for the co-creators. "There is still no emphasis on a single center or command position, still no place of total control. Instead, there are multiple narratives that thread through the city, intersecting and bypassing each other as the case may be." And this is done by how the co-creators chose to lay out the roads, attached the buildings and where the cars should start.

Interface, Interaction, and Iteration.

The elements we use in the isle of tune modification consists of ;

- Forty-four one second long audio tracks. Five provided from recordings from the actual locations where the fragments of the city part was originally captured Thirty-nine one second long audio tracks provided by the invited co-creators
- Twelve graphical 3d rendered elements to represent the city part's architecture
- Three vehicles

Figure 5. Abadyl of tunes map, show a collaboration with musician Johan Salo

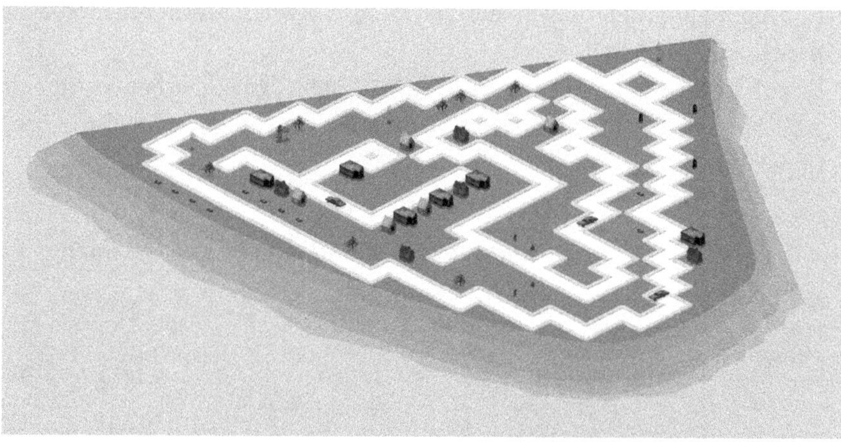

The structure for this setup we use the original Isle of tune 2d grid that matches the different city-parts that constitute how the co-creator can layout the graphical elements, with connected sounds. The graphical elements are divided into six groups were four of them contains the sound database and the other two is used for layout the roads and the cars.

The importance of this project is to let the co-creators have full access to both the media database and the layout of the sounds and player objects on the grid. This establishes a practice for the invited co-creator that make use of the geometrical or geographical space where interface and visual expression are inseparable units. The idea here is to reveal the certain levels of the programming structure of both the visual and soundscape construction for the co-creators so they can easily understand its constraints.

Worldmaking

Quite late in this project we recognized Nelson Goodman´s Ways of worldmaking (1978) and recognized how we over the years in a similar way worked with his proposed activities of worldmaking that is:

Composition and Decomposition Weighting Ordering Deletion and Supplementation Deformation

Layout, ordering and editing the roads includes periodicity and proximity; and the standard editing and ordering of the tunes in the database as well as adjusting the rhythm and tempo of the different tunes that are played by the cars.

Taking apart and putting together, analyzing the components at the same time composing wholes, by connecting features into complexes, by adding, breaking up, and restructuring. The importance here is to let the co-creators have full access to both the media database and the layout of the sounds and player objects on the grid. Establish a practice for the invited co-creator that make use of the geometrical or geographical space where interface and visual expression are inseparable units in their making of their world. The idea here is to reveal the certain levels of the scripting structure of both the visual and soundscape construction itself.

The art and act of letting the cars drive around in the constructed space are to the urban system what the pickup is to the record player. Where the three cars follow the thick and thin of an urban tune. And by letting the cars drive through the streets of the city one shapes, plays and share the soundscape of Abadyl.

CONCLUSION

The now four projects we have conducted so far have all proven that this limited approach challenges the co-creators presumptions about how to create soundscapes, play music and construct city-spaces, even though all the different parts are already known beforehand.

This way of letting the co-creators redefine a world in a situation where information is both fixed and limited has again proven both for us who are working with the actual framework and writing the scenarios but also for the co-creators themselves, were they both de/reconstruct their music making capabilities through limitations and constraints. So by providing ambiguous fragments (content/constraints/interface) as a starting point, we managed to challenge the co-creators assumptions in order to produce soundscapes and dwell the city of Abadyl at the same time. Where the question for us is how to create an awareness the underlying programming structure that makes the interaction and execution of the music possible and at the same time make it visible interesting to look and listening upon. By the interface revealing some scripting structures for co-creators with no or limited scripting knowledge, it adds a positive friction in relations to the co-creators own professional knowledge and support them in bringing the noise to Abadyl.

FUTURE WORK

In a time of big data, filter bubbles and data mining and the idea that the flow of information could be seen as an ever-changing landscape or even a continent, that is explored and discovered at the same time. The question for me in this line of work is how we could use the set of rules and procedures for further visual and artifactual explorations. Where one would like to probe that landscape/continent (Johansson, 2013) through different data-sets, by mixing real time data with digital 3d generated imagery and/or physical objects and installations based on similar principles described in this paper. Using dynamic visual expressions and real time simulations as the main surface of generating expressions that constantly, discrete and slowly influence imagery/artifact on different digital surfaces and artifacts. We hope to discover new relationships between our actual world and the data that represent it and maybe reveal something new in the already known.

ACKNOWLEDGMENT

Thanks to Jim Hall letting us use Isle of tune and all the modifications he implemented for making our explorations possible. Also, thanks to Johan Salo (Abadyl of tunes, Jon Ovander (Formula one story engine), Lukas, Isak & Noak Jönsson (Hit strip music) and Kevin Kerney (3D soundscape)

REFERENCES

Alexander, C., & Ishikawa, S. (1977). *A pattern language: Towns, buildings, construction*. New York: Oxford University Press.

Bardzell, J. (2009). Interaction Criticism and Aesthetics. In *Proceedings of CHI'09: World Conference on Human factors in Computing Systems*. ACM. doi:10.1145/1518701.1519063

Boden, M. (1997). *The Creative Mind: Myths and Mechanism*. London: Weidenfeld &Nicholson.

De Monchaux, N. (2016). Local code: 3,659 proposals about data, design & the nature of cities. New York: Princeton Architectural Press.

Goodman, N. (1978). *Ways of Worldmaking*. Indianapolis, IN: Hackett Publishing.

Greenaway, P. (1992). *Hundert Objekte zeigen die welt* [Hundred objects to represent the world]. Verlag erd Hatje.

Hoberg, C. (2006). *Komplexitets Max*. Stockholm, Sweden: Dialoger Förlag.

Johansson, M., & Linde, P. (2004). *Fieldasy*. Sheffield, UK: Fieldasy. Pixel Raiders.

Nelson, H. G., & Stolterman, E. (2003). The design way: Intentional change in an unpredictable world: foundations and fundamentals of design competence. Englewood Cliffs, NJ: Educational Technology Publications.

Poggenpohl, S., & Sato, K. (2009). *Design Integration*. Bristol, UK: Intellect.

Sharples, M. (1996). An Account of Writing as Creative Design. In A Writer's Reference (6th ed.). Academic Press.

Tschumi, B. (1994). *Architecture, and disjunction*. Cambridge, MA: MIT Press.

Wardrip-Fruin, N. (2009). *Expressive processing: digital fictions, computer games, and software studies*. Cambridge, MA: The MIT Press.

Chapter 10
Experiencing Commercial Videos for Online Shopping:
A Cross-Cultural User's Design Approach

Jose Cañas-Bajo
Aalto University, Finland

Johanna Silvennoinen
University of Jyväskylä, Finland

ABSTRACT

Although online shopping has become a popular and convenient instrument for companies to buy and sell products, the design of these web-shops does not always offer the rich multisensory experiences that physical retailing offers. In the chapter, the authors argue that introducing audio-visual contents in the design could provide dynamic multisensory information to offer more engaging experiences to the consumer. Despite existing controversies regarding universalism of the emotional experiences induced by perceptual processes, the authors present evidence that suggests cultural modulations of videos experiences. Spanish and Finnish participants interacted with audiovisual products depicting videos of a culturally loaded brand design. Content analysis of participants' verbalizations helped to identify categories and subcategories that defined the representation of the video elements and their relative weight depending on the cultural background of the viewer. Although results indicate common elements affecting viewers of the two countries, they differ in the relative weight to global aesthetics features.

DOI: 10.4018/978-1-5225-5023-5.ch010

INTRODUCTION

For some years now, Internet is considered a useful tool that helps companies to rapidly spread information about their products to very diverse geographical location at a very low cost. Internet use many times reduces the cost for marketing and also permits companies to offer their customers the possibility of buying from home independently of where the product or the company is located. The reduction of the product cost and the possibility of buying from any location make Internet shopping an attractive instrument for companies and customers. However, displaying the product on internet for online shopping greatly differ from the displays of the product in traditional physical shops where the costumers can not only look at the products from fixed angles, but also move and look at them from different distances and perspectives, touch them and even try them, Hence, internet does not usually offer the rich multimodal experiences than physical shopping offers. For these reason, because online shopping usually display the products in two dimensional surfaces and lacks the richer sensory experiences than the physical shops offer, the design of the online shop needs to take special care of the arrangement of the visual elements, the format of presentation, the combination of colours, etc. Usually the products are introduced by presenting pictures and verbal descriptions of their features, but the designer's job is to try to induce positive impressions and good experiences in the users by distributing the element of the web in particular ways. The choice of colours, sounds, and images seeks to influence the users and induce them to by the product (Childers et al. 2002). Despite these efforts, many aspects of the real shopping environment are missing in the static web displays, and this might introduce communication barriers between companies and customers. One way of reducing these possible problems is to introduce dynamic elements by adding video contents to the online shops. Videos can be useful devices to induce emotions and change consumers' perceptions of an object. The mental representations of a product can be influenced by the emotions attached to the audiovisual contents where the product is inserted (Moran, 1981). By exposing an individual to a video-based content about a product, emotions toward the product can be created, which might lead to a positive tendency to consume it (Wang & Cheong, 2006). However, to ensure that these audio-visual elements induce positive experiential, and informational representations of the products, we need research that helps to provide guidelines for enhancing visual communication with the users. As suggested by Childers, et al. (2002) costumers of online-shops do not only use the web to buy in a more efficient manner, but they also look for entertainment. Adding video contents to the shop web design may introduce this important hedonic dimension by making the presentation of the product richer and more attractive. Videos may provide attractive contexts where to display the products, to enhance their aesthetic features or to describe the quality of their manufactured elements. An aesthetically enhanced video might increase the probability that the consumer buy the product, but also may induce him/her to visit the online shop in the future, and to forward the link to other prospective shoppers (Reinecke & Gajos 2014). Recently, Ecklers and Rogers (2014) also suggested that audio-visual contents form part of the viral phenomena, highlighting the influences that videos might have as strategies for marketing (Purcell 2010).

As we mentioned, however, there is not enough research directed to investigate the impact that different elements have on the consumer experience. In addition, sometimes companies have made ill use of videos by applying emotional sensationalism and try to catch the audience attention through unconventional designs and ideas with the risk of triggering a negative reaction on the potential consumer (Ecklers & Rogers 2014) instead of the desired positive experience. But what element catches attention and interest in the consumers when exploring an online-shop? Interest is an important emotion for online shopping that has been explored in the context of film studies. According to Silvia (2005) interest involves evaluation of the degree to which an idea or event is complex, unexpected and difficult to process, but at the same time is evaluated as possible to cope and to understand. Thus, people find interesting complex or unusual events that can be understood after some effort (Silvia 2005a; 2005b; 2005c; 2006).

For the theory, interest is the more central emotion that filmmakers and scriptwriters should seek in their audiences. Interest keeps the viewers in their seats, with the motivation of spending their mental resources on the audiovisual sequences and with high expectations for the future. Interest gives an emotional structure to the video and provides the motivation to keep watching it (Tan, 1996). Although films and commercial videos differ in many aspects, the need to rise interest and to be able to influence the emotional experience of the viewer is common to both. The use of videos poses interesting psychological problems that are also common to films because they both integrate visual elements with music, sounds and narrative structure to convey meaning and emotions (Gross & Levenson 1995) and this combination can be used to rise interest for the presented products (in the case of commercial videos) or in the visual narrative and plot (in the case of films). Different aspects of the video design such as the shot composition, the movements of the camera, the music and sounds, the script for particular scenes interact to produce emotional reaction toward the product, and to influence the consumers' behaviour. Similarly, all these very different aspects of the video interact to conforming the users' experience of the product and capture their interest. Identifying the elements of videos that capture interest, and induce positive emotions, and understanding the way in which the audience experience them, is critical for the companies using videos with commercial goals (Kaikati & Kaikati 2004). Research should, therefore, provide answer to the question of how video contents can be used to enhance customers' experiences. Two recent studies (Silvia & Berg, 2011; Tarvainen et al., 2014) have tried to identify these elements by presenting people with short clips of pictures and asking viewers to rate the novelty, complexity, interest, and emotions elicited by the clips. Although the participants vary in their interest ratings, diverse combinations of novelty, complexity, and comprehensibility seem to predict interest. Similarly, in a recent study by Cañas-Bajo et al. (2017), participants were asked to signal the more interesting elements of a film and to indicate their emotional valence while watching the film, The results of this study showed that complex sequences that combined positive and negative emotions were rated as the more interesting parts of the film. In addition, studies should also be directed to identify elements that might differ for people with different cultural background.

Viewers' Experiences of Audio-Visual Contents

Experiencing a product involves creating a mental representation of the cognitive and emotional contents associated to the product. When researchers and designers discuss about users' (or consumers') experiences, they mean the conscious part of the symbolic representation of reality that we have in the mind (Newel & Simon, 1972). This means people have mental representations of their actions, their environment, their own internal states and their emotions, and that experience can be seen as the conscious part of mental representations (Saariluoma & Jokinen 2014). Consciousness is the result of appraisal processes involving three different information sources: the perceptual stimuli, the associative process in memory, and conscious reasoning processes (Roseman & Smith, 2001). These three processes proceed sequentially so that appraisal starts from the sensory and the perceptual processing of the stimuli whose aspects are detected and perceived by our senses. Information is then recovered from memory through fast and automatic associative processes, from the representation of the stimuli to the memory contents that start providing meaning to the events. Finally, the conscious reasoning process encodes the meaning and provides the bases of cognitive and emotional experiences that will endure in time and might influence behavior (Saariluoma & Jokinen, 2014). In this sense, appraisal processes define the cognitive aspect, which produces a specific emotional state. The result of the appraisal process determines whether an individual consciously likes or dislikes the stimuli (Scherer 2009; Saariluoma & Jokinen 2014; Eckler & Bolls 2011). Appraisal theory can also be applied to emotional states coming from audiovisual stimuli. Experiencing an audiovisual product involves creating a mental representation of the cognitive and the emotional contents associated with the product. It means that people have representations of the actions and the environment depicted in the product that combine with their own internal mental states. Videos might be useful tools to induce emotions and change consumers' perception of an object. The mental representations of a product can be influenced by the emotions attached to the audiovisual contents where the product is inserted (Moran, 1981). By exposing an individual to a video-based content about a product, external emotions will be created in consciousness. The information will be presented with sensitive additions that are implicit in the image and the sound, leading to a posterior appraisal when the viewer faces the product (Wang & Cheong, 2006).

Videos and other forms of audiovisual presentations could provide contexts that may influence appraisal processes by providing associations with the presented information and influencing the apperception of the product, in turn changing the viewer's emotional experience (Moggridge, 2010). Thus, the user experience includes thoughts, perceptions, or feelings when using a certain product, and the way that the product is presented may increase or decrease the quality of the experience (Albert, 2010; Albert & Tullis, 2013). Taking this approach might be critical for increasing the video's effectiveness in facilitating knowledge about the product and the customers' intention to buy it and other possible products from the same brand. Similarly, the impact of and interest of a video may be facilitated by understanding the factors affecting the audience's emotional and cognitive experiences of the video. Designing video for advertisements or to implement on-line shopping web sites should then start from understanding the different elements that influence the viewers' cognitive and emotional aspects (Garrett, 2010; Helfenstein, 2012; Norman, 2005).

Taking the users' experiences as a starting point is essential for increasing the effectiveness of the video in facilitating knowledge of the product. User-centred design aims to create products that meet the specific needs of the users in order to provide greater satisfaction and good experiences by reducing effort when interacting with the products (Norman 2005). As mentioned, identifying the elements that enhance users' experiences might be critical since some elements might not produce the desired experience for the users. For example, in the field of web design some studies have shown that adding features such as colours to facilitate the understanding or increasing the aesthetic of a product, could produce the opposite effect on the user experience by distracting users' attention and hinder understanding (Jiang & Benbasat 2004; Norman 2005). In audio-visual products, we might also find that introducing some technical or narrative features may have the opposite effect that the one intended. This is why focusing on the users' experiences might also be crucial for creating industrial applications that really meet the goals of the companies. However, this approach has not very frequently been used in studying online shopping. Users' approaches many times involves the use of methods that try to extract the mental representation of the user or consumer, and this in many occasions requires time consuming qualitative methods to assess mental contents. The study that we describe below represents one attempt to characterise the consumer's mental representation of the video contents and to identify the elements that capture the attention and interest of the viewer.

CROSSCULTURAL APPROACHES IN DESIGNING AUDIO-VISUAL CONTENTS

Understanding users' experience is even more important when shopping websites are aimed at a multicultural audience. For audio-visual elements to be successful, they need to adjust their contents to the type of product and to the cultural characteristics of the user. These cultural elements might influence users' experiences of the product. An obvious advantage of online shopping is that costumers can be reached from many different geographical and cultural backgrounds, but this advantage would only materialize if the features of the online shop match the needs of the costumers. For this reason, cross-cultural studies from a users' experience perspective are needed for successful and efficient design of online shopping interfaces. To the extent that competition between companies has become globalized, it is important to understand the users' experiences and the impact of their products across cultures. Creating cross-cultural attractive interfaces where consumers from different cultural backgrounds can experience pleasant thoughts and feelings is especially critical (Van Slyke 2010). In this context, it is possible that an audiovisual product developed in a country and a cultural context does not have the same impact on the user from another country or culture. Although, it might be possible to find some features that users may generally consider usable and attractive (e.g., designs that follow the laws of Gestalt), many factors are strongly influenced by cultural values (Ito & Nakakoji 1996). Thus, the success of a video design may depend not only on the coordination of the quality of individual technical or narrative elements comprising it, but also on external factors such as historical and cultural aspects, which might influence the consumer (Dempster, 2006). Audience reception studies (Hall, 1966; Morley, 2006; Tomlinson 1999) have stressed the idea that the messages conveyed through the media are not passively accepted by the audience, and that people interpret their meaning based on their cultural background. For audio-visual elements to be successful, they need to be adjusted to the type of a product and to the cultural characteristics of the users. Audio-visual elements of products displayed in online shopping web sites convey culturally differing perceptions, which influence users' experiences of the products, and thus affects the adoption of e-commerce (Pavlou & Chai 2002).

Cross-cultural studies have also tried to identify a number of dimensions in which cultural differences affect the type of relationships and communication of companies and organizations (Hofstede 2010; Trompennaars 2010; Kasemsap 2016).

For example, Hostfede identified six cultural dimensions in which countries differ (Power distance; Individualism vs. Collectivism; Uncertainty avoidance; Masculinity vs. Femininity; Long-term vs. Short-term orientation; Indulgence versus Restraint). Obviously these dimensions may have an effect on how people experience and react to audiovisual products.

One important dimension in online shopping web design is aesthetics since many aesthetics elements have been shown to positively influence purchase intentions (Bloch 1995). The appeal of the product in first impression formation seems to influence behaviour before consciously considering of buying a product (Lindgaard et al. 2006). However, it is not clear whether there are universal features for successful design and many studies have shown that aesthetic impressions may vary across individual personalities, genders, and ages (Moss & Gunn 2009; Cyr et al. 2010; Reinecke & Bernstein 2011; Hsiu-Feng 2014). Thus, aesthetic experiences involve interpretative interaction with several states and actions and thus cannot be considered merely as an immediate response (Carroll 2001). Additionally, important for the design of online shops is the many demonstrations that aesthetics vary with geographical location and culture (O'Connell 2014). For example, in their studies Reinecke and Gajos (Reinecke & Gajos 2014) observed that geographically close countries share similar visual preferences regarding color and visual complexity. For example, Finland and Russia preferred the lowest visual complexity and colorfulness of all countries they study. Similarly, Northern European countries seemed to prefer lower colorfulness than Southern European countries. In a similar vein, Marcus & Gould (2000) showed that Asians prefer brighter, more colourful and animated web sites than Westerners and that these more colourful web sites are sometimes perceived as information overload by the Westerners. In general, all these studies suggest that the countries with a regular exchange of cultural values share similar website preferences. Hence, in designing video contents to convey positive emotions toward a product or a brand, it is critical to consider the cultural dimensions in which people differ, not only when making aesthetic judgments, but also when considering other aspects of the products.

The idea of culturability introduced by Barber and Badre (1998) and extended by Cyr and Trevor-Smith (2004) summarizes the importance of cultural design. Culturability represents the mixture of design elements and culture to achieve usability in webs

...color, spatial organization, fonts, shapes, icons, and metaphors, geography, language, flags, sounds, and motion contribute to the design and content of a Web page, which directly affects the way that a user interacts with the site. (Badre 2001).

To address culturability in web and video design attention needs to be paid to many features (visual complexity, music, animation, narrative features, etc.) that can be culturally determined. Decision about these elements from the users' perspective will have an impact on the representation that people make of the product and this mental representation would impact apperception of the product. Thus, the use of videos as a possible solution for enhancing emotional experiences and interest in web shopping calls for research intending to understand the multiple elements of videos that can impact these representations.

Thus, differences in national and cultural dimensions may have an effect on how people experience and react to audiovisual products For example, in a recent cross-cultural study (Cañas-Bajo el al. 2015) images and videos of Iittala products (Finnish design) were compared with the goal of finding whether participants from Finland and Spain, appraised the designs similarly. In the study participants interacted with different audiovisual products depicting images/videos of the brand design. Through content analysis, four main categories defining the representation of the product were identified. Interestingly, the relative weight of these categories depended on nationality of the participants and on the type of visual presentation (image/video). The results indicated cultural differences in some dimensions (memory, usefulness) regarding the products, however the differences regarding these dimension were reduced when participants were shown videos instead of static images. These results suggest that videos are effective in inducing emotions and in changing consumers' perception of an object. Therefore, when designing web interfaces for internationally oriented web shops, audio-visual content may be used to reduce cultural differences and have more global impact.

This proposal can influence how we think of videos and use them. As we argued, videos can be used in online shopping as a way of offering customers a rich multimodal experience that can be closer to the experience of a physical shop (Childers et al., 2002; Reinecke & Gajos, 2014), but also audiovisual presentations of the products can be used to reduce the communication limits due to cultural differences by providing richer perceptual, multisensory experiences of the products.

Some theories propose (Barrat, 2012) that perceptual and cognitive processes are independent modules with different properties: perceptual processes are universal and independent of cultural factors, while cognition is prone to be influenced by culture. Accordingly, to the extent to which videos depend on perceptual processing, cultural and contextual factors should play a smaller role than for other type of object representations. Hence, videos can be used to reduce cultural differences and represent an useful strategy to become more global (Purcell, 2010). However, more evidence is needed to support these claims.

Are Users' Experiences of Videos Culturally Mediated?

Studies in this research area lack knowledge for identifying the factors that increase consumers' interest and the aspects of the video contents that help in creating pleasant users' experiences across different cultures (Cañas-Bajo et al. 2015). The aim of empirical case study that we describe below is to identify elements of audio-visual contents that enrich the product experience. Because the idea is to enhance cognitive and emotional mental contents of the consumer, the approach that we have taken in our research is to try to understand mental contents, and very specifically, what aspects of the displays capture interest in the viewers. The studies reviewed in the previous section also suggest that the answer to these questions might be culturally mediated. Hence, in our case study, (Cañas-Bajo & Silvennoinen, 2017) we have taken a cross-cultural approach and explored the mental contents and experiences of people from two countries, such as Finland and Spain, that are geographically distant (see Reinecke & Gajos 2014, for data showing differences in aesthetics judgment depending on geographical distance) and that also differ according to some Hostftede's cultural dimension. For instance, in the Hosftede's dimensions regarding "uncertainty avoidance", "collectivism" or "restrain" both countries present different results (Hofstede 2010). These differences has been found relevant when people from those countries buy new technological product (Cañas-Bajo et al., 2015).

In our study Spanish and Finnish participants were exposed to three videos representing Scandinavian design products and we investigated whether participants' mental representations of the videos include elements related to aesthetics qualities of the videos (e.g. the video is interesting, the video is colourful), but also if we could identify aesthetic emotions (Scherer 2009) and intentional aesthetics elements (Carroll 2001). According to Scherer an aesthetic experience is a mixture of interacting cognitive and emotional states when appraised produced aesthetic emotions (e.g. the video makes me feel nostalgic). Similarly, we tried to identify if intentional dimensions of aesthetic also impacted the viewers' experiences. Recent philosophical accounts of aesthetics in arts proposed that aesthetic experiences might also include appraisal of the aesthetics intention of specific element (e.g. the color is trying to capture my attention). From this view, the study not only assessed the elements of the video influencing the viewers' experience, but also aimed to provide support to the psychological reality of some aesthetics dimensions. In addition, we aimed to seek evidence of the importance of culture in modulating users' experiences of audiovisual products. Although appraisal theory and audience studies seem to suggest that culture plays a role in how we perceive and represent the external world, he question of the universality of emotions elicited by videos and films is subject to controversy. Thus, some cognitive theories assume that the capacity of making sense of a projected video is universal and not culturally determined (Barrat,

2012). According to this idea, a functioning perceptual system can visually allow any person to understand a film. According to the modularity hypothesis (Barrat, 2012), perceptual and cognitive processes are independent modules with different properties. As we mentioned, this hypothesis suggests that perceptual processes are universal and independent of cultural factors, while cognition is prone to be influenced by culture. Hence, to the extent that the presented stimuli depend on these modules, culture might have a larger or a smaller effect.

The universality of emotional processing is a classic question in the field of psychology (Darwin, 1872; Ekman & Friesen, 1971). Universalists have argued that basic emotions, such as hunger, fear, sadness, and so on, are universal and independent of culture. In their research and attempt to identify basic, culturally independent emotions, Ekman and Friesen (1971) provide evidence that facial emotional expressions are interpreted in similar ways by people from different cultures. The question of the universality of emotions is important for the video industry because if emotions are universal, video producers can use a culturally independent language (images) to convey such emotions. From this perspective, the visual language involved in video production (cuts, shots, angles, shadows, and so on) can be considered universal resources to convey culturally independent meanings and emotional experiences (Brown, 2003; Nielsen, 2005). From this standpoint, efficiently managing audiovisual resources should result in producing the desired emotional impact on the viewers.-

Against the idea of universality, the research on web and visual design has shown that aesthetic emotions may vary with culture. Additionally, communication studies have stressed that research should be targeted to features of the audience and that individual differences in culture and education may influence aesthetic emotions and the appreciation of cultural products. Films and other audiovisual products (series, clips, short films, etc.) are cultural items that can be influenced by individual factors regarding knowledge and orientation toward cultural products. Research on the so-called *cosmopolitan orientation* (Cleveland, et al., 2009; Hannerz, 1990; Skrbis & Woodward, 2007; Szerszynski & Urry, 2002; Woodward et a., 2008) has shown that knowledge and interest in cultural products, as well cultural habits of individuals, play a major role in the consumption of foreign products. *Cosmopolitan-oriented* people are identified by their interest in other cultures and geographically distant places and activities, as well as by their preference to consume products with different cultural combinations. A cosmopolitan orientation is usually associated with openness and compromise with social matters (Meuleman, & Savage, 2013).

In technology and design research, sociocultural differences and tendencies have been approached from the life-based design (LBD) paradigm (Leikas et al., 2013). This approach argues that individuals differ in their *forms of life* (Wittgenstein, 1958), constituting the habits, interests, and activities that they show in a given culture and context. Religious beliefs, support for a football team, and being part of a choir constitute forms of life. Cultural orientations are also considered aspects that comprise different forms of life (Rössel & Schroedter, 2015). In this theoretical context, designers have shown that people with different forms of life might vary in the types of activities and technological devices that are useful to them. From this perspective, designers need to pay attention to the aspects of individuals' lives and the contexts in which technology fits. Similarly, if cultural orientation and cosmopolitanism are considered forms of life, the design of cultural products should take into account the cultural habits of prospective consumers. People with interest in and openness to other cultures and consumers of cultural products may differ in many aspects that may impact their cultural experiences (Geertz, 1973). From this viewpoint, researchers and creators of audiovisual contents must consider different cultures and forms of life when targeting broad audiences. Thus, its approach is that culture may have an effect on how people experience and react to audiovisual products and that audiovisual companies should understand these different reactions when creating videos conforming to the cultural characteristics of its members. An audiovisual product created in a specific cultural context may not have the same impact on a viewer with a different cultural background. Thus, it is critical for designing video advertisements, industrial videos, or feature length films to understand viewers' experiences across cultures.

To sum up, different orientations and theoretical approaches assign varying roles and importance to cultural factors, and clearly more empirical studies are needed to try to understand the relative weight of universality and culturalism in the appreciation of video contents.

AN EMPIRICAL STUDY ON CULTURAL INFLUENCES ON VIDEO EXPERIENCES OF VIDEO PRESENTATION OF CULTURALLY LOADED PRODUCTS

Method

In the study described in, (Cañas-Bajo & Silvennoinen, 2017). Spanish (N = 20) and Finnish (N = 20) participants were exposed to three videos representing Scandinavian design products (Videos of Iitala products) with audio-dubbings presented in Finnish or in Spanish depending on the participants' nationality (See Appendix A). Both Finnish and Spanish participants were presented each of the videos and then asked: what thoughts and emotions does the video suggest to you? What video elements would you highlight?). Participants were tested individually in a quiet at the University of Granada, Spain (for the Spanish participants) or at the University of Jyväskylä in Finland (for the Finnish participants). We performed content analyses (Weber 1990) to extract the basic dimensions of the participants' mental representations of the videos. The approach for the content analysis was data-driven because the focus was on analyzing the type of feelings, emotions and impressions attached to different properties of the videos by the participants. Semantic units were transformed into content categories and then the frequencies of the semantic units within the categories were recorded in the following manner. Through several iterations of categorisations and mutual agreement between the authors, the categories and sub-categories were identified. First, we identified single words, phrases, and idioms that constituted semantic units for describing cognitive and emotional experiences produced by features of the videos. The second phase of the analysis was to critically evaluate the semantic units into initial categories. This classification phase of the coded semantic units resulted in 30 categories (see Appendix B). In the third analysis phase, second level categorisation was conducted to construct higher-level main categories from the initial sub-categories. These higher-level main categories were related to overall impression of the video (aesthetics, emotion, memory, desirability, etc.), narrative structures (plot, characters, and scenario), visual design (light, colour, shapes) and audio design. The frequency of semantic units within each category was counted for statistical analyses. These analyses were conducted to detect what aspects of the videos raised the users' interest and how these aspects affected the emotional and cognitive experiences of the viewers.

Results[1]

As mentioned, semantic units referring to video contents and emotional experiences elicited by them were classified according to four main categories. The first category was "Overall impression of the video" and contained subcategories related to overall features of the video. Three of these features referred to aesthetics dimensions: 1) Aesthetic persuasion: represented aesthetic evaluations regarding the intentionality of the video. Examples: *"They are trying to make you think about the shapes of everything"* *"to sell these objects is fine"; 2)* Aesthetic emotion: represents aesthetic evaluations of the general emotions experienced by the viewers. Examples of this subcategory are: *"I liked it a lot" "It did not make me feel anything" "The video makes me feel calm"*. 3) Aesthetic Quality: Represents quality attributes attached to the video by the viewers. Examples of this subcategory might be *"The video gives warmth" "I think it is interesting"*. In addition to aesthetics some units referred to what we termed: 4) Memory matches & Associations: represented mental associations such as memories and familiarity with the context. For example verbalization such as *"you wake up in the morning and breakfast is made"*, *"this suggests memories of Finland"*. 5) Desirability: represented enhanced experience of the product through it presentation in the video. Examples: "*Makes you want to buy*"; and 6) Rhythm and length representing evaluation of the dynamics and rhythm of the footage. Examples: "*it felt like too long", "it moved forward all the time with a calm phase"*

The second main category referred to the "Narrative Structure of the video". This broad category also contained aesthetic element corresponding to the three aesthetics subcategories: persuasion, emotion *and* quality, but this time referring to narrative elements of the video. Examples of 1) aesthetic persuasion are "*that are tried to invent to represent some forms, how it can move and so on*", "*it didn't try to explain anything just showed*" 2) aesthetic emotion are "*Well sometimes analogies and metaphors annoy* me", "*this story didn't get me on its side concerning my emotions*" and 3) aesthetic quality are "*comparison between older and newer version was kind of useless*", "*There was interesting information that it takes 30 hours of work per one Aalto vase*". In addition, we identified units referring to the general description of the videos regarding the main idea perceived by the viewer. We termed this subcategory "Overall idea" and examples are "*they are talking about lamps and lights" "they are talking about design*" "*video featuring household items*". Some other subcategories within "Narrative structure" referred to evaluation of specific aspects of the storyline such as "*It explains how the design of curves and forms emerged" "It is interesting to see how it's done"(*Plots aspects subcategory); Evaluation of Character featured in the story line and their role: "*two artists involved in his work" "Then turn to the figure of the designer*"(Character subcategory); Evaluation of the Scenarios appearing in videos such as "*It seems the house of a couple" "what it does is use the Scandinavian landscape*" (Context subcategory); and Product presentation: evaluations on how the product is presented in the storyline. Examples: "*they see that each piece is unique" "Products were presented like in a catalogue*"

The third main category referred to elements of "Visual Design". Again some unit referred to the three aspects of aesthetics, persuasion, emotion, quality, as they related to visual elements. Examples of 1) aesthetic persuasion are "*all the elements a cozy atmosphere was tried to be created*", "*They try some kind of unity in here with the distribution of the elements*" 2) aesthetic emotion are "*peaceful feeling, because of the happy, bright images*", "*stillness, and safety inside the home walls*" and 3) aesthetic quality are "*the quality of the image was horrible and it disturbed me a lot", "visually really stimulant*" In addition some units can be classified within the Composition subcategory that included evaluation of the visual composition of the element in the image. Examples are "*Quite a lot of graphic lines and details*", "*the video plays whit the camera angle of the view*"; some units were related to the Text subcategory that represented evaluation of the subtitles appearing in one of the videos. Example "*text, high-spaced video image were the core elements*"; Product presentation that included evaluations the way in which the product was visually and the fit within the general visual design. Examples: "*products were presented in their natural environment*". In addition, some subcategories referred to perceptual impressions; Materials referred to valuations regarding the material of the product and in other visual objects in the video. Examples: "*Soft wool and hard metal wire in that self, so that in here was lots of balancing elements*", "*there are a lot of wooden surfaces*"; Shapes referred to evaluation and comments on the visual shapes of object in the videos. Example: "*The shape of the vase was clearly the most important element here*"; Light referring to evaluations regarding the disposition of lights and lightning effects on the visual design. Examples: "*At the end of the video the room's lighting gets darker*", "*With the light and the glass, shadows came which looked nice*"; Colors included evaluation of the colors in the visual display and its integration into the overall visual design. Examples: "*there are lots of interesting colors in these*", "*Quite simple colors*"

The fourth main category referred to the units regarding "Audio Design". Similarly, to the previous categories the Aesthetics categories regarding persuasion *(e.g. "Music was taken into account, and to all the parts that created the space),* emotion *(e.g. I like the music; The music wasn't supposed to disturb so much)* and quality were included as they applied to audio elements. (*e.g. the music suggests tranquility; the music influenced me).* In addition, some units referred to the quality of the Narrator/off voice (e.g.*the voice was relaxing*); and to the ·Congruency of the music with other elements of the video (*e.g. Music and the forms were unified*; *I think the music wasn't the right one for this video*)

Figure 4 (see Appendix C) shows the mean number of verbalizations for each group (Finnish/Spanish) in each of the four higher-level categories. An analysis of variance (ANOVA) was carried out including Category as within-subjects variable and Nationality as between-subjects variable. Both main effects were significant (Category: $F(2, 81) = 41$; p<.05; $\eta_p = .52$; Nationality: $F(1, 38) = 8{,}38$; $p<.05$; $\eta = .12$). *"Overall impression of the video"* produced more verbalizations than the other categories (mean = 9.67). Moreover, Spanish reported more verbalizations (mean = 7.45) than Finish participants (mean = 5.24). Importantly, there was also a significant interaction, $F(2, 81) = 3{,}22$; $p<.05$; $\eta_p = .08$. Bonferroni corrected pairwise comparisons showed that there were significant differences between Spanish and Finnish participants in the first category: *"Overall impression of the product"* ($p<.01$) and a marginal difference in the fourth category *"audio design"* ($p= .056$). Thus, Spanish participants produced more verbalizations related to overall impression with the product and audio design (mean = 12.15) than Finnish participants (mean = 7.20).

When we analysed the specific subcategories within the overall categories, we observed that the differences in *"Overall impression"* between Spanish and Finish participants came from their verbalization regarding *Aesthetic quality* ($p < .05$)

and *Memory matches* ($p < .05$). Similarly, we analysed the broader *"Audio design"* category and observed significant differences in Aesthetics persuasion ($p < .05$) with Spanish participants producing more verbalizations than Finnish participants, all other comparison within this category were not significant ($p > .05$).

Although the overall *"video design"* category did not reach significance, two of the subcategories within it did: *"Composition"* and *"Shapes"* were significant ($p < .05$), both showing more verbalizations by Spanish participants than by Finnish participants. Furthermore, the subcategory *"Plot"* was also significant ($p < .05$), although its higher-level category *"Narrative Structure"* was not. Again, Spanish made more verbalizations than Finnish.

DISCUSSION

The study reported here was designed to study possible differences between Spanish and Finnish participants in the emotional experiences elicited by videos in the context of online shopping. As we mentioned in the introduction, although design studies suggest that videos experiences might be culturally mediated, some psychological and film studies theories proposed that some emotional experiences are universal and that may be universal features that attract the viewers' interest. The results of the content analysis performed in the data form our empirical case study indicated that Finish and Spanish participants focused in different aspects of the videos. In general the Spanish participants produced more verbalizations than the Finnish, but importantly they also differed from the Finnish participants in their greater reliance in aspects related to the overall impression of the videos and in the audio categories. Regarding "overall impressions", the differences between the two groups came from the subcategory "aesthetic quality" where the Spanish participants verbalised more global perceptual and emotional aspects of the video (*"the video is colourful, it suggests calm, it is interesting"*. etc.) than the Finnish participants. However, these differences did not appear when local and concrete visual features such as shapes, colours or lighting of specific objects or parts of the videos were considered. Similarly, Spanish participants represented mental associations and global contextual features (e.g. *you wake up in the morning and breakfast is made; reminds me of my town, etc."*). Hence, the pattern of the verbalizations suggests that there are qualitative differences in the elements of the video contents that capture attention and interest for the two groups. Thus, the Spanish participants seem to pay more attention than the Finnish to contextual factors and memories reflecting an overall, global approach to the video contents, while Finnish participants seen to pay more attention to local functional features. It is also interesting that the only differences found in the category of visual design referred to more global features, such as composition or shapes and differences in the narrative structure dimension, also came from the Spanish mentioning more features related to the overall plot.

Although these differences might be due to a number of factors, this global approach to perception by the Spanish participants could also be associated to higher scores in Hostftede "collectivism" dimension and lower in "Restraint" of countries such as Spain. Obviously, this hypothesis is speculative and should be the focus of further research, but is lo consistent with cognitive research suggesting that cultural factors influence the focus with which people perceive the world. For example, Colzato et al (2008) compared neo-calvinist and atheist Dutch participants in the global-local task (Navon 1977) and showed that neo-Calvinists focused more on the local features composing global figures than the atheists.

Similarly, the dimension of audio design also showed differences between the Spanish and Finnish participants. Thus, the Spanish participants also paid more overall attention to the audio design of the video, especially to the emotional aspects composing aesthetics elements of the audio, than the Finnish participants. This difference may also be related to the differences in the "collectivism" dimension since cultural studies in emotion perception (Massuda et al. 2008) suggest the less individualistic societies pay more attention to contextual cues (e.g. music) to interpret emotions. These cultural differences in appraising the elements of the video that are brought into consciousness (overt verbalizations) by Finnish and Spanish viewers have implications for the design of the contents and aesthetics elements of the videos. Thus, our results suggest that videos designed for Spanish audiences should enhance global contextual and emotional features (global plot, global associations, global emotions suggested by light, color and music in the video), whereas audiovisual contents for Finnish viewers should also enhance local features of specific parts of the videos and individual objects. Because we have compared only two countries, we are not at the moment able to dissociate whether the obtained differences are due to geographical distance as suggested by the research by Reinecke & Gajos (2014), or to cultural dimension such "uncertainty avoidance", "collectivism" or "restrain" (Hofstede 2010). Further research should help to understand the underlying reasons for these differences and to generalise some of our conclusions to countries varying in geographical locations and cultural values.

Note that the pattern of results goes against theoretical positions suggesting that many aspects of videos and films evoke universal emotions and experiences. Modular theories (Barrat, 2012) assume that perceptual processes are universal and independent of cultural factors, while cognition is prone to be influenced by culture. From this assumption, videos that contain more perceptual features should be less prone to cultural modulation that other type of stimuli. However, our results seem to indicate that despite the rich perceptual features of the presented videos, they were represented and experienced differently by our Finish and Spanish participants. This suggests that video design should pay attention to cultural factors when aiming to reach an international audience, and there is a need to adjust the video contents to the cultural characteristics of the viewers since the cognitive module may convey culturally different perceptions, which influence the viewers' experiences that may lead them to the adoption of e-commerce (Pavlou & Chai, 2002).

Although identifying aesthetic dimension was not the main goal of the study, it is important for aesthetics theories that we were able to identify in people mental representation of the video different aesthetic dimensions. Thus, we were able to classify units that reflected aesthetic contents of some visual, audio or narrative qualities of the video, separately from units that represented emotional aspects raised by the aesthetic quality of the video. But most interesting was the fact that we were able to identify in both Spanish and Finnish participants, what we called aesthetic persuasion that identifies elements clearly containing an interpretative dimension. Thus, within this dimension, mental contents as reflected by the participants' verbalizations referred to units where participants verbalized their interpretation of the intended meaning of different elements of the video (Carroll 2001).

These results indicate that audiovisual experiences involve evaluations of aesthetic persuasion regarding the overall impression of the video, and in more detail, concerning the narrative structure, visual design, and audio design. The elicitation of impression of aesthetic persuasion as a cognitive process can be explained with appraisal theory of emotion (Smith & Kirby 2001; Jokinen et al. 2015; Silvennoinen et al. 2015). Appraising aesthetic persuasion of an audiovisual representation requires information derived from all three-appraisal sources of information (Scherer 2009; Saariluoma & Jokinen 2014; Eckler & Bolls 2011). Thus, the evaluation of the aesthetic persuasion of a video includes perceiving the audiovisual stimuli, associations withdrawn from memories, and critical reasoning considering the intents of the aesthetics of the stimuli. In addition, aesthetic persuasion appraisals include visual literacy processes in detecting design methods and choices through which the aesthetic strives to create meanings, attitudes, and experiences and elicit emotions in the viewer.

Furthermore, overall impressions, narrative structure, visual and audio design involve aesthetic emotions and evaluations of aesthetic qualities. In appraisal theory of emotion, aesthetic emotion is conceptualized as a separate emotion typology from utilitarian emotions (Mastandrea 2014; Scherer 2009). Utilitarian emotions focus on their adaptive function, which help in adapting to certain stimuli or event. Aesthetic emotions are subjective affective responses to artifacts and to their intrinsic qualities (Mastandrea 2014). Thus, aesthetic emotion and aesthetic quality are intertwined in audiovisual experiences. Aesthetic emotion is a result of the cognitive evaluation of the significance of the event (in this case the audiovisual representation), this process also involves aesthetic qualities that are attributed to mentally represented information contents of the encountered audiovisual representation.

The categories presented in this paper illustrate the mentally represented contents of audiovisual experiences, and can be utilized to inform experience-driven design processes for online videos, and to understand the cognitive process of audiovisual experience from cross-cultural perspective. In addition, our study provides support to aesthetics theories that defend the multidimensional aspects of aesthetics including cognitive, emotional and intentional/interpretative aspects of it. Similarly, our results support and extend previous research suggesting that videos are effective in inducing emotions and in modulating consumers' perceptions. Thus, the mental representations of the viewers were affected by the properties associated to the audiovisual context.

A possible limitation of the study is the association of nationality to familiarity with the product and video presentation styles. Although, the use of familiar Finnish items had the advantage that allowed us to observe the influence of cultural aesthetics (for the Finnish participants the product aesthetics matched their aesthetics preferences) in the mental representations, it is possible that the differences that we obtained between Finnish and Spanish participants might be due to familiarity with the product more than with other cultural dimensions. However, the fact that familiarity was only important for images but not when video was presented, suggests that there is more than familiarity in the different ways that Finnish and Spanish participants appraised and mentally represented the objects. Further research with unfamiliar items should try to dissociate the role of familiarity form other cultural dimensions.

SOLUTIONS, RECOMMENDATIONS AND FUTURE DIRECTIONS

The main focus of our study was to answer the question of whether the way of experiencing commercial videos depended of the cultural background of the viewers. Our results seem to suggest that it does. However, some limitations of our study force us to be cautious with our conclusions. First, although we have selected culturally different countries (according to Hofstede's [2010] dimensions) to conduct our observations, these countries come from occidental cultures where many values and forms of life are similar. Hence, an obvious line of research from our data is to directly compare emotions from many more different cultural orientations and larger variations in values, habits, and forms of life. Second, a similar reasoning can be applied to the selected videos since they all come from the same brand, with highly shared values and aesthetics. Additionally, in our attempt to control the selection of videos, we have picked videos depicting similar objects and using similar aesthetic elements and it is possible other styles and products are more open to cultural differences. Third, individual differences in cultural background, age, and expertise might provide better insights into individual and social factors influencing emotional experiences with films. Expert studies in many fields have been highly influential in revealing mental representations that differ for knowledgeable individuals. Reports involving radiologists or chess experts show that perceptions (apperceptions) of objects change with experience. Similarly, cultural background, openness to new experiences, educational level, and age yield different results (Saariluoma et al., 2016). Cultural globalization is a relatively new phenomenon that unevenly impregnates people's lives. Future research should explore the roles of these individual factors in film experience.

CONCLUSION

The results of our study suggest that cultural factors related to global versus local perception and emotions modulate the mental contents of video viewers, and therefore influence how users experience culturally designed products. Different elements composing the overall impression of the video, and the emotional perceptions of music aesthetics seem to impact differently Spanish and Finnish viewers. Thus, the elements of the medium in which the product is presented affects viewers in different ways and could be utilized in providing guidelines in designing product advertisements or web applications. An essential contribution of this paper to user psychology is that results clearly show that there are relevant cultural differences in the way the video elements are apperceived. Thus, this study builds a bridge between the field of user experience, user psychology, and cross-cultural psychology.

ACKNOWLEDGMENT

This research was partially supported by The Doctoral School in the Faculty of information Technology (University of Jyvaskyla, Finland) and by a postdoctoral grant from the project "Elements of Success" by TEKES Foundation. We are grateful to Iittala for providing the video materials for the experiment.

REFERENCES

Albert, W. (2010). *Measuring the user experience*. Morgan Kaufmann.

Albert, W., & Tullis, T. (2013). *Measuring the user experience: collecting, analyzing, and presenting usability metrics*. Newnes.

Badre, A. (2001). *The effects of cross cultural interface design orientation on world wide web user performance*. Academic Press.

Barber, D. B., & Badre, A. (1998). A Year It Was! Culturability: The merging of culture and usability, 1-14.

Barratt, D. (2012). The Geography of Film Viewing: What Are the Implications of Cultural-Cognitive Differences for Cognitive Film Theory? In 2012 Conference for the Society for Cognitive Studies of the Moving Image. SCSMI.

Bloch, P. H. (1995). Seeking the ideal form: Product design and consumer response. *Journal of Marketing*, *59*(3), 16–29. doi:10.2307/1252116

Brown, G. (2003) The Moving Camera. Part I. *Zerb*, 58.

Cañas-Bajo, J., Perälä, P. M., Silvennoinen, J. M., & Saariluoma, P. (2015, September). Affective contents of cross-cultural audiovisual experience. In *Proceedings of the 19th International Academic Mindtrek Conference* (pp. 82-88). ACM. doi:10.1145/2818187.2818294

Cañas-Bajo, J., & Silvennoinen, J. M. (2017). Cross-Cultural Factors in Experiencing Online Video Contents in Product Marketing. *International Journal of Art, Culture and Design Technologies*, *6*(1), 40–56. doi:10.4018/IJACDT.2017010103

Carroll, N. (2001). *Beyond aesthetics: Philosophical essays*. Cambridge University Press. doi:10.1017/CBO9780511605970

Childers, T. L., Carr, C. L., Peck, J., & Carson, S. (2002). Hedonic and utilitarian motivations for online retail shopping behavior. *Journal of Retailing*, *77*(4), 511–535. doi:10.1016/S0022-4359(01)00056-2

Cleveland, M., Laroche, M., & Papadopoulos, N. (2009). Cosmopolitanism, consumer ethnocentrism, and materialism: An eight-country study of antecedents and outcomes. *Journal of International Marketing*, *17*(1), 116–146. doi:10.1509/jimk.17.1.116

Colzato, L. S., van den Wildenberg, W. P., & Hommel, B. (2008). Losing the big picture: How religion may control visual attention. *PLoS One*, *3*(11), e3679. doi:10.1371/journal.pone.0003679 PMID:19002253

Cyr, D., Head, M., & Larios, H. (2010). Colour appeal in website design within and across cultures: A multi-method evaluation. *International Journal of Human-Computer Studies*, *68*(1), 1–21. doi:10.1016/j.ijhcs.2009.08.005

Cyr, D., & Trevor-Smith, H. (2004). Localization of Web design: An empirical comparison of German, Japanese, and United States Web site characteristics. *Journal of the American Society for Information Science and Technology*, *55*(13), 1199–1208. doi:10.1002/asi.20075

Darwin, C. (1872). The Descent of Man and Selection in Relation to Sex. D. Appleton.

Dempster, A. (2006). Risky business: The art of managing creative ventures. In *Cultural Industries: The British Experience in International Perspective.* Humboldt-Universität zu Berlin.

Eckler, P., & Bolls, P. (2011). Spreading the virus: Emotional tone of viral advertising and its effect on forwarding intentions and attitudes. *Journal of Interactive Advertising*, *11*(2), 1–11. doi:10.1080/15252019.2011.10722180

Eckler, P., & Rodgers, S. (2014). Viral Advertising. The handbook of international advertising research, 184-202.

Ekman, P., & Friesen, W. V. (1971). Constants across cultures in the face and emotion. *Journal of Personality and Social Psychology*, *17*(2), 124–129. doi:10.1037/h0030377 PMID:5542557

Garrett, J. J. (2010). *Elements of user experience, the: user-centered design for the web and beyond.* Pearson Education.

Geertz, C. (1973). *The interpretation of cultures* (Vol. 5019). Basic Books.

Gross, J. J., & Levenson, R. W. (1995). Emotion elicitation using films. *Cognition and Emotion*, *9*(1), 87–108. doi:10.1080/02699939508408966

Hall Edward, T. (1966). *The hidden dimension.* Garden City, NY: Anchor.

Hannerz, U. (1990). Cosmopolitans and locals in world culture. *Theory, Culture & Society*, *7*(2), 237–251. doi:10.1177/026327690007002014

Helfenstein, S. (2012). Increasingly emotional design for growingly pragmatic users? A report from Finland. *Behaviour & Information Technology*, *31*(2), 185–204. doi:10.1080/01449291003793777

Hofstede, G. (2010). *National cultural dimensions.* Academic Press.

Hsiu-Feng, W. (2014). Picture perfect: Girls' and boys' preferences towards visual complexity in children's websites. *Computers in Human Behavior*, *31*, 551–557. doi:10.1016/j.chb.2013.05.033

Ito, M., & Nakakoji, K. (1996, August). Impact of culture on user interface design. In *International users interface* (pp. 105–126). John Wiley & Sons, Inc.

Jiang, Z., & Benbasat, I. (2004). Virtual product experience: Effects of visual and functional control of products on perceived diagnosticity and flow in electronic shopping. *Journal of Management Information Systems, 21*(3), 111–147. doi:10.1080/07421222.2004.11045817

Jokinen, J. P., Silvennoinen, J. M., Perälä, P. M., & Saariluoma, P. (2015, April). Quick affective judgments: Validation of a method for primed product comparisons. In *Proceedings of the 33rd Annual ACM Conference on Human Factors in Computing Systems* (pp. 2221-2230). ACM. doi:10.1145/2702123.2702422

Kaikati, A. M., & Kaikati, J. G. (2004). Stealth marketing: How to reach consumers surreptitiously. *California Management Review, 46*(4), 6–22. doi:10.2307/41166272

Kasemsap, K. (2016). Cultural Perspectives and Cultural Dynamics: Advanced Issues and Approaches. *International Journal of Art, Culture and Design Technologies, 5*(1), 35–47. doi:10.4018/IJACDT.2016010103

Leikas, J., Saariluoma, P., Heinilä, J., & Ylikauppila, M. (2013). A methodological model for life-based design. *International Review of Social Sciences and Humanities, 4*(2), 118–136.

Lindgaard, G., Fernandes, G., Dudek, C., & Brown, J. (2006). Attention web designers: You have 50 milliseconds to make a good first impression! *Behaviour & Information Technology, 25*(2), 115–126. doi:10.1080/01449290500330448

Marcus, A., & Gould, E. W. (2000). Crosscurrents: cultural dimensions and global Web user-interface design. *Interactions, 7*(4), 32-46.

Mastandrea, S. (2014). How emotions shape aesthetic experiences. In P. P. L. Tinio & J. K. Smith (Eds.), *The Cambridge Handbook of the Psychology of Aesthetics and the Arts* (pp. 500–518). Cambridge, UK: Cambridge University Press, UK.

Masuda, T., Ellsworth, P. C., Mesquita, B., Leu, J., Tanida, S., & Van de Veerdonk, E. (2008). Placing the face in context: Cultural differences in the perception of facial emotion. *Journal of Personality and Social Psychology, 94*(3), 365–381. doi:10.1037/0022-3514.94.3.365 PMID:18284287

Meuleman, R., & Savage, M. (2013). A field analysis of cosmopolitan taste: Lessons from the Netherlands. *Cultural Sociology, 7*(2), 230–256. doi:10.1177/1749975512473991

Moggridge, B. (2010). *Designing media.* MIT Press.

Moran, T. P. (1981). The command language grammar: A representation for the user interface of interactive computer systems. *International Journal of Man-Machine Studies, 15*(1), 3–50. doi:10.1016/S0020-7373(81)80022-3

Morley, D. (2006). Unanswered questions in audience research. *Communication Review, 9*(2), 101–121. doi:10.1080/10714420600663286

Moss, G. A., & Gunn, R. W. (2009). Gender differences in website production and preference aesthetics: Preliminary implications for ICT in education and beyond. *Behaviour & Information Technology, 28*(5), 447–460. doi:10.1080/01449290802332662

Navon, D. (1977). Forest before trees: The precedence of global features in visual perception. *Cognitive Psychology, 9*(3), 353–383. doi:10.1016/0010-0285(77)90012-3

Newell, A., & Simon, H. A. (1972). Human problem solving: Vol. 104. *No. 9*. Englewood Cliffs, NJ: Prentice-Hall.

Nielsen, J. I. (2005). Bordwell on Bordwell. *Part IV-Levels of Engagement, 16*, 9.

Norman, D. A. (2005). *Emotional design: Why we love (or hate) everyday things*. Basic books.

O'Connell, R. M. (2014). *Visualizing Culture: Analyzing the Cultural Aesthetics of the Web*. Peter Lang Publishing.

Pavlou, P. A., & Chai, L. (2002). What Drives Electronic Commerce across Cultures? Across-Cultural Empirical Investigation of the Theory of Planned Behavior. *Journal of Electronic Commerce Research, 3*(4), 240–253.

Purcell, K. (2010). *The state of online video*. Washington, DC: Pew Internet & American Life Project.

Reinecke, K., & Bernstein, A. (2011). Improving performance, perceived usability, and aesthetics with culturally adaptive user interfaces. *ACM Transactions on Computer-Human Interaction, 18*(2), 8. doi:10.1145/1970378.1970382

Reinecke, K., & Gajos, K. Z. (2014, April). Quantifying visual preferences around the world. In *Proceedings of the SIGCHI Conference on Human Factors in Computing Systems* (pp. 11-20). ACM.

Roseman, I. J., & Smith, C. A. (2001). Appraisal theory. In K. R. Scherer, A. Schorr, & T. Johnstone (Eds.), *Appraisal processes in emotion: theory, methods, research* (pp. 3–19). Oxford, UK: Oxford University Press.

Rössel, J., & Schroedter, J. H. (2015). Cosmopolitan cultural consumption: Preferences and practices in a heterogenous, urban population in Switzerland. *Poetics*, *50*, 80–95. doi:10.1016/j.poetic.2015.02.009

Saariluoma, P., Cañas, J. J., & Leikas, J. (2016). *Designing for Life: A Human Perspective on Technology Development*. Springer. doi:10.1057/978-1-137-53047-9

Saariluoma, P., & Jokinen, J. P. (2014). Emotional dimensions of user experience: A user psychological analysis. *International Journal of Human-Computer Interaction*, *30*(4), 303–320. doi:10.1080/10447318.2013.858460

Scherer, K. R. (2009). The dynamic architecture of emotion: Evidence for the component process model. *Cognition and Emotion*, *23*(7), 1307–1351. doi:10.1080/02699930902928969

Silvennoinen, J. M., Rousi, R., Jokinen, J. P., & Perälä, P. M. (2015, September). Apperception as a multisensory process in material experience. In *Proceedings of the 19th International Academic Mindtrek Conference* (pp. 144-151). ACM. doi:10.1145/2818187.2818285

Silvia, P. J. (2005). Cognitive appraisals and interest in visual art: Exploring an appraisal theory of aesthetic emotions. *Empirical Studies of the Arts*, *23*(2), 119–133. doi:10.2190/12AV-AH2P-MCEH-289E

Silvia, P. J. (2005). Emotional responses to art: From collation and arousal to cognition and emotion. *Review of General Psychology*, *9*(4), 342–357. doi:10.1037/1089-2680.9.4.342

Silvia, P. J. (2005). What is interesting? Exploring the appraisal structure of interest. *Emotion (Washington, D.C.)*, *5*(1), 89–102. doi:10.1037/1528-3542.5.1.89 PMID:15755222

Silvia, P. J. (2006). *Exploring the psychology of interest*. Oxford University Press. doi:10.1093/acprof:oso/9780195158557.001.0001

Silvia, P. J., & Berg, C. (2011). Finding movies interesting: How appraisals and expertise influence the aesthetic experience of film. *Empirical Studies of the Arts*, *29*(1), 73–88. doi:10.2190/EM.29.1.e

Skrbis, Z., & Woodward, I. (2007). The ambivalence of ordinary cosmopolitanism: Investigating the limits of cosmopolitan openness. *The Sociological Review*, *55*(4), 730–747. doi:10.1111/j.1467-954X.2007.00750.x

Smith, C. A., & Kirby, L. D. (2001). *Toward delivering on the promise of appraisal theory*. Academic Press.

Szerszynski, B., & Urry, J. (2002). Cultures of cosmopolitanism. *The Sociological Review*, *50*(4), 461–481. doi:10.1111/1467-954X.00394

Tan, E. S. (1996). *Emotion ond the Strtucture of Narrative film: Film as an Emotion Machine, prijev*. B. Fasting.

Tarvainen, J., Sjöberg, M., Westman, S., Laaksonen, J., & Oittinen, P. (2014). Content-based prediction of movie style, aesthetics, and affect: Data set and baseline experiments. *IEEE Transactions on Multimedia*, *16*(8), 2085–2098. doi:10.1109/TMM.2014.2357688

Tomlinson, J. (1999). *Globalization and culture*. University of Chicago Press.

Trompenaars, F., & Hampden-Turner, C. (2010). *Riding the waves of innovation: Harness the power of global culture to drive creativity and growth*. McGraw Hill Professional.

Van Slyke, C., Lou, H., Belanger, F., & Sridhar, V. (2010). The influence of culture on consumer-oriented electronic commerce adoption. *Journal of Electronic Commerce Research*, *11*(1), 30.

Wang, H. L., & Cheong, L. F. (2006). Affective understanding in film. *IEEE Transactions on Circuits and Systems for Video Technology*, *16*(6), 689–704. doi:10.1109/TCSVT.2006.873781

Weber, R. P. (1990). *Basic content analysis (No. 49)*. Sage. doi:10.4135/9781412983488

Wittgenstein, L. (1958). *Preliminary Studies for the "Philosophical Investigations"*. Generally Known as The Blue and the Brown Books.

Woodward, I., Skrbis, Z., & Bean, C. (2008). Attitudes towards globalization and cosmopolitanism: Cultural diversity, personal consumption and the national economy. *The British Journal of Sociology*, *59*(2), 207–226. doi:10.1111/j.1468-4446.2008.00190.x PMID:18498592

ADDITIONAL READING

Cañas-Bajo, J., Leikas, J., Jokinen, J., Cañas, J. J., & Saariluoma, P. (2016). How older and younger people see technology in Northern and Southern Europe: Closing the generation gap. *Gerontechnology (Valkenswaard), 14*(2), 110–117. doi:10.4017/gt.2016.14.2.010.00

Cañas-Bajo, J., Perälä, P. M., Silvennoinen, J. M., & Saariluoma, P. (2015, September). Affective contents of cross-cultural audiovisual experience. In *Proceedings of the 19th International Academic Mindtrek Conference* (pp. 82–88). ACM. doi:10.1145/2818187.2818294

Cañas-Bajo, J., Santamaría-Muñoz, L., Valtanen, J. P., Berki, E., Ross, M., & Staples, G. (2015). Multicultural edutainment role models for women IT professionals – an innovative proposal to creative industries. *Twentieth International Conference on Outlook on Computing Education.*

Cañas-Bajo, J., & Silvennoinen, J. M. (2017). Cross-cultural factors in experiencing online video contents in product marketing. *International Journal of Art, Culture and Design Technologies, 6*(1), 40–56. doi:10.4018/IJACDT.2017010103

Silvennoinen, J., Rousi, R., & Mononen, L. (2017). Creative Interpretation in Web Design Experience. *The Design Journal, 20*(sup1), S134-S145.

Silvennoinen, J., Vogel, M., & Kujala, S. (2014). Experiencing visual usability and aesthetics in two mobile application contexts. *Journal of Usability Studies, 10*(1), 46–62.

Silvennoinen, J. M., & Jokinen, J. P. (2016). Appraisals of salient visual elements in web page design. *Advances in Human-Computer Interaction.*

Silvennoinen, J. M., & Jokinen, J. P. (2016, May). Aesthetic Appeal and Visual Usability in Four Icon Design Eras. In *Proceedings of the 2016 CHI Conference on Human Factors in Computing Systems* (pp. 4390-4400). ACM. doi:10.1145/2858036.2858462

Silvennoinen, J. M., Rousi, R., Jokinen, J. P., & Perälä, P. M. (2015, September). Apperception as a multisensory process in material experience. In *Proceedings of the 19th International Academic Mindtrek Conference* (pp. 144-151). ACM. doi:10.1145/2818187.2818285

KEY TERMS AND DEFINITIONS

Audiovisual Experience: People's cognitive and emotional impression of audiovisually presented products and materials.

Audiovisual Representation: Mental representations that incorporate visual and audio elements of the perceptual stimuli.

Cross-Cultural Factors: Dimensions in which cultural differences impact the type of relations and the types of communication. In this context, these dimensions may influence the way usability and aesthetics are perceived.

Global Aesthetics: Subjective quality of objects, related to subjective sensori-emotional values and judgements of sentiment and taste.

Mental Representation: Symbolic mental contents that people have of their actions, environment, and their own internal states.

User-Centered Design: A model that seeks to satisfy the user's needs when interacting with artifacts by providing a good experience of this interaction without investing larger efforts.

ENDNOTE

1 Results are all described in "Cross-Cultural Factors in Experiencing Online Video Contents in Product Marketing" (Cañas-Bajo & Silvennoinen, 2017). They are reproduced here with permission of IGI Global.

APPENDIX A

Still-images of the videos used in the experiment and links to videos on YouTube.

Figure 1. YouTube link: https://www.youtube.com/watch?v=gCqw-NTVjT4

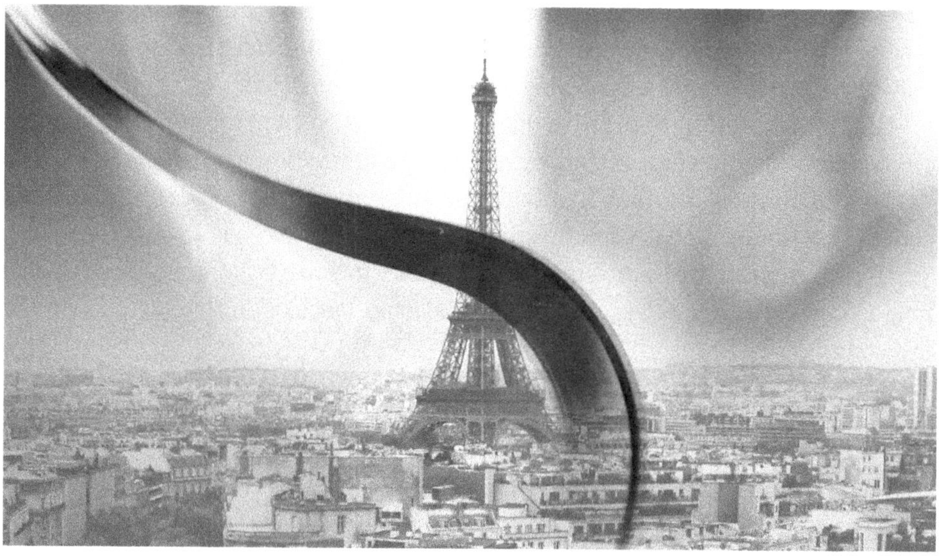

Figure 2. YouTube link: https://www.youtube.com/watch?v=YXiKgElEulg

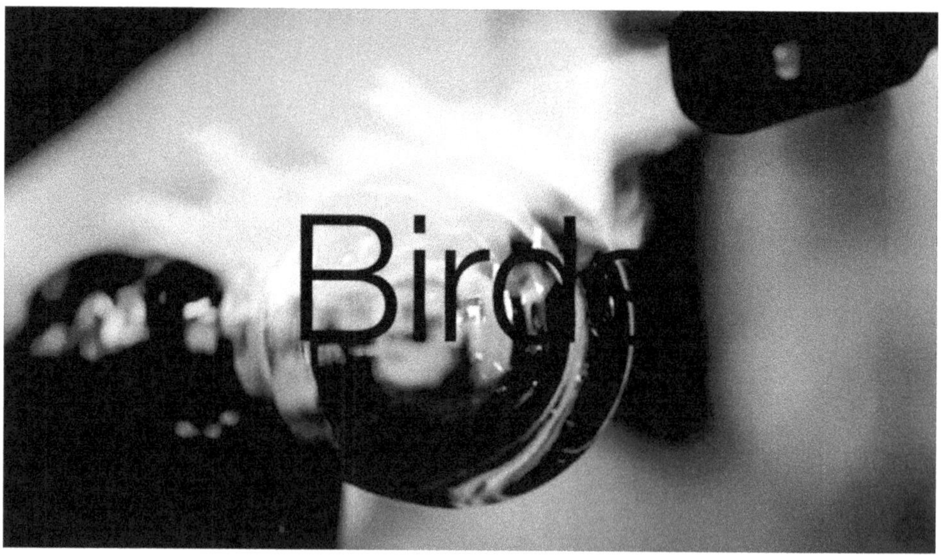

Figure 3. YouTube link: https://www.youtube.com/watch?v=bl9LsiQpcl4

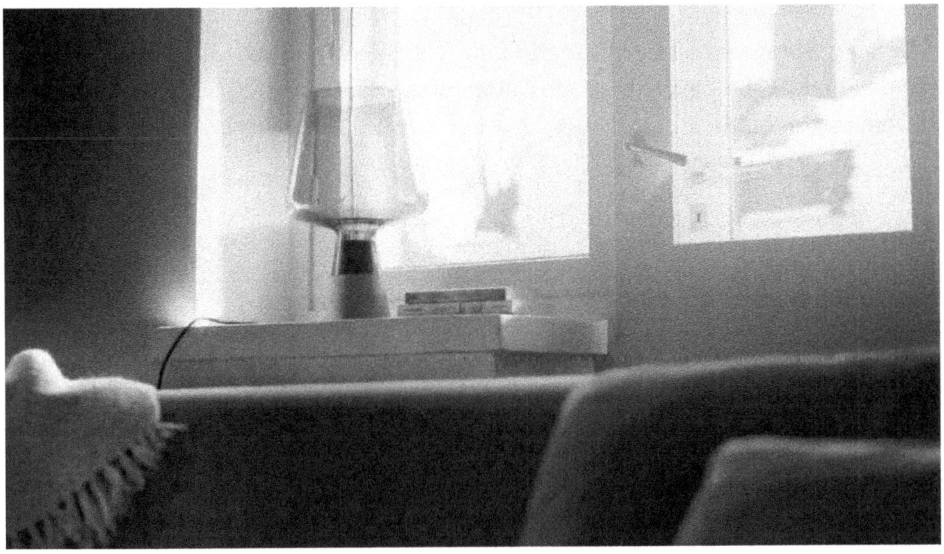

APPENDIX B

Mean percentage semantic units within the categories and sub-categories produced by the Finish and Spanish groups after presentation of the videos (significant differences in bold).

Table 1.

Category and Sub-Categories	Finnish	Spanish
Overall Impression of the Video		
Aesthetic persuasion	1,05	1,05
Aesthetic emotion	2,55	3,40
Aesthetic Quality	2,00	4,20
Memories & Associations	0,9	2,70
Desirability	0,40	0,20
Rhythm & Length	0,30	0,60
Total	7,20	12,15
Narrative Structure		
Aesthetic persuasion	0,95	0,85
Aesthetic emotion	0,10	0,00
Aesthetic quality	0,30	0
Overall idea	1,45	1,7
Plots aspects	2,70	4,25
Characters	0,60	1,35
Context	0,20	0,45
Product presentation	1,50	0,85
Total	7,85	9,30
Visual Design		
Aesthetic persuasion	0,20	0,30
Aesthetic emotion	0,10	0,25
Aesthetic Quality	0,75	0,25
Product presentation	9	0,45
Composition	0,70	1,15
Text	0,20	,10
Shapes	0,60	1,70
Materials	0,20	,70

continued on following page

Table 1. Continued

Category and Sub-Categories	Finnish	Spanish
Light	0,60	,75
Colors	0,30	,55
Total	1,80	2,15
Audio Design		
Aesthetic persuasion	0,20	0,10
Aesthetic emotion	0,10	0,55
Aesthetic quality	0,70	0,65
Narrator / off voice	0,55	0,30
Congruency with video	0,25	0,55
Total	4,1	6,2

APPENDIX C

Figure 4. Number of verbalizations as a function of Nationality, Category.

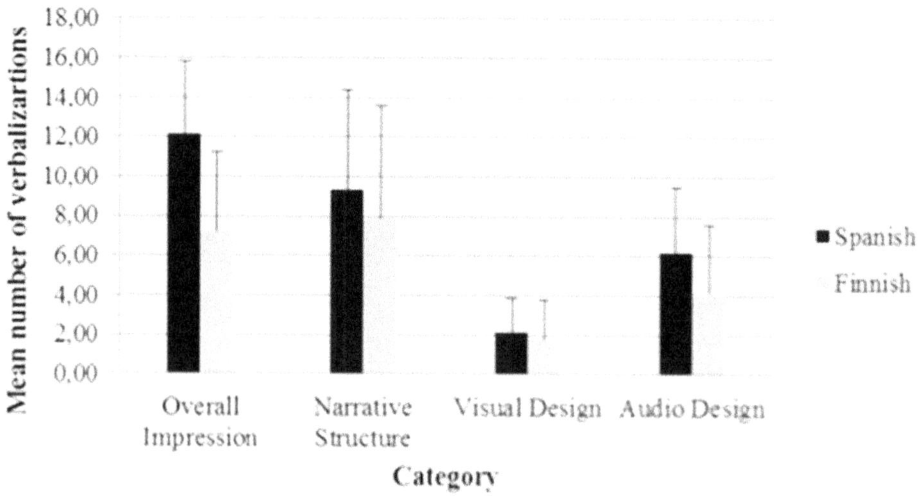

Chapter 11
Rapid Product Development and Application in Ceramics Production

Folasayo Enoch Olalere
Vaal University of Technology, South Africa

ABSTRACT

This chapter practically investigates how rapid prototyping technology can be effectively adopted to enhance ceramics production. The researcher used an experimental approach that analysed the conventional methods used in ceramics production and then introduced computer-aided design tools (CAD) and rapid prototyping technology into the development process. After that, four ceramic products were developed to test the viability of the rapid production process. Finally, the process was evaluated and compared to the conventional methods. The findings revealed that the introduction of CAD and rapid prototyping technology into the development process reduces the development time and also helps to ensure the accuracy of the prototypes produced. Besides, visualising and validating the design digitally (CAD models) also helps to identify possible faults at the early stage of the development process. Therefore, the ideas explored in this chapter will give insight to ceramics designers and artisans on how they can effectively eliminate bottlenecks in the production process.

DOI: 10.4018/978-1-5225-5023-5.ch011

INTRODUCTION

Physical prototyping can be a major bottleneck, slowing down the product development process and seriously constraining the number of design alternatives that can be examined. According to Gary Hawley, the conventional process of physical prototyping (Hand carving and turning) is time consuming and never produced 100 percent, accurate models. As a result, it is hard for the clients to understand the concepts being proposed fully. Also, ceramics companies are too careful about introducing new products because it takes as long as four weeks to create prototypes (Gary 2012). Hence, many manufacturers easily accept that eliminating physical prototypes equates to shorter or shortened product development cycle time and also a competitive strategy for reducing development cycles and getting products to market faster. However, the reality is that product development teams are still forced to include physical prototyping in the design loop to verify proper functionality, ease of assembly and also use it to create mould for mass production

Lately, the evolution of rapid product development has provided extraordinary opportunities for new product development and manufacturing improvement. Research from the Aberdeen Group shows that manufacturers that use rapid product development build half of the number of physical prototypes as the average manufacturer, get to market 58days faster than average, and experience 48 percent lower prototyping cost (Aberdeen 2006). Instead of building multiple physical prototypes and then testing them to see if they'll work, companies can conduct testing digitally throughout the process by using digital prototyping to catch design problems up front. Therefore manufacturers can reduce the number of physical prototypes they need to create before manufacturing a product, which also reduces the cost and time needed for physical prototyping (Aberdeen 2006). Against this backdrop, this chapter seeks to explore the effectiveness of the rapid product development in ceramics production.

RAPID PROTOTYPING TECHNOLOGY

Prototyping is a fundamental design initiation, which involves the construction of working models of conceived products for mass production. A prototype is the first or original example of a product that has been or will be copied or developed; it is a model or preliminary version (Chua, Leong, and Lim 2010). According to Soegaard (2010), a prototype is often used as part of product design process to allow engineers and designer the ability to explore design alternatives, test theories and confirm performances before starting production of a new product. For example, some prototypes are used to confirm and verify consumer interest in a proposed design whereas other prototypes will attempt to verify the performance or suitability of a specific design. Prototypes are also used to revise the design to reduce cost through optimisations and refinement. There are two main modes of prototyping: Visual/ Virtual Prototyping and Physical Prototyping. Visual Prototyping is the process of simulating the appearance; colour, size or shape of the intended design visually while physical prototyping is the building of a real model.

The application of CAD tools in the two modes of prototyping has resulted in digital and rapid prototyping. Digital prototyping helps product developers to design, iterate, optimise, validate and visualise their products digitally throughout the product development process (Autodesk 2013). Rapid prototyping is the automatic construction/fabrication of physical objects directly from computer aided design (CAD) data sources. The technologies are being increasingly used also in non-prototyping applications; thus, the techniques are often collectively referred to as solid free-form fabrication; computer automated manufacturing or layered manufacturing (William 1998). These systems add and bond materials in layers to form objects. With this additive technologies, objects can be formed with any geometric complexity or intricacy without the need for elaborate machine setup or final assembly (Venuvinod and Ma 2004).

The first commercially available machine based on this principle was Stereo Lithography Apparatus-1 (SLA-1) released by 3D Systems, Inc. in 1984 (Hul 1984). There are now numbers of rapid technologies that uses additive principles; these include Selective Laser Sintering (SLS), Fused Deposition Modelling (FDM), Layered Object Manufacturing (LOM) and 3D Printer (3DP). Each of these technologies has their unique strengths and varying base materials, as shown in Table. 1.

The advantage of rapid prototyping is that any solid 3D freeform can be generated without the aid of external tooling, so most of the problems associated with the subtractive jiggering and jollying, Computer Numeric Control and Lathe are sidestepped. Just as shown in Table 2, rapid prototyping technologies have several

Table 1. Rapid prototyping technologies and their base materials (Olalere and Ab. Aziz 2014)

Rapid Prototyping Technologies	Base Materials
Selective Laser Sintering (SLS)	Thermoplastic & metal powders
Fused Deposition Modelling (FDM)	Thermoplastic, eutectic metals
Stereo lithography (SLA)	Photopolymer
Laminated Object Manufacturing (LOM)	Paper
Electron Beam Melting (EBM)	Titanium alloys
Direct Metal Laser Sintering (DMLS)	Alloy metals
3D Printing (3DP)	High performance composite

advantages compared to CNC and conventional manufacturing processes. These include lower risk factor and nominal time to process the CAD data during pattern building, the absence of secondary or nominal post operation, and the ability to integrate multiple operations in the building process.

Research by Chua, Leong, and Lim (2010) shows that the integration of rapid prototyping technology in the product lifecycle can reduce the time and cost significantly to 10-50%. As illustrated in Figure 1, when rapid prototyping technology is integrated into the product life cycle, the assembling and function testing can be done simultaneously during manufacturing. This is due to the ability of RP technology (especially high-end FDM) to integrate multiple operations in manufacturing, which saves time and production cost. Due to the advantages of rapid prototyping, the technology has been employed in many advanced ceramic industries, right from the conceptualization of ideas to realisation of processes (Woodward 1996).

Its application has been motivated by the advances in engineering ceramics where methods of creating complex shapes are limited (Cawley 1997). For example, research by Huson (2012) and Walters (2012) on 3D printing of ceramics for design concept modelling at the Centre for Fine Print Research, the University of the West of England confirms that rapid prototyping technology is viable tools for developing designs with complex shapes. Similarly, Gary and Thomas (2012) also confirms that rapid technology makes a concept to manufacturing process much faster, compared with the traditional techniques (such as hand throwing, turning and hand modelling).

Yue and Zhi (2012) also researched on rapid development of new sanitary ware products of ceramics and application study with the computer. The researchers

Table 2. Comparison of Conventional, CNC, and RP construction methods

Variables	Conventional	CNC Machining	Rapid Prototyping
Operations	Single Operation	Simultaneous operations	Multiple operations
Post-Operations	Secondary and tertiary machining operations	Secondary machining operations	No or minimal post operation
Tools	Different types of cutting and finishing tools	Different types of cutting & finishing tools	No tools required
CAD data	CAD model not required	CAD data required	CAD data (STL format) required
Risk factor/error	Very high risk factor	High risk factor	Lower risk factor
Intricate Features	Intricate features within the pattern can't be built	Forming Intricate features are limited	Intricate features are easily built
End product Operator	Single pattern at a time	Single pattern at a time	Possibility of multiple patterns
Nature of the product	Generally solid pattern	Solid & hollow pattern	Solid and hollow pattern

Figure 1. Results of integration of Rapid Prototyping technology (Chua, Leong, and Lim 2010)

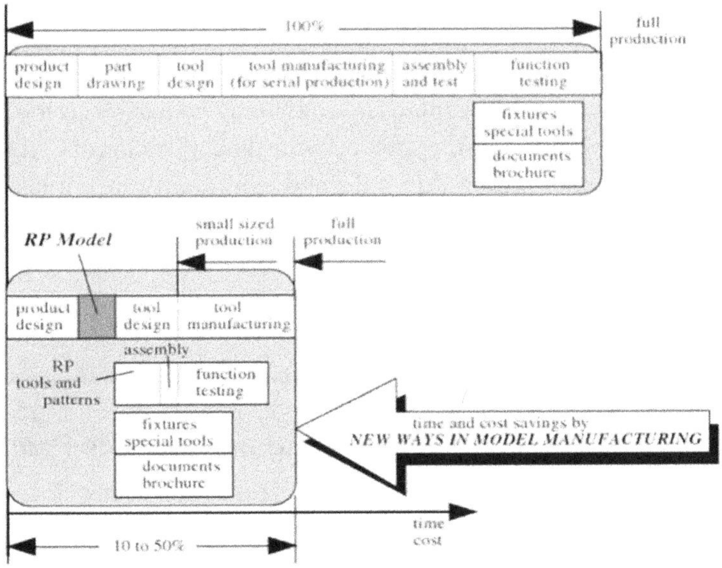

analysed how product development process can be simplified to reduce development cost and time. This was achieved by proposing the use of some rapid development process such as "Hole Radius of Ray light Technology", 3D point cloud graph, 3D scanning algorithm and 3D lofting cloud graph. In relation to this, Jorge et al. (2004) also research on rapid prototyping & tooling applied in product development of ceramic components. The researchers reviewed some of the works done at INEGI2 and also seek to introduce new methods for product development of decorative ceramic components using CAD 3D design or engineering tools. With respect to this, the researchers used an experimental approach by designing new ceramic components using different software such as Allias, Solid Works, Solid Edge and Unigraphics and then used Laminated Object Manufacturing (LOM) technology to generate a rapid prototyping models.

Also, Jan and Konrad (2006) worked on rapid prototyping but focused on the manufacturing of ceramic components for medical and technical applications via Selective Laser Melting (SLM). The work focused on the use of Zirconia and tricalcium-phosphate (TCP) ceramics as the materials for rapid manufacturing. Zirconia ceramics was used in manufacturing customized dental restoration and also technical applications where high strength, wear resistance and heat resistance are required. TCP ceramics was used in the rapid manufacturing of customised bio-restorable bone substitute implants.

Even though most of the previous researches on rapid technology focus on large-scale engineering ceramics, research by Jennifer (2010) shows that rapid product development could also help to reduce drafters significantly in small to mid-size companies. Therefore, this chapter seeks to explore the potential of rapid technology in enhancing prototyping in small and medium scale ceramic industries. The study used an experimental approach, combining both engineering and artistic perspective in developing ceramic prototypes with the application of CAD tools (such as Autodesk Sketchbook, AutoCAD and Solid works software). It also generates rapid prototypes using rapid prototyping machine (uPrint 3D Printer) and finally compared the digital method with the conventional method.

STUDY

Conventional Methods of Prototyping in Ceramics Production

The conventional methods of prototyping are the manual/traditional methods used in ceramic product development. Detailed data on these methods was collected through primary and secondary data. Primary data was collected from three cottage ceramics firm using interview, participation and observation. The data collection gives the general overview of the conventional method used by potters around the world, which reveals that the three major materials used for building prototypes in ceramic production are wood, clay and plaster of Paris (POP).

- **Using Wood for Building Prototypes:** According to Andrew (2011), each new ceramic piece begins with a prototype made of wood. The advantage of using wood is its durability compared to clay and POP. Besides, wood is also great for achieving sharp, transitional lines and edges. The production process begins with sketches that illustrate the side view or profile and the top view. With reference to the side view drawing as a blueprint, the wood is shaped using a surform tool or turned on a lathe to make a solid round form whose shape is close to the side profile of the finished piece. Then, the top-view drawing is used as a cutting template, which is glued to the top of the form (Figure 2). Using a band saw, the model is cut into its shape following the outside edges of the glued-on template and the sides of the form are then sanded smooth to erase any irregularities from sawing (Figure 2)

Before using the wooden prototypes to create the plaster mould, seam lines are first drawn on the prototypes to know how many parts will make up the mould. Then, a clay slab is added to the top of the prototype for a pouring gate. After that, the mould

Figure 2. Finished wooden prototypes of various vessels (Andrew 2011)

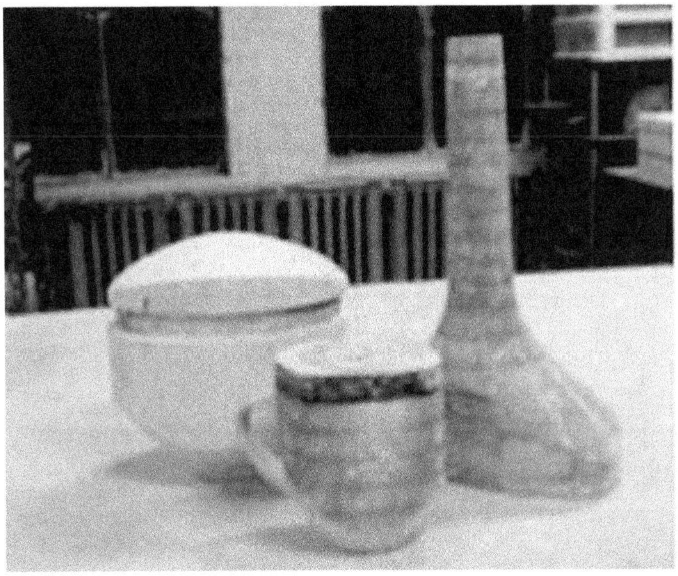

is created following the seam lines (as shown in Figure 3), which is later used for slip casting. This process of caving the model/prototype from wood takes between 3-5days depending on the intricacy and size of the models. Therefore, a complete process of product development using wood takes between 2-4weeks (14-28 days).

- **Using Clay for Building Prototypes:** The property of clay such as plasticity as made it a very useful material in building the object. This is because it can be used to form any shape. Therefore, some designers find it easier to use for building prototypes. Building models with clay can be achieved using wheel throwing or hand building. Throwing is done on a potter's wheel; has the wheel-head rotates rapidly, the solid ball of soft clay on the wheel-head is pressed, squeezed and pulled gently upwards and outwards into a hollow shape. Unfortunately, this method is only appropriate for round shape objects (Figure 4).

Hand building process is a method of constructing pieces by hand from coils of clay, combining flat slabs of clay, or pinching solid balls of clay or some combination of these. This method is mostly used for intricate geometrical shapes (Figure 5).

After the clay model/prototype has been produced, it is allowed to dry before using it to create the plaster mould (Figure 6), which is then used for slip casting. According to Woodward (1996), the process of using clay for building prototype in

Figure 3. Mould halfway through the casting process (Andrew, 2011)

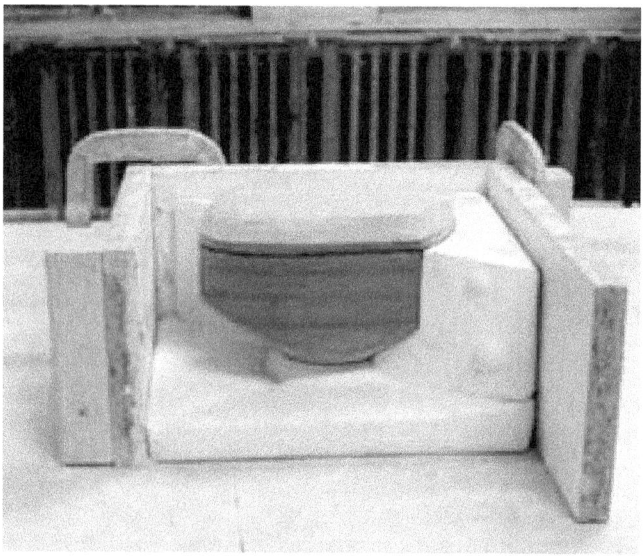

Figure 4. Throwing on a potter's wheel

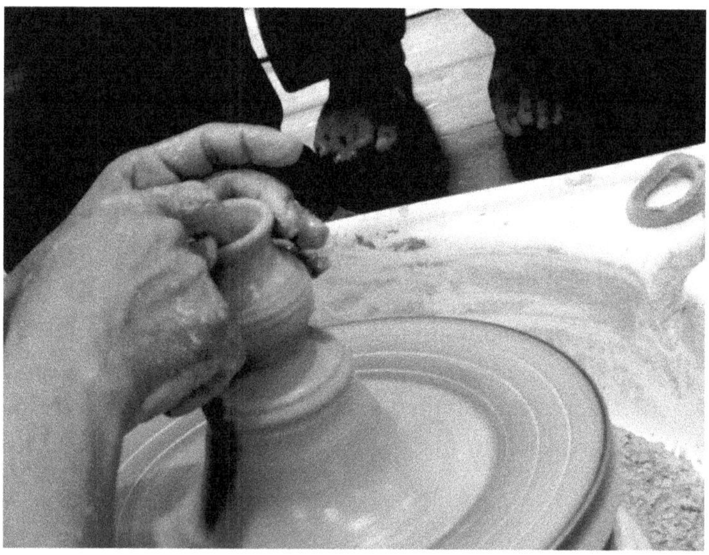

the conventional ceramic product development takes between 4-6weeks (28-42 days) depending on the size and form of the model and also the weather condition. Some of the disadvantages of using clay are shrinkage, longer processing time due to the drying time and also clay models are not durable unless fired which will increase the development time and cost.

Figure 5. A clay model produced by hand building method

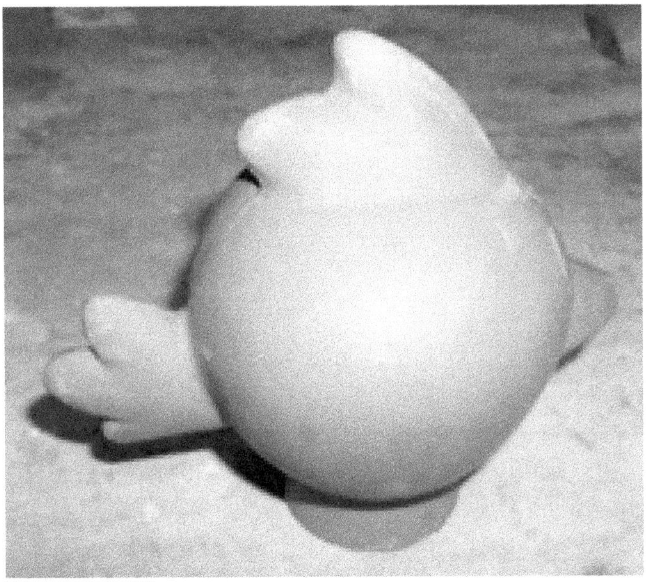

Figure 6. Creating mould with clay model

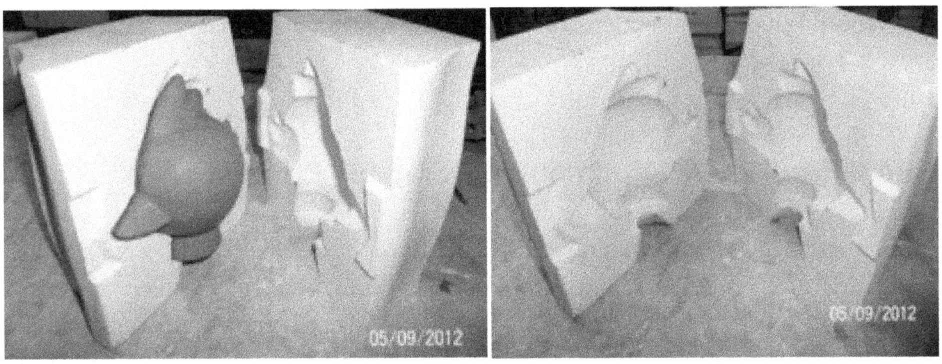

- **Using Plaster of Paris for building Prototypes:** Plaster of Paris is a plaster made by calcining gypsum. It is a white powder that forms a paste when mixing with water. Its application in ceramic production includes the building of models and mould. Before creating the plaster models, design ideas are first developed into detailed two-dimensional drawings (Fig. 7). The drawings are used as a guide to creating the prototypes using handing tuning technique. The material (POP) is first mixed in appropriate proportion with water and left to solidify. After that, the solidified plaster is mounted

Figure 7. Two-dimensional drawings of design ideas

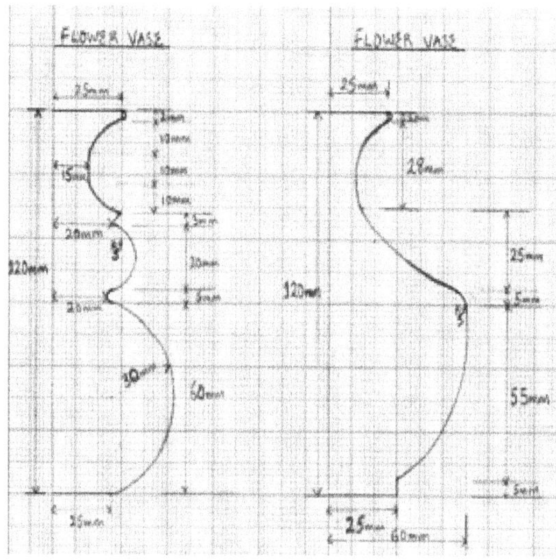

on the turning wheel and shaped out to the desired form as the wheel-head rotates (Fig. 8). The mock-up models (Fig. 9) were then used to create plaster mould. According to Gary Hawley (a designer at Denby Pottery Company), the product development process, using plaster prototypes takes between 2-4weeks (Gary 2012).

Rapid Product Development in Ceramics Production

Rapid product development integrates CAD tools (Autodesk Sketchbook Designer, AutoCAD, and Solidworks) and rapid technology into the development process, and four products were produced to test its viability. The experimental process includes three stages: concept development, the creation of prototypes, and the creation of plaster moulds (Figure 10).

- **Concept Development:** The conceptual designs were sketched at this first stage using Autodesk Sketchbook Designer software (Figure 11), and then developed into detailed two-dimensional drawing with the aid of AutoCAD (Figure 12). After which the detailed 2D drawings were used as a guide to building the 3D CAD models (digital prototypes) using Solidworks software (Figure 13). The part drawings and sectional views of each product were also generated [Figure 14(A-D)] to check the accuracy of the digital models.

Figure 8. Turning wheel with a prototype and the shaping tools used

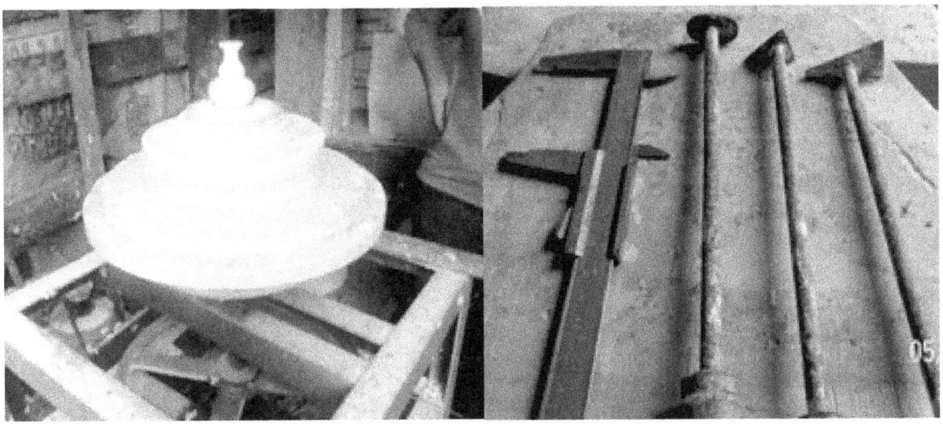

Figure 9. Prototypes created with plaster of Paris (POP)

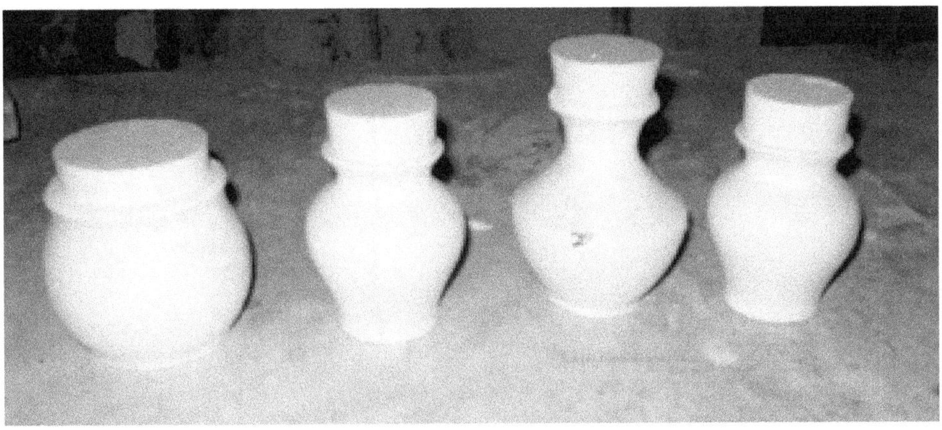

- **Creation of Prototypes:** After the concept development, the 3D CAD models were used to generate rapid prototypes using uPrint 3D Printer. The printer is a Fused Deposition Modelling (FDM) technology, which builds accurate and stable rapid prototypes in ABS-plus (Acrylonitrile Butadiene Styrene) plastic (GXSC 2011; uPrintSE 2012). ABS is a thermoplastic with glass transition temperature of approximately 105°C (221°F) (Stratasys 2007). The printer uses two materials (support & modelling materials), and the maximum build size of the machine is 203×203×152mm (8×8×6inchs).

Figure 10. Experimental process

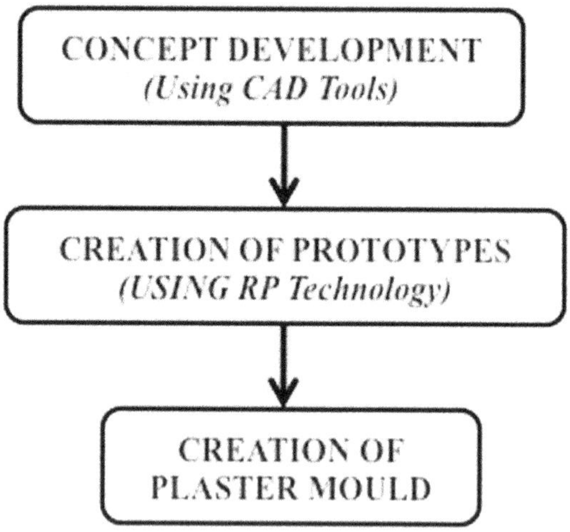

Figure 11. Conceptualized digital sketches of the product

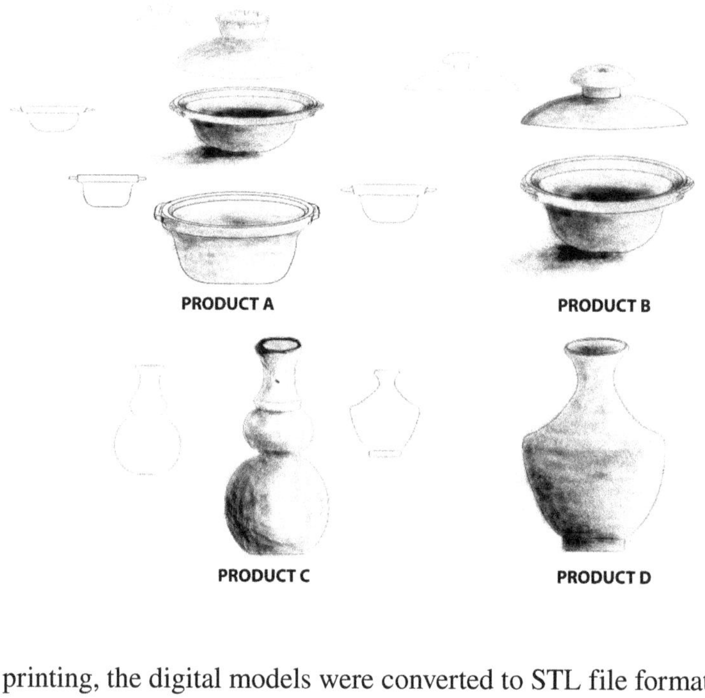

Before printing, the digital models were converted to STL file format. The file is accessed by the printer's software (Catalyst EX) and converts the CAD system's STL output into 3D model print paths. The Catalyst EX software can also be used to scale, arrange and place the CAD model. Once the CAD model is sent to the printer, the printer builds the model alongside with a support (Figure 15). The

Figure 12. Digital 2-dimensional drawings (Product A, B, C & D)

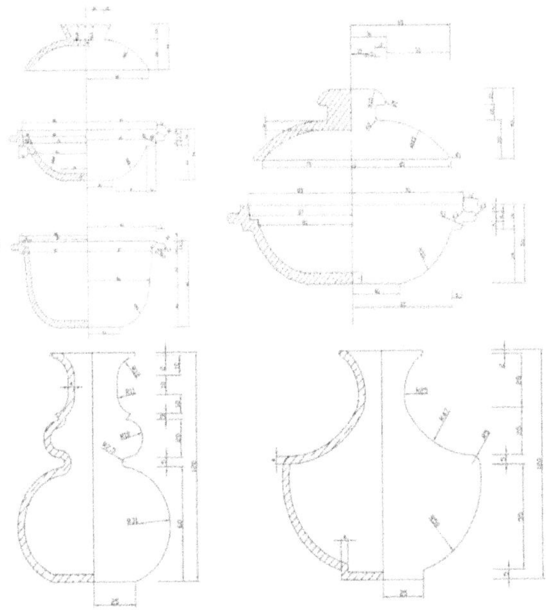

Figure 13. 3D CAD models of the products built with Solidworks

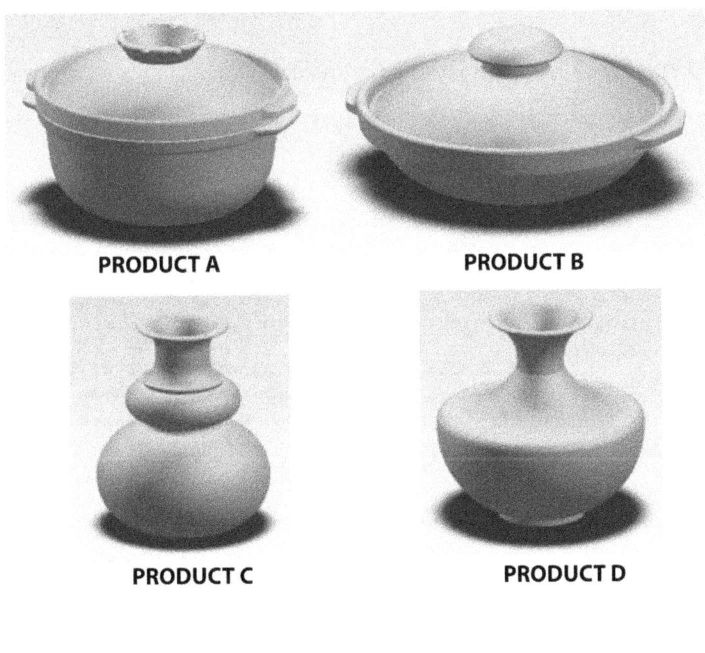

Figure 14. Part drawings and sectional views of the models

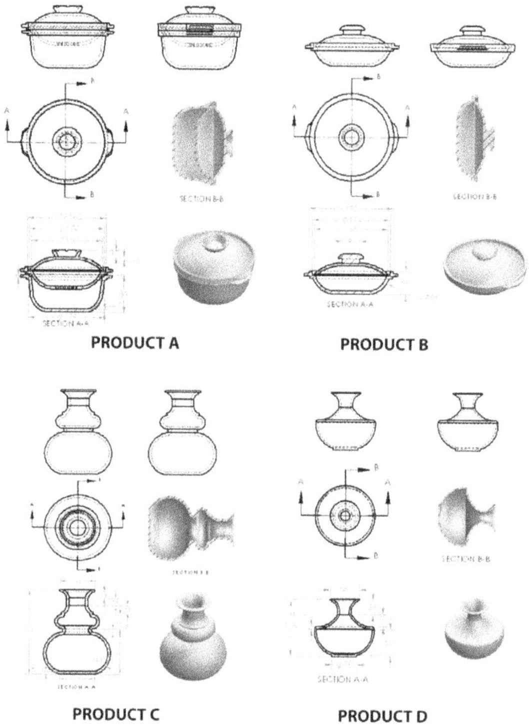

Figure 15. Prototypes with the support material

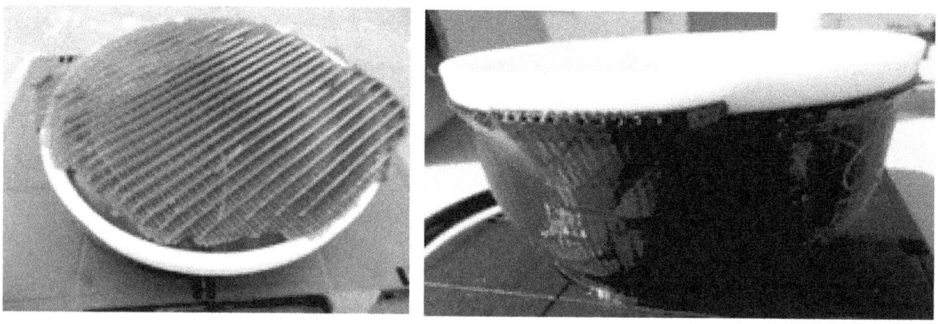

support material is a soluble support technology (SSF), which can be dissolved from the model in a wave wash support cleaning system (ultrasonic) that maintains the right temperature and agitation for efficient support removal from the model's body (uPrintSE 2012). The dissolving of the support material takes between 30-60 minutes depending on the shape and size of the model. Figure 16 shows the rapid prototypes built with the uPrint 3D printer.

Figure 16. (A-D) Images of models built by RP-Machine

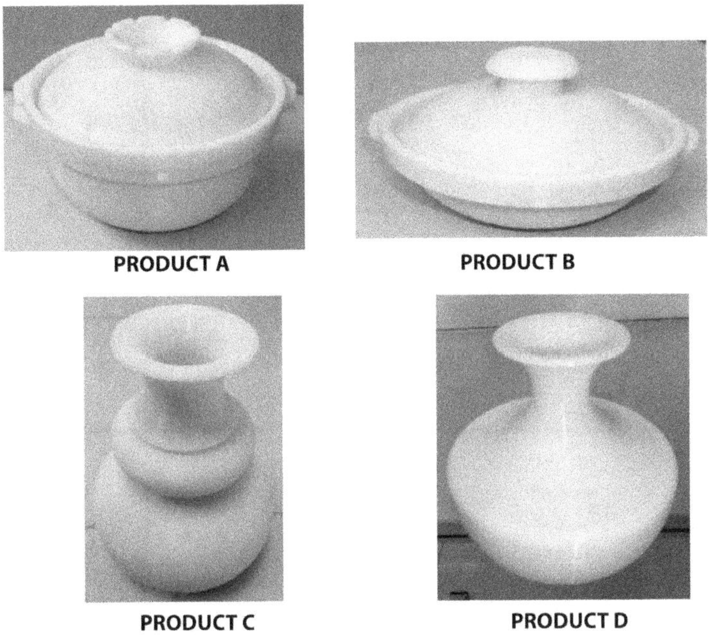

Figure 17. Rapid prototypes and the plaster moulds

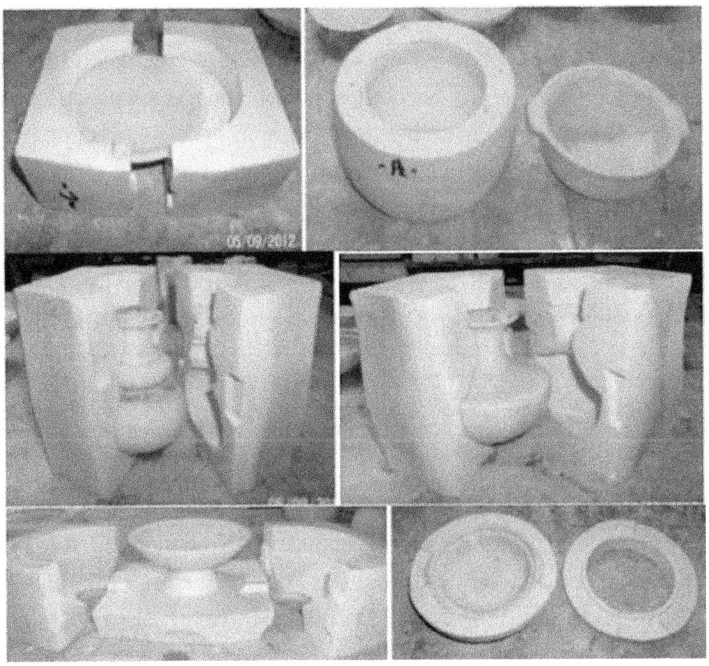

Figure 18. Final ceramic products

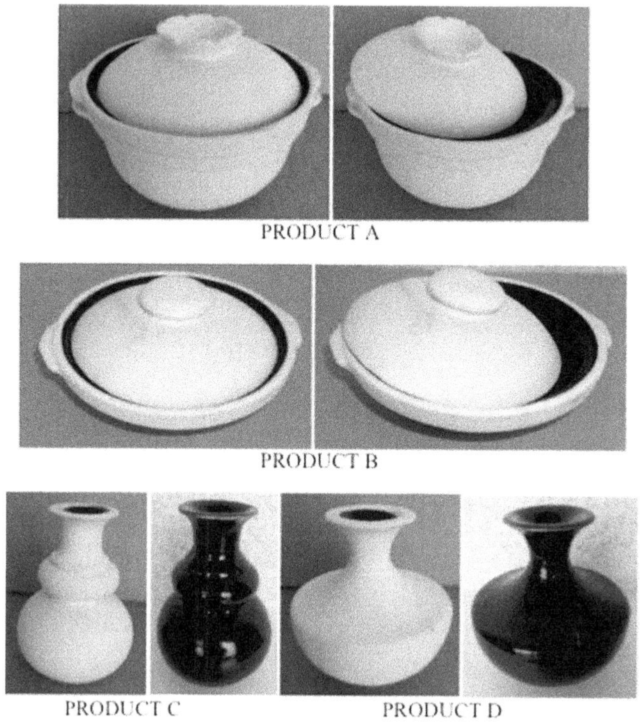

- **Creation of Mould:** The RP-models were used at this stage to produce plaster moulds (Figure 17). Since the models have been digitally checked and analysed at the early stage of the development process (concept development stage), the possible faults have been discovered and fixed thereby reducing the risk of prototype failing at this stage. Therefore, the created moulds were further used for slip casting, and the casted pieces were allowed to dry, and then fired (bisque and gloss firing), as shown in Figure 18. The overall time for the complete product development process is shown in Table 3.

CONCLUSION

There are three basic ways of creating prototypes: conservative, subtractive, and additive. Conservative approach involves applying the needed forces to deform material to the required shape, without either adding or removing material, i.e., the material is conserved (Venuvinod and Ma 2004). However, the disadvantages of this approach are that it consumes a lot of energy. Also, it focuses on form generation

Table 3. The overall time and cost of building the ceramic prototypes

PRODUCTS	TIME/DAYS
A (130×180mm; 420g)	13
B (85×190mm; 235g)	10
C (120×55mm; 115g)	8
D (115×60mm; 116g)	8

Table 4. Comparison of the conventional and rapid product development based on time

Methods		Development Time/Days
Conventional Methods	Wooden Prototype	14-28 days
	Clay Prototypes	28 – 48 days
	POP Prototypes	14 – 28 days
Rapid Prototyping Technology		8 – 13 days

without providing means for controlling material composition. In subtractive approach, unwanted segments are chipped away from a block of material. This approach is used in a conventional method for building models from plaster. Some of the challenges of this approach are that; it waste materials, only those form features accessible by the subtractive tools can be created and lastly, like the conservative approach, it focuses on the form generation without providing means for controlling material composition (Venuvinod & Ma, 2004).

The additive approach starts with nothing and builds an object incrementally by adding materials. The materials added each time can be the same or different. Thus, problems of form generation and composition can be addressed at once through the same process. The advantage of this approach is that any solid 3D freeform can be generated without the aid of external tooling, so most of the problems associated with the conservative and subtractive methods are sidestepped (Venuvinod & Ma, 2004). This additive approach was practically used in this study to build RP-models from CAD data. When compared to the three conventional methods use in ceramics production, the study shows that rapid product development reduces development time (Table 4) and also helps to ensure the accuracy of the prototypes produced. Besides, visualising the design digitally (CAD models) also helps to Identify possible faults at the early stage of the development process.

REFERENCES

Aberdeen. (2006, September). The Transition from 2D Drafting to 3D Modelling Benchmark Report. *Design Bulletin.*

Andrew, G. (2011). Ceramic Mold Making Techniques (2nd ed.). Retrieved from www.ceramicartsdaily.org

Autodesk. (2013). Digital Prototyping. *Autodesk*. Retrieved from http://usa.autodesk.com/digital-prototyping/

Cawley, J. D. (Ed.). (1997). *Proc. Int. Gas Turbine and Aeroengine Congress and Exhibition, Orlando, FA, 1-6*. American Society of Mechanical Engineering.

Chua, C. K., Leong, K. F., & Lim, C. C. S. (2010). *Rapid Prototyping: Principles and Applications*. World Scientific Publishing Company. doi:10.1142/6665

Gary, H. (2012). *The Denby Pottery Company Denby Pottery Extracts Unanticipated Benefits from 3D Printing*. Retrieved from www.printin3d.com

Gary, H., & Thomas, A. (2012). 3D Printing at Denby Pottery. In *Towards a New Ceramic Future: CFPR Symposium*. Victoria and Albert Museum, Hochhauser Auditorium.

GXSC. (2011). *uPrintSE 3d Print Pack*. Retrieved from http://store.gxsc.com/s.nl/it.A/id.1040/.f

Hul, C.H. (1984). *Apparatus for Production of 3D Objects by Stereolithography*. Academic Press.

Huson, D. (2012). 3D Printing of Ceramics for Design Concept Modelling CFPR 3D Lab. In *CFPR Symposium*. London: Arts and Humanities Research Council (AHRC).

Jan, W., & Konrad, W. (2006). Rapid Manufacturing of Ceramic Components for Medical and Technical Applications via Selective Laser Melting. Steinbachstr. Germany: Frauhofer Institute for Laser Technology (ILT).

Jennifer, H. (2010). 3D Model-Based Design: Setting the Definitions Straight. *MCADCafe*. Retrieved from http://www10.mcadcafe.com/nbc/articles/2/867959/3D-model-based-design-setting-definitions-straight

Jorge, F., Rui, J., Ricardo, P., & Ana, M. (2004). Rapid Prototyping and Rapid Tooling Applied in Product Development of Ceramic Components. *Materials Science Forum II, 455–456*, 835–38.

Olalere, F. E., & Ab, S. A. (2014). Computer-Aided Ceramic Design: Its Viability for Building User-Centered Design. In Contemporary Advancements in Information Technology Development in Dynamic Environments. Hershey, PA: IGI Global. doi:10.4018/978-1-4666-6252-0

Soegaard, M. (2010). *Prototyping*. Retrieved from http://www.interaction-design.org/encyclopedia/prototyping.html

Stratasys. (2007). ABS. *Stratasys*. Retrieved from www.nrri.umn.edu/NLTC/ABS07.pdf

uPrintSE. (2012). *uPrint 3d Printer*. Author.

Venuvinod, P. K., & Ma, W. (2004). *Rapid Prototyping: Laser-Based & Other Technologies*. Boston, MA: Kluwer Academic. doi:10.1007/978-1-4757-6361-4

Walters, P. (2012). Ceramic 3D Printing : A Design Case Study. In *CFPR Symposium*. London: Arts and Humanities Research Council (AHRC).

William, P. (1998). *Rapid Prototyping Primer*. Retrieved from http://www.me.psu.edu/lamancusa/rapidpro/primer/chapter2.htm

Woodward, C. (1996). *Computer Aided Industrial Design for Ceramics and Glass Industries*. Dest Artes Oy.

Yue, W. Y., & Zhi, F. Z. (2012). Rapid Development of New Sanitary Ware Products of Ceramics and Application Study with Computer. *Advanced Materials Research*, *466–467*, 201–206.

Chapter 12
Mobile Learning:
Benefits of Augmented Reality in Geometry Teaching

Rui Leitão
Loughborough University, UK

Joao M. F. Rodrigues
University of the Algarve, Portugal

Adérito Fernandes Marcos
Universidade Aberta, Portugal

ABSTRACT

As a consequence of the technological advances and the widespread use of mobile devices to access information and communication in the last decades, mobile learning has become a spontaneous learning model, providing a more flexible and collaborative technology-based learning. Thus, mobile technologies can create new opportunities for enhancing the pupils' learning experiences. This chapter presents the development of a game to assist teaching and learning, aiming to help students acquire knowledge in the field of geometry. The game was intended to develop the following competences in primary school learners (8-10 years): a better visualization of geometric objects on a plane and in space; understanding of the properties of geometric solids; and familiarization with the vocabulary of geometry. Findings show that by using the game, students have improved around 35% the hits of correct responses to the classification and differentiation between edge, vertex, and face in 3D solids.

DOI: 10.4018/978-1-5225-5023-5.ch012

INTRODUCTION

Mobile platforms are now part of young people's life. Their pervasiveness makes them an ideal vehicle for the development of educational content both for classroom activities and informally. The possibility of learning anywhere and at any time is one of the most remarkable features of mobile learning. The past years have seen a great increase in mobile learning, alongside a larger offer of contents and technological possibilities and a greater interest among people in buying and using mobile platforms. Mobile technologies are becoming more embedded, ubiquitous and networked, with enhanced capabilities for rich social interactions, context awareness and internet connectivity. Such technologies could have a great impact on learning (Naismith, Lonsdale, Vavoula, & Sharples, 2004). As a research field, mobile learning is relatively new. According to Catenazzi & Sommaruga (2013, p. 9), 'the first initiatives date back to the last decades of the 1900s, but the wide diffusion of mobile learning took place starting from 2000 as a result of the large availability of mobile technologies'.

Mayer (2005) describes the cognitive theory of multimedia learning as several cognitive processes that include the selection of relevant visual and verbal materials, organization of these visual and verbal mental representations in coherent structures in working memory (short-term memory), and integrating the representations among themselves and with prior knowledge. The author also refers to the importance of design to prime these different cognitive processes in order to create multimedia instructional messages, in other words, to communicate containing images and words intended to foster learning (Mayer, 2005). Furthermore, the *Cognitive Load Theory* complements this process model by describing three types of cognitive load: intrinsic cognitive load, which describes the natural complexity of the information, germane cognitive load, which describes the amount of mental effort in the acquisition of knowledge by the learner in comprehending the materials, and extraneous cognitive load, which describes processing demands of information that is not directly related to the learning task, but to the manner in which instruction materials are designed. (Sweller, 2010). Therefore, in this cognitive perspective, the design of educational materials has effects on the cognitive process.

Pupils' motivation is probably one of the most important factors for teachers in the enhancement of the learning process (Williams & Williams, 2011). Motivation is an inner state which stimulates pupils to engage in a certain task (Lei, 2010). In this context, by promoting entertainment and voluntary performance, games are related to a cognitive process of intrinsic motivation (Dichev, Dicheva, Angelova, & Agre, 2014).

Given the current technological possibilities and the popularity of games, the interest in games in a learning context has been on the rise (Bie & Lipman, 2012;

Razak, Connolly, Baxter, Hainey, & Wilson, 2012; Hamari, Koivisto, & Sarsa, 2014; Schlagenhaufer & Amberg, 2015; Cheng, 2017). Sales, user statistics and public opinion polls are all evidence that computer games have become a dominant entertainment. Tobias and Fletcher (2011) state that, for the same reason, it is hardly surprising that teachers and trainers in different sectors consider the use of games in the teaching and learning process. Games and cognition are deeply intertwined and, with the advent of computer games, new scientific interests in this relationship have emerged (Gamberini, Barresi, Majer, & Scarpetta, 2008).

Primary-school learners experience difficulty in developing and consolidating their spatial sense, namely the visualization and the understanding of the properties of geometric figures on a plane and in space. They also present difficulties in acquiring the vocabulary associated to geometrical concepts such as face, edge, vertex, plane, prism, pyramid, cylinder, cone, sphere, among others.

In educational practice, physical objects are commonly used to establish a socially shared meaning. 'Physical objects support collaboration both by their appearance, the physical affordances they have, their use as semantic representations, their spatial relationships, and their ability to help focus attention' (Billinghurst, 2002, p. 3). The development of a game to be used in the teaching of primary-school geometry, whose main purpose was to develop learners' spatial sense, with emphasis on visualization could be of major importance to improve these pupils' competences, in addition, the union of the game with augmented reality could bring same advantages. Augmented reality (AR) offers functionalities which improve immersion, interaction and imagination, experiences connected to the constructivist principles of learning. Moreover, AR offers an intimate relationship between virtual and physical objects in real time.

This paper presents an AR game for primary-school learners (8-10 years old), aimed to facilitate the appropriation and the interpretation of geometric solids and 3D objects. The main objective of the game was to help students develop a better visualization ability; a better understanding of the properties of geometric solids in space; and familiarization with the vocabulary of geometry. The technology in this project was used to "augment" the visual field of the user with the information necessary in the performance of the tasks. The camera detected position (AR) markers which then generated virtual 3D objects, where the student was taken to learn the concepts of edge, vertex and face. The game allowed the teacher and the learner to engage in experimental operations in real time and at the same time. Learners adopted a natural interactive method and enjoyed the experience in a real environment, without resorting to the mouse or the keyboard. The teaching context was assumed to be favourable to the experimentation with new teaching/learning models given the low cost of AR technology as well as the presence of necessary infrastructure within schools.

The next section will present the state of the art. Then, the steps and the process of the conceptual design involved in the development of the game will be described. Afterwards, the evaluation of the prototype and further research avenues will be presented. Finally, the last section contains a discussion and proposals for future work.

STATE OF THE ART

Connections between play and education for both adults and children have already existed for a long time. According to A. Krentz (Krentz, 1998), 'etymologically in Greek the terms "paideia" the word for education/culture, "paida" the word for play/pastime/sport and "paides" the word for children, have the same root, and the three terms often show up in the same context' (Krentz, 1998, p. 5). Although the relationship between the words *play* and *children* may suggest that play is more connected with children's activities instead of adults', the term extends to activities we might not consider as laborious, serious, or solemn. The term *play*, in the context of the ancient Greek texts, is built on the opposite of the term *work*, which in the mainly agrarian context of Greek life had connotations of agricultural labour (D'Angour, 2013).

In Plato's *Republic*, the connection between play and education is central. Krentz (1998) states that in order to understand Plato's philosophical message it is necessary to pay close attention to the 'connection between education/culture (paideia), and the pedagogical approach (paidagogia) to teaching and learning that are to be carried out in the community. The central aim of pedagogy (paidagogia) is to encourage learning as a form of play (paidia), which is the most persuasive and effective approach to learning for free citizens in a society which honours philosophers' (Krentz, 1998, p. 6). According to Jane McGonigal, the ancient Greek Herodotus reported the use of game-playing in society by a king to distract his citizens from a famine (McGonigal, 2011). More recently, in the early Soviet era, game elements were used by the Soviet Union leaders as a substitute for monetary incentives for performing at work (Dicheva, Dichev, Agre, & Angelova, 2015).

One of the early and best-known projects in the field which combines games and AR in an education context is Construct3D (Hannes, 2004). Construct3D is a major project that resulted in a geometric construction tool especially designed for the teaching of mathematics and geometry in secondary and tertiary education. It allows the visualization of three-dimensional (3D) objects which so far had to be calculated and constructed by means of traditional methods. AR is used as a way of improving interfaces for a future generation, allowing students to work directly in a 3D space. Despite the advantage of free hand movements, in this project the use of a Head Mounted Display (HMD) was a necessary tool, and the possibility of mobility in other spaces was not accessible. In the opinion of Akçayir and Akçayir (2017), one of the principal reasons why AR technology is so widely used at present is that it no longer requires expensive hardware and sophisticated equipment, such as HMD. The authors state that 'the technology now can be used with computers or mobile devices. Thus, using AR technology is not as difficult as it was in the past. It is used today in every level of schooling, from K-12 to the University level' (Akçayır & Akçayır, 2017, p. 1).

A study by Lin et al. (2015) who integrated AR into teaching activities to assist K-12 students in learning solid geometry revealed, among other findings, that AR-assisted teaching has a favourable effect on students with low academic achievement. Perceptions indicated a low feeling of task load experienced by most users and, after AR-assisted teaching, student performances from the experimental groups improved, although the experimental and control groups did not differ significantly from each other.

In a previous study by these authors, the findings were that teaching geometry by means of information technology enhances students' comprehension of abstract mathematical concepts. They present as a possible reason that image-based teaching can help students to focus their attention (Lin et al., 2015). In addition to this, previous studies also confirmed that the use of three-dimensional dynamic multimedia in teaching geometry may encourage the curiosity of students, support their active learning and increase their achievement in geometry (Erbas & Yenmez, 2011).

In the last four years the number of AR studies in the field of education has increased (Cheng, 2017) and one the most reported advantage is that it promotes enhanced learning achievements (Akçayır & Akçayır, 2017). In science learning, studies have shown that students often hold robust misconceptions about several scientific ideas and that digital simulations and dynamic visualization tools have helped to increase the learning outcomes by providing scaffolding to understand different aspects of the contents (Yoon, Anderson, Lin, & Elinich, 2017). According to the authors, AR has the potential to augment users' interactions, engagement, and experiences and has revealed numerous affordances for science learning, including 'supporting students' scientific spatial ability, by (a) allowing them to manipulate and learn content in three-dimensional perspectives; (b) engaging them in scientific inquiry by encouraging them to make observations, ask questions, collaborate with others, and investigate and interpret data; and (c) enhancing their conceptual understanding by enabling them to visualize invisible or abstract concepts or events' (Yoon et al., 2017, p. 158). In a project that investigated optimal uses of AR in science museums, Yoon et al. (2017) found that students who interact with an exhibit using AR are better able to understand the science concepts than students in a non-AR condition. Furthermore, an experimental study by Lin et al. (2013), which intended to validate the correlation between spatial ability and mathematics performance, revealed that spatial ability and mathematics scores are highly correlated.

Moreover, with the increasing demand for Science, Technology, Engineering, and Mathematics (STEM) professionals in a worldwide innovation and global economics, STEM disciplines are viewed as a means of developing citizens' scientific literacy. Despite this panorama, Hsu et al. (2017) state that readiness and motivation on the part of students to pursue STEM majors and careers seem to be decreasing. Through a study developed with high school students in which they had learning experiences with AR lessons, the authors reported students' feedback that they felt inspired to deepen their knowledge, understood the importance of a transdisciplinary work for the innovation and creativity needed in futures places, and felt motivated to choose STEM-related majors at university (Hsu, Lin, & Yang, 2017). In other words, in addition to all the advantages mentioned by different authors about the use of AR in the context of teaching, the use of this technology also seems to have the capacity to motivate students both to feel more interest in deepening their knowledge about the contexts in question as well as in the field of technologies where AR is inserted. According to Squire and Jan (2007)

science education needs to prepare students for a future world in which multiple representations are the norm and adults are required to "think like scientists" (...) and AR games offer an opportunity to create a "post-progressive" pedagogy in which students are not only immersed in authentic scientific inquiry, but also required to perform in adult scientific discourses. (Squire & Jan, 2007, p. 5).

Despite the various advantages, technical limitations associated with AR can also be found in the literature. According to Akçayır and Akçayır (2017), without a well-designed interface and guidance for the students, AR technology can be too complicated for them to use. The authors also found that the various devices that deliver AR applications may cause additional technical problems as well that bulky AR technologies such as HMDs are not easy to handle, therefore AR technologies should be developed to be smaller, lighter, more portable, and fast enough to display graphics.

DESIGN PROCESS

With the complexity of design problems increasing dramatically in the last decades, the need to use a method has become more important in the development of a design project. Although we can find design process tendencies in very different places and in different tasks, they all end up having in common the inclusion of the end user

(Foo & Mårtensson, 2016). Therefore, there are several methods to achieve goals in design projects bearing in mind end users, deadlines etc.

Design thinking is a process of addressing problems in the development of a product, which can facilitate innovation and open up alternatives, not only lead to artefact creation. According to Liedtka (2013), when the individual elements of design thinking are combined and viewed together as a system from the beginning to the end aimed at solving problems, it emerges as a clearly distinctive way of thinking. The concept offers a unique integration framework that brings together both the creative and the analytical reasoning modes. Moreover, it is accompanied by a process and a set of tools and techniques (Liedtka, 2013). Brown (2008) reinforces the idea that design thinking must be present from the very beginning in order to be part of the creation process and increase the possibility to generate more ideas. Through this process, the author claims that the development is faster than by any other means (Brown, 2008).

In addition to technological considerations, the process should consider human behaviour, its needs and preferences. A human-centred project, especially when it includes observation-based research, will reflect more precisely people's needs.

The first step in any process should be understanding what project is being developed and why. In this stage, a mode of participatory design known as *Bodystorming* was adopted. *Bodystorming* emerged in the 1950s (Osborn, 1963) and is often considered a way to generate ideas and a form of start prototyping in context (Oulasvirta, Kurvinen, & Kankainen, 2003; Oppegaard & Still, 2013; Foo & Mårtensson, 2016;). According to Maguire (2001a), the awareness of context issues grew in the mid-1980s, promoted by the work of Whiteside and his colleagues who 'found that although many products performed well in their laboratory experiments, they did not work when transferred to the real world. They put this down to the fact that the research often overlooked something crucial to the context in which the product would be used' (Maguire, 2001a, p. 454). Thus, design sessions were carried out in the natural context, in a classroom, instead of being lab-based.

In the case of the project/application described in this paper, since one of the authors taught in this field, it was possible to carry out these sessions in two classes and understand the problems in a learning and teaching context. He observed that a substantial part of pupils revealed difficulty in understanding the third dimension of geometric solids, namely depth. One of the found difficulties was related to the explanatory process since two-dimensional drawings on the blackboard were used to represent three-dimensional objects. During these sessions, in randomly chosen classes, the pupils were asked about their use of smartphones, if they used to play games and what kind of games they used to play, as well questions related with geometric solids and the third dimension. The observation of the pupils' discussion and activities provided meanings and interpretations which were introduced in the interaction design process of the game.

The next step was the definition of objectives, the identification of expected behaviours and the description of players. The game "vertice" (see Figure 1) aimed to facilitate learning about and interpreting geometric solids and 3D objects; namely, to enable learners to develop their capacity for visualization, their understanding of the properties of geometric solids in space and their familiarization with the vocabulary of geometry. At the same time, fun was expected to be an essential component of the experience. The expected behaviour was that the player (learner) would play the game and continue to play it via a user-centred, highly engaging interface.

In this context, the name for the game and the first sketches (Figure 1) emerged. Several studies

implied that different aesthetic designs can induce emotions and that these emotions affect users' performance and cognitive process' and, moreover, the users positive perceptions suggested 'that positive emotions were produced by the different design of multimedia elements such as visual design principles, design layout, colour, and sound. (Um & Plass, 2012, p. 7).

Sketch

The design process ordinarily starts with a conceptual idea of the envisaged product when its design parameters are still vague and incomplete. During the working process, the final product can be represented concretely in the form of drawings or computer models (Bayer et al., 2015). Prototyping is about the visual representations of complex systems and interaction models that meet high levels of design and usability. Although we can find several definitions of prototype in software engineering and human-computer interaction (Alperowitz, Weintraud, Kofler, & Bruegge, 2017), most practitioners will agree on the general meaning of a dynamic prototype as a

Figure 1. Conceptual design, different phases of the process

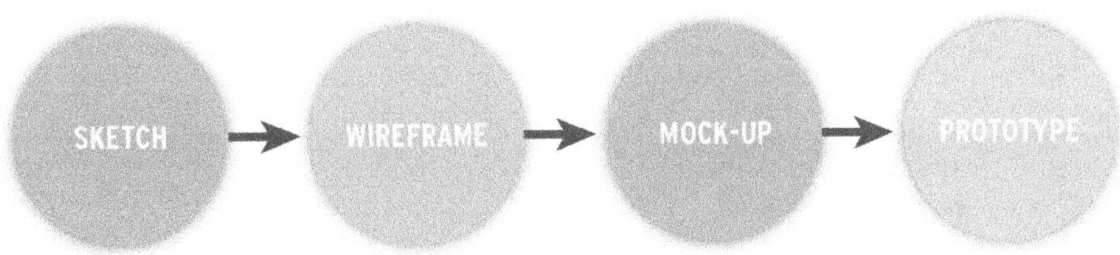

working model under construction. Despite the importance of drawing to complete this process successfully, prototyping is not about improving the ability to draw, but about the intellectual and physical freedom to express ideas easily and spontaneously. Different forms of prototyping can help narrow the search for a solution in different phases of the process and these were also employed in the reported project (Figure 2).

In early stages of designing the user interface of applications, sketching and wireframing are useful as a communication medium to explore, express, detect usability issues and communicate the design intentions (James & Brad, 2001; Murugappan, Piya, Yang, & Ramani, 2017; Sanchez Ramon, Garcia Molina, Sachez Cuadrado, & Vanderdonckt, 2013). This flexibility is particularly important to sketch rough design ideas quickly, to test designs by interacting with them and to fill in the design details as it becomes necessary to make choices. Sketching is basically a freehand drawing that gives a low-fidelity representation of the application. It is also considered a fast way to generate an idea for brainstorming. Several kinds of tools are used as part of the design process for expressive design representations (Coyette, Kieffer, & Vanderdonckt, 2007). Since conceptual sketching is characterized by ambiguity and the need to create several design variations quickly, these tools must support flexibility. According to James and Brad (2001) sketching with a pen is a mode of informal and perceptual interaction that has been shown to be especially valuable for creative design tasks.

For designers, the ability to rapidly sketch objects with uncertain types, sizes, shapes, and positions is important to the creative process. This uncertainty, or ambiguity, encourages the designer to explore more ideas without being burdened by concern for inappropriate details such as colours, fonts, and precise alignment. (James & Brad, 2001, p. 57).

Figure 2. Sketching phase

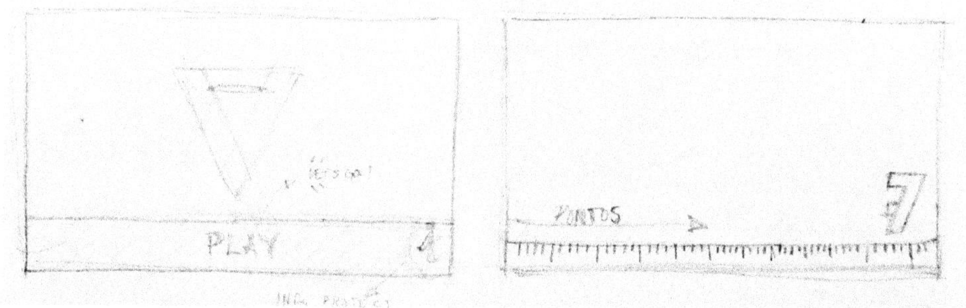

Wireframe

As the design process evolves, the used representations move towards a higher level of detail and fidelity. Although a wireframe is still considered a low fidelity representation, wireframing tools are used to refine the concepts that come out of sketching and they frequently offer some facilities to generate mock-ups and prototypes (Sanchez Ramon et al., 2013). Without the distraction of graphic design, wireframes show layout, content and functionalities and sizes are almost pixel perfect. Wireframes can be also hand-drawn. In this phase, the representations of every important piece of the product are focused on basic structural issues and usually they have the appearance of a set of boxes.

Mock-Up

Although sketches and wireframes as low-fidelity models are very helpful in the early stages of the design process and offer a clear set of advantages compared to high-fidelity prototyping (Coyette et al., 2007), they are insufficient in validating ideas and concepts. Thus, mock-ups help to define and refine the requirements by providing certain functionalities and contribute to new features in later phases of the project. They are very useful to define the overall interaction of the user with the application (Alperowitz et al., 2017) and to test. They are much easier, cheaper and quicker to make than a functional prototype in which studies and experiments are often more complex and less clean (Schmidt & Albrecht, 2017). Thus, this step is important for getting to the prototype phase. There are several kinds of tools to create mock-ups regarding the form, precision, interactivity and feasibility (e.g. Axure, Balsamiq, Sketch for interactive mock-ups; Illustrator, Photoshop and paper for a different interaction approach) and some services offer the possibility to execute the mock-up on a mobile platform (Alperowitz et al., 2017). Figure 3 shows part

Figure 3. An example of a part of a mock-up of two mobile screens with some interaction information, produced with the Illustrator software

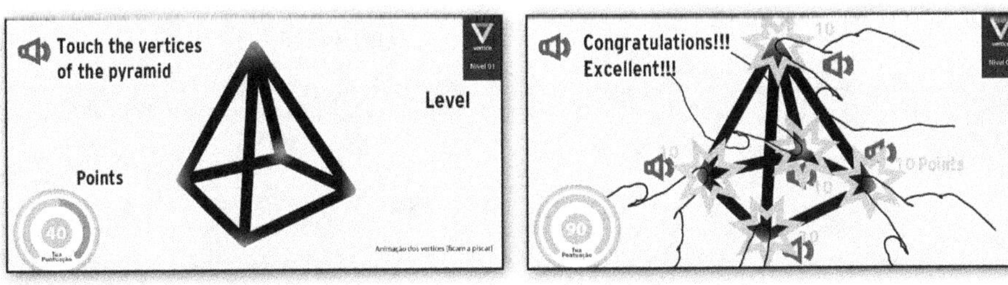

of the mock-up of two mobile screens with some interaction information for the developed application, produced with vector graphics software.

Since engagement was expected to be an essential component of the experience for the pupils interacting with the AR application, some game elements and game mechanics were incorporated: engagement cycles to familiarize the learner with the game environment, and progression cycles which represent the different levels of difficulty and competences. Regarding the engagement cycles, an initial, very simple scenario was created, consisting of only one area of a plane which presented the player with only one vertex. For a better engagement with this cycle and to minimize the rejection by the player, the area which he/she was expected to touch was highlighted by means of a red flashing light, as shown in Fig. 4 left. At the same time, the player heard an instruction: "Touch the vertex". After players' interaction with the vertex, feedback on their action was immediate, rendering them capable of interpreting their choice against the objective. A score was attributed and an encouragement was heard: "Very well, congratulations!". This step was repeated with the introduction of new objects. The continuous iteration of this cycle – action-feedback-interpretation – aimed to help learners to develop their cognitive abilities gradually and to get used to the interface.

The end of these three engagement cycles also marked the end of the first level which introduced the concept of *vertex*. The second level maintained the same approach, but this new progression cycle introduced the concept of *edge*. Again, players were presented with three engagement cycles which increased progressively in their degree of difficulty. The third level brought the last of these engagement cycles, meant to allow learners to experience the game and at the same time acquire knowledge on basic concepts. This third level was another progression cycle which transferred knowledge on the concept of *face* by means of the same approach which assisted the player through the signalling of the area which he/she was expected to indicate. After these first cycles, the signalling was eliminated and new objects were

Figure 4. Construction of solid geometric figures and exportation from blender

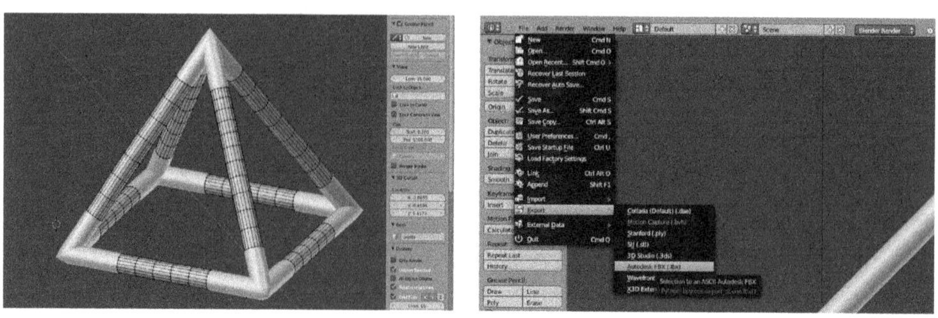

presented, the first ones always with very simple structures, followed by increasingly complex structures.

Prototype

Finally, to test and to demonstrate the feasibility of the functionality of the product while it is still possible to take risks, the use of prototypes provides several benefits and a basis for discussion in order to explore the product requirements (Schmidt & Albrecht, 2017). Hence prototypes allow to explore how the technology works and what it feels like to use it. Tools to produce high-fidelity models must enable building a user interface that looks complete and equipped with a wide range of editing functions for all graphic user interface (GUI) and heads-up displays (HUD) that should be usable.

In an initial stage, pre-production, the first components to be developed were the graphical user interfaces (GUIs). The initial panels, the score indicators, the level in which the player finds him/herself, the buttons and their behaviours, and the data introduction toolbars were built in vector graphics format. Next, these were exported to a .png format for a potential use of the images with transparency features and without a background. It was then necessary to draw the 3D objects, which were created in Blender ("Blender," 2017) given its ease of usage and its free availability. All the drawings, already with the division of objects according to the interaction needs for the vertex, edge and face elements, were exported to the Autodesk FBX format (.fbx). The design of the recognition pattern (marker) represented the next step. A drawing was developed in A4 size, with some images of geometric figures and with the image which represented the game icon. Care was taken to draw an image which contained the necessary information, but which, at the same time, contemplated economical printing in greyscale. The marker was prepared using the Vuforia developer portal ("Vuforia," 2017). The instructions on the expected actions as well as the feedback phrases to confirm or question the player's choices were recorded, for which Audacity 2.0.4 ("Audacity," 2017) was used. More detailed information can be found in Leitão, Rodrigues and Marcos (2014, p. 69).

In the second stage of the implementation, all the developed components referred to above were imported into Unity3D ("Unity 3D," 2017). The Unity scenes corresponded to the levels. Therefore, in the first scene, which was named "start" and which corresponded to level 0, the icon for the game was introduced, as well as the two buttons to start or to quit the game. This stage worked with the Main Camera, present by default in Unity, since there were no immersive stages yet. The game objects named GUI Textures served to introduce the images and then to position, rescale or turn them as desired. A folder called Resources was created to which all the designed GUI were imported in .png format in order to take advantage of the transparencies. The importance of the transparencies resided in the fact that sometimes the interfaces would superimpose themselves onto the images viewed by the camera. Similar steps were followed for the sound. Once the sound was selected, it was indicated into the Inspector panel that it should be heard on starting the game ('Play on Awake') and that it should stay in constant rotation ('Loop'). For the two buttons, it was first necessary to create two 3D Text game objects, and then to apply to each of them a script developed in Java in order to obtain the interaction.

In level 1, where the game effectively began, the marker was first integrated with the Vuforia extension. After making the registration in the application's online platform, it was necessary to create a database where the marker image was processed to immediately after download a file and import it into Unity 3D. The Vuforia application offers two types of databases. The Devise Database, was chosen as being the most appropriate for the simplicity of the project. Inside Unity, the installation of the extension Vuforia for Android and iOS was necessary. After this step, it was necessary to delete the Main Camera which comes by default with Unity 3D and replace it with the AR Camera which Vuforia provides for RA applications. Afterwards, the marker (Image Target Prefab) had to be added to the stage and configured. The Image Target (images which Vuforia SDK can detect and follow) as well as the AR Camera are located in the folder Qualcomm Augmented reality/ Prefabs. All that was needed was drag and drop inside Hierarchy, to introduce the marker which would determine the position of the objects.

In the Inspector panel, with the AR Camera selected, it was required to add the script 'Data Set Load Behaviour' and activate it. Similarly, with the image target selected, it was necessary, in the script 'Image Target Behaviour', to configure the Data Set component with the marker to be used and, automatically, in the stage view, it was possible to view the marker. It was also essential to create a light point for a better visualization. In this case, a directional light was applied in order to illuminate the whole scene. The stage was now set for the introduction of the 3D models. The important point here was to maintain the model subordinated to the Image Target and, afterwards, to position and scale it as desired in the scene itself.

At this level, two GUI Textures were introduced for the .png images, one corresponding to the information on the current level and another to the score background. A GUI Text was also added to the score background for the updating of the score points. In this case, since the information was not an image but a text, a semi-bold sans serif font was chosen for better legibility. The colour was chosen to create a good contrast with the background. It was important at this stage to subordinate these GUI to the AR Camera.

The animations were tackled next. In this application, they consisted of flashing which indicated the point of contact. Previously in the development of 3D models, care was taken to separate the contact points as different objects, which facilitated the animations and the action attributions. With only the vertex selected, in *animation* (in a process similar to other animation and video software) the colour was changed and its constant rotation (loop) was activated in a space of 30 frames. In this way, the flashing effect was obtained by means of a gradient towards red and then with a gradual return to its original colour. A Java script was also attributed, similar to the one attributed to the buttons in the initial stage, which this time indicated the passage to the next level and the attribution to the *val* GUI (GUI text) of the first points. Finally, the sound was added: the initial sound corresponded to the expected action and then the feedback to the performed action.

As initially envisaged, in a process of simplification of all the structures, all the levels were developed in the same way. It was only necessary to change the models in *Image Targets*, position, scale and turn them as desired, and reproduce the animations. Additionally, the Android Developer Tools had to be installed in order to allow the connectivity of the game with an Android device. For this purpose, it was necessary to indicate its localization in the Unity 3D options in *external tools*. The Unity Remote application allows the Android device to function as a remote controller of the Unity3D editor project. In this way, the development turns to be faster since one does not have to constantly compile and implement in the Android device every time the project needs to be visualized following further modifications (see Unity, 2017).

PROTOTYPE EVALUATION AND FURTHER RESEARCH AVENUES

The evaluation of the prototype was carried out in a public primary school in Portugal, which was receptive to the project since one of the authors had been teaching there for several years. The evaluation took place at the beginning of the 2013/2014 school year and involved two classes (53 pupils, 30 girls and 23 boys, aged 8-10). In the first class, with 27 pupils, the geometric concepts of vertex, edge and face were explained verbally and by means of drawings on the whiteboard (Class 1) – control class. In the second class (Class 2), with 26 pupils, in addition to a short, simplified explanation like that provided to Class 1, the game was presented to the children. It was also explained to them how the game worked with AR, similar to the process employed in more popular games (such as Invizimals of PlayStation Portable or more recently Pokemon Go of Niantic, Inc.). The children were also told that they were going to hear instructions during the game and were shown the image that served as marker, towards which they had to direct the device.

Children were given the opportunity to interact with the game. Initially, each pupil interacted with the game individually, but after a while more and more pupils

Figure 5. Experimental group in interaction with the game

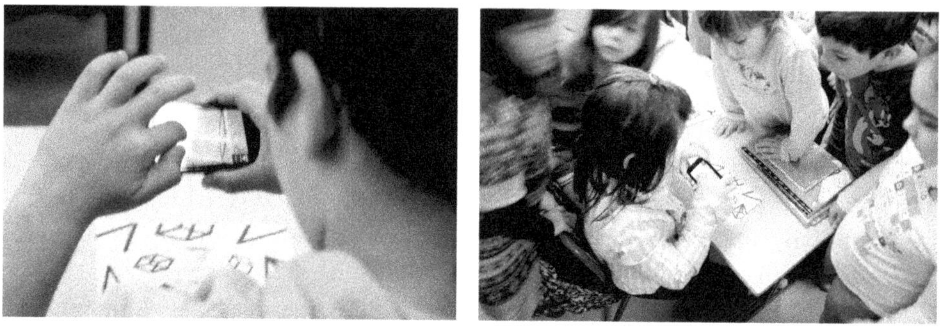

gathered around the player: some who had not had their turn yet to satisfy their curiosity, and others, who had already played, to offer help. The game ran generally smoothly and pupils encountered little difficulty. Although the sound was not always at its best, which could prejudice the understanding of the action required, it did not affect the course of the game thanks to the visual signals present in the first levels, which were meant to create greater involvement with the game. The tendency to use the smartphone horizontally was noted in all children, probably related to their use of PlayStation Portable. More detailed information can be found in Leitão, Rodrigues and Marcos (2014).

After these sessions, the students in both classes were asked to complete a test according to the learning outcomes in order to get information on their learning. It consisted of several figures in black with the edge, or the vertex or the face highlighted. For each figure a form with a checkbox was presented, allowing the student to select one of the options – edge/vertex/face. The results for Class 1 were as follows: 33.3% pupils identified the edge, 55.6% pupils identified the vertex and 59.3% identified the face. In Class 2, they were the following: 84.6% pupils identified the edge, 88.5% the vertex and 88.5% the face. Overall, in Class 1 (control class) after the presentation and work done by the teacher, the students hit correctly 49.4% of the questions, and in Class 2 the students hit correctly 87.2% of the questions (Figure 6).

As noticed, there was a great difference in the learning of the concepts between the pupils in Class 1 and Class 2. One of the reasons could result from the pupils' level of concentration during the explanation of the concepts, without the game, in Class 1. Explanation is a traditional method for knowledge transmission, implying a passive attitude and little engagement for the pupils. Additionally, in Class 2, the game served to repeat various times the concepts through interaction and experimentation, which contributed to the consolidation of knowledge.

The small sample size employed in the study represented one limitation. Carrying out the experiment in a larger number of classes and schools would be important to

Figure 6. Left: Percentages of correct answers by evaluated item. Right: overall percentage of correct answers

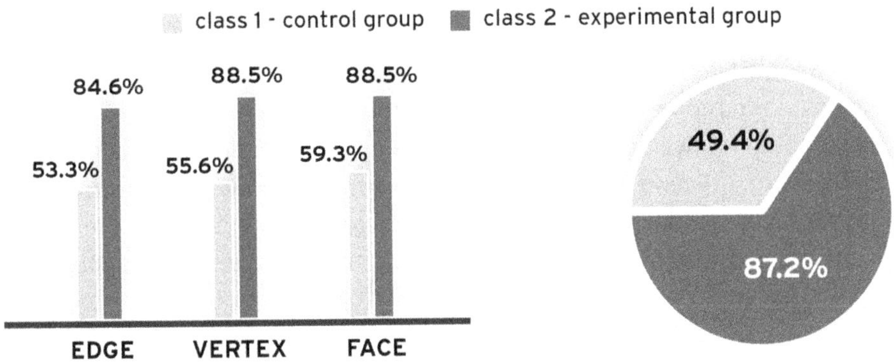

get a better 'validation' of how much the game can improve learning. In this case, the experiment led to an increase of almost 35% of correct answers when playing the game when compared with traditional teaching. In the future, the game can also be extended to other geometrical concepts such as plane, prism, pyramid, cylinder, cone, sphere, etc. Another limitation was not to have done a prior test in order to assess the children's knowledge of geometric concepts before the explanation or the interaction with the game. This meant that it was not possible to determine whether the learning occurred only during the class in which the concepts of vertex, face and edge were presented to them (by one method or the other), or whether the children already had some previous knowledge. A third area which would have deserved more exploration was the children's degree of contact and interaction with technology and whether this had any influence on the results of the learning through the game. A plausible hypothesis would be that children already familiar with technology (computers, smartphones, tablets, etc.) find it easier to interact with the game and, consequently, derive greater benefit from it. Similarly, future work could also include a study to establish the relationship between learning outcomes and the children's degree of familiarity with new technologies (types, usage frequency, level of difficulty, purposes etc.). This would allow better precision in the evaluation of the game's efficiency in the attainment of learning outcomes.

CONCLUSION

Game-based learning (Muñoz, Lunney, Mc Kevitt, Noguez, & Neri, 2013; Razak et al., 2012) takes children to immersive environments where they learn to use an impressive range of tools and complex machines. Many hours are dedicated to the memorization of scenarios, during which they develop sophisticated tactics to reach their objectives and win the game. In many cases, games are developed in collaboration with children all over the world in an exchange of solutions. Eventually games can contribute to emotional and social development, including different forms of cooperation and competition. Moreover, they allow children to discover why rules are important and which ones work best. While playing, children direct their attention to details in order to have the best possible performance in the game. In fact, these experiences help children to develop more mature thinking and problem-solving abilities (Yatim & Masuch, 2007). However, creating engaging game applications can be very difficult. As stated by Maguire (2001b),

human-centred system development is a collaborative process which benefits from the active involvement of various parties, each of whom have insights and expertise to share. It is therefore important that the development team be made up of experts with technical skills and those with a "stake" in the proposed software. (Maguire, 2001b, p. 589).

Furthermore, is not a simple step to bring together a team.

The elaboration of the game "vertice" aimed to provide pupils with a learning tool which would facilitate knowledge acquisition according to constructivist principles. Constructivism advocates that learners should, through own experience (Huang, Rauch, & Liaw, 2010) build new knowledge upon the knowledge they already possess, a process which allows each learner to have their own idiosyncratic version of a specific piece of knowledge. In this way, knowledge is actively constructed by the learner, and not passively absorbed from books or verbal explanations. The game "vertice" validated once again the statements of several authors, that by playing games students learn to act, their curiosity is stimulated, they acquire initiative and self-confidence and they develop better language skills, reasoning and concentration (Przybylski, Rigby, & Ryan, 2010). The improvement of cognitive capacities as well as knowledge acquisition are facilitated by exploration, action and experimentation. When associated to social interaction, these methods, in fact, encourage more that just individual knowledge acquisition, but a collective construction of knowledge and a negotiation of meanings (Silva & Silveira, 2012). The learning process becomes more pleasant, different from a traditional class which implies passive knowledge reception. Such methods promote active learning by means of active problem-solving (Boyle, Connolly, & Hainey, 2011). According to Prensky (2005), when playing, individuals interact with the game environment and receive immediate feedback on their actions, thus being able to interpret their choices in accordance with their objectives. The continuous repetition of this cycle (action-feedback-interaction) allows players to develop their cognitive capacities gradually. The teachers and learners (as also shown in our experiment) have recognized that games can help the development of strategic thinking, planning, communication, numeracy, negotiation skills, group decision-making and information processing (Kirriemuir & McFarlane, 2004).

Augmented reality is a growing field that can be explored for different applications adapted to every knowledge area, presenting a great potential in education with the development of educational games. As already mentioned, the game presented here demonstrated a capacity to awaken curiosity, motivation and initiative among the children due to its engaging and interactive nature. Thanks to its novelty, the use of AR stirred great interest on the part of the children during the game. From a pedagogical point of view, it helped them attain the competences that the game aimed to develop: ability of visualization, the understanding of the properties of geometric solids and familiarization with the vocabulary of geometry. The pupils' receptivity to the game was, as expected, very positive, since they came across curricular content in a different and entertaining manner. The 3D model, especially given its powerful capacity of representation, proved to be an efficient means to convey the concept of three-dimensionality, compared to the bi-dimensional drawings on the whiteboard.

Design thinking was the problem-solving approach adopted in the development of the game. It implied constant attention to the target audience, i.e., children/pupils. It was also an approach which sought to achieve a balance between the objectives (improvement of pupils' visualization abilities, their understanding of the properties of geometric solids in space, familiarization with the vocabulary of geometry) and the image and presentation, bearing in mind popular games in this age group. As a result, the interfaces, which represent the information and the instructions for the pupil/user, were designed as secondary elements in order to simplify usability and to interfere as little as possible with the camera vision. Another constant concern was user-centeredness, meant to facilitate pupils' learning and interaction with the game, while not limiting their performance. The use of gamification techniques (Deterding, Dixon, Khaled, & Nacke, 2011) – whereby games and game elements were used in an educational context – facilitated the process of behaviour definition, which influenced interaction and motivation. In addition to a scoring system that increased pupils' motivation, the game dynamics allowed integrating educational content with the game narrative.

A further development of the application could envisage responsive design which allows content to adapt to screens of different sizes. The introduction of additional levels could also be an aspect worth exploring, to include for instance more complex geometric solids and other types of difficulty (such as movement and time). The sound, too, presents potential for further development, not only to improve its quality, but also by employing a voice more suitable to the target group, by creating new soundtracks to accompany the storyboard and by diversifying the interaction prompts and responses. Additionally, the introduction of leaderboards would not only enable players to measure their progress, but also to compare it with that of other players. And, finally, game analytics would be an easy means of tracking the players' behaviour. Through the collection and analysis of data, it would be possible to obtain feedback to improve design, quality, and to identify errors and correct them.

The presented paper and application reinforces that which has already been referred to by other authors (e.g. Kaufmann, 2004), namely that learning by means of games can significantly benefit learning outcomes and the development of more mature thinking. The results suggest that the game-based learning with AR has a positive effect on pupils' content knowledge. The capacities of new technologies, the miniaturization of devices, the universal access, the handy analysis of digital data, the ease of diffusion and update, the bandwidth and, not least, the financial accessibility of means and resources, ensure good conditions for the use of this type of games in teaching.

ACKNOWLEDGMENT

This research was supported by the Arts and Humanities Research Council Design Star CDT (AH/L503770/1), the Portuguese Foundation for Science and Technology (FCT) projects LARSyS (UID/EEA/50009/2013) and CIAC-Research Centre for Arts and Communication.

REFERENCES

Akçayır, M., & Akçayır, G. (2017). Advantages and challenges associated with augmented reality for education: A systematic review of the literature. *Educational Research Review*, *20*, 1–11. doi:10.1016/j.edurev.2016.11.002

Alperowitz, L., Weintraud, A. M., Kofler, S. C., & Bruegge, B. (2017). Continuous prototyping: Unified application delivery from early design to code. *CEUR Workshop Proceedings*, *1806*, 36–43.

Audacity. (2017). Retrieved June 15, 2017, from http://www.audacityteam.org

Bayer, J., Bukhari, S. S., Langenhan, C., Liwicki, M., Althoff, K.-D., Petzold, F., & Dengel, A. (2015). Migrating the Classical Pen-and-Paper based Conceptual Sketching of Architecture Plans Towards Computer Tools - Prototype Design and Evaluation. *The Eleventh IAPR International Workshop on Graphics Recognition. IAPR International Workshop on Graphics Recognition (GREC-2015), 11th, Located at 13th International Conference on Document Analysis and Recognition (ICDAR), August 22-23, Nancy, France*. Retrieved from http://www.dfki.de/web/forschung/publikationen/renameFileForDownload?filename=bare_conf.pdf&file_id=uploads_2934

Billinghurst, M. (2002). Augmented Reality and Education. *New Horizons for Learning, 21*(3), 195-209. <ALIGNMENT.qj></ALIGNMENT>10.4018/jgcms.2011010108

Blender. (2017). Retrieved June 15, 2017, from https://www.blender.org

Boyle, E., Connolly, T. M., & Hainey, T. (2011). The role of psychology in understanding the impact of computer games. *Entertainment Computing, 2*(2), 69–74. doi:10.1016/j.entcom.2010.12.002

Brown, T. (2008). Design thinking. *Harvard Business Review, 86*(6), 84–92, 141. doi:10.5437/08956308X5503003 PMID:18605031

Catenazzi, N., & Sommaruga, L. (2013). Mobile Learning and Augmented Reality: New Learning Opportunities. *Social Media: Challenges and Opportunities*, 9–13.

Cheng, K.-H. (2017). Exploring Parents' Conceptions of Augmented Reality Learning and Approaches to Learning by Augmented Reality With Their Children. *Journal of Educational Computing Research*, (1): 1–24. doi:10.1177/0735633116686082

Coyette, A., Kieffer, S., & Vanderdonckt, J. (2007). *Multi-fidelity Prototyping of User Interfaces*. Academic Press.

D'Angour, A. (2013). Plato and Play: Taking Education Seriously in Ancient Greece. *American Journal of Play, 5*(3), 293–307.

de Bie, M. H., & Lipman, L. J. A. (2012). The use of digital games and simulators in veterinary education: An overview with examples. *Journal of Veterinary Medical Education, 39*(1), 13–20. doi:10.3138/jvme.0511.055R PMID:22433739

Deterding, S., Dixon, D., Khaled, R., & Nacke, L. (2011). From game design elements to gamefulness: defining gamification. *Proceedings of the 15th International Academic MindTrek Conference on Envisioning Future Media Environments - MindTrek '11*, 9–15. doi:10.1145/2181037.2181040

Dichev, C., Dicheva, D., Angelova, G., & Agre, G. (2014). From Gamification to Gameful Design and Gameful Experience in Learning. *Cybernetics and Information Technologies*, *14*(4), 80–100. doi:10.1515/cait-2014-0007

Dicheva, D., Dichev, C., Agre, G., & Angelova, G. (2015). Gamification in Education : A Systematic Mapping Study Gamification in Education : A Systematic Mapping Study. *Journal of Educational Technology & Society*, *18*, 75–88.

Erbas, A. K., & Yenmez, A. A. (2011). The effect of inquiry-based explorations in a dynamic geometry environment on sixth grade students' achievements in polygons. *Computers & Education*, *57*(4), 2462–2475. doi:10.1016/j.compedu.2011.07.002

Foo, H., & Mårtensson, O. (2016). *Designing for engagement using gamification in mobile applications*. Academic Press.

Gamberini, L., Barresi, G., Majer, A., & Scarpetta, F. (2008, Summer). A game a day keeps the doctor away: A short review of computer games in mental healthcare. *Journal of Cyber Therapy and Rehabilitation*, *1*(2), 127–145.

Hamari, J., Koivisto, J., & Sarsa, H. (2014). Does gamification work? - A literature review of empirical studies on gamification. *Proceedings of the Annual Hawaii International Conference on System Sciences*, 3025–3034. doi:10.1109/HICSS.2014.377

Hannes, K. (2004). *Geometry Education with Augmented Reality*. Retrieved from http://gretchen.ims.tuwien.ac.at/media/documents/publications/kaufmann_diss.pdf

Hsu, Y.-S., Lin, Y.-H., & Yang, B. (2017). Impact of augmented reality lessons on students' STEM interest. *Research and Practice in Technology Enhanced Learning*, *12*(1), 2. doi:10.1186/s41039-016-0039-z

Huang, H.-M., Rauch, U., & Liaw, S.-S. (2010). Investigating learners' attitudes toward virtual reality learning environments: Based on a constructivist approach. *Computers & Education*, *55*(3), 1171–1182. doi:10.1016/j.compedu.2010.05.014

James, A., & Brad, A. (2001). *Sketching Interfaces: Toward More Human Interface Design*. Academic Press.

Kirriemuir, J., & McFarlane, A. (2004). Literature Review in Games and learning. *Futurelab Series*, (8), 39. Retrieved from https://telearn.archives-ouvertes.fr/hal-00190453%0Ahttp://telearn.archives-ouvertes.fr/hal-00190453/

Krentz, A. (1998). Play and Education in Plato's Republic. *Twentieth World Congress of Philosophy*, (2). Retrieved from https://www.bu.edu/wcp/Papers/Educ/EducKren.htm

Lei, S. A. (2010). Intrinsic and Extrinsic Motivation : Evaluating Benefits and Drawbacks from College Instructors '. *Perspectives*, *37*(2), 153–160. Retrieved from http://web.a.ebscohost.com/ehost/pdfviewer/pdfviewer?sid=1b4f8e6a-7ae0-444d-9d6c-b0e343f3df1f%40sessionmgr4003&vid=0&hid=4106

Leitão, R., Rodrigues, J. M. F., & Marcos, A. F. (2014). Game-Based Learning: Augmented Reality in the Teaching of Geometric Solids. *International Journal of Art, Culture and Design Technologies*, *4*(1), 63–75. doi:10.4018/ijacdt.2014010105

Liedtka, J. (2013). Design Thinking: What it is and why it works. *Design at Darden Working Paper Series*. Retrieved from http://batten.squarespace.com/storage/books-journals-articles/DesignThinking_WhatitIs_Liedtka.pdf

Lin, H.-C. K., Chen, M.-C., & Chang, C.-K. (2015). Assessing the effectiveness of learning solid geometry by using an augmented reality-assisted learning system. *Interactive Learning Environments*, *23*(6), 799–810. doi:10.1080/10494820.2013.817435

Maguire, M. (2001a). Context of Use within usability activities. *International Journal of Human-Computer Studies*, *55*(4), 453–483. doi:10.1006/ijhc.2001.0486

Maguire, M. (2001b). Methods to support human-centred design. *International Journal of Human-Computer Studies*, *55*(4), 587–634. doi:10.1006/ijhc.2001.0503

Mayer, R. E. (2005). Cognitive Theory of Multimedia Learning. In *The Cambridge handbook of multimedia learning* (pp. 31–48). Cambridge University Press. doi:10.1017/CBO9780511816819.004

McGonigal, J. (2011). *Reality is broken: Why games make us better and how they can change the world*. Penguin.

Muñoz, K., Lunney, T., Mc Kevitt, P., Noguez, J., & Neri, L. (2013). An Emotional Student Model for Game-Based Learning. In *Technologies for Inclusive Education. Beyond Traditional Integration Approaches*. doi:10.4018/978-1-4666-2530-3.ch009

Murugappan, S., Piya, C., Yang, M. C., & Ramani, K. (2017). FEAsy : A Sketch-Based Tool for Finite Element Analysis. *Journal of Computing and Information Science in Engineering*, *17*(3), 031009. doi:10.1115/1.4034387

Naismith, L., Lonsdale, P., Vavoula, G., & Sharples, M. (2004). *Literature Review in Mobile Technologies and Learning Literature Review in Mobile Technologies and Learning*. Slough: FutureLab. Retrieved from http://www.futurelab.org.uk/resources/documents/lit_reviews/Mobile_Review.pdf

Oppegaard, B., & Still, B. (2013). Bodystorming with Hawkins's block: Toward a new methodology for mobile media design. *Mobile Media & Communication, 1*(3), 356–372. doi:10.1177/2050157913496421

Osborn, A. F. (1963). Applied imagination: Principles and procedures of creative problem-solving (3rd ed.). New York : Scribner's.

Oulasvirta, A., Kurvinen, E., & Kankainen, T. (2003). Understanding contexts by being there: Case studies in bodystorming. *Personal and Ubiquitous Computing, 7*(2), 125–134. doi:10.1007/s00779-003-0238-7

Prensky, M. (2005, October). Engage Me or Enrage Me: What Today's Learners Demand. *EDUCAUSE Review*, 60–64. Retrieved from http://cff.wiki.elanco.net/file/view/Engage+Me+or+Enrage+Me.pdf/84842561/Engage+Me+or+Enrage+Me.pdf

Przybylski, A. K., Rigby, C. S., & Ryan, R. M. (2010). A motivational model of video game engagement. *Review of General Psychology, 14*(2), 154–166. doi:10.1037/a0019440

Razak, A. A., Connolly, T., Baxter, G., Hainey, T., & Wilson, A. (2012). The use of Games-Based Learning at Primary Education Level Within the Curriculum for Excellence: A Combined Result of two Regional Teacher Surveys. In *Proceedings of the 6th European Conference on Games Based Learning* (p. 52). Academic Press.

Sanchez Ramon, O., Garcia Molina, J., Sachez Cuadrado, J., & Vanderdonckt, J. (2013). GUI Generation from Wireframes. In *Actas del XIV Congreso Internacional de Interacción Persona-Ordenador*. Madrid: Academic Press. Retrieved from http://hdl.handle.net/2078/153911

Schlagenhaufer, C., & Amberg, M. (2015). A Descriptive Literature Review and Classification Framework for Gamification in Information Systems. *ECIS 2015 Completed Research*, 0–15. <ALIGNMENT.qj></ALIGNMENT>10.18151/7217466

Schmidt, A. (2017). Understanding and researching through making-A Plea for Functional Prototypes. *Interaction, 24*(3), 78–81. doi:10.1145/3058498

Silva, D., & Silveira, I. (2012). Jogos Educacionais com Realidade Aumentada: Desafios de Integração e Possibilidades na Construção de Objetos de Aprendizagem. In Conferencias LACLO 3.1. São Paulo: Academic Press.

Squire, K. D., & Jan, M. (2007). Mad city mystery: Developing scientific argumentation skills with a place-based augmented reality game on handheld computers. *Journal of Science Education and Technology*, *16*(1), 5–29. doi:10.1007/s10956-006-9037-z

Sweller, J. (2010). Element interactivity and intrinsic, extraneous, and germane cognitive load. *Educational Psychology Review*, *22*(2), 123–138. doi:10.1007/s10648-010-9128-5

Tobias, S., & Fletcher, J. D. (2011). *Computer games and instruction*. Charlotte, NC: Information Age Publishing.

Um, E., Plass, J. L., Hayward, E. O., & Homer, B. D. (2012). Emotional design in multimedia learning. *Journal of Educational Psychology*, *104*(2), 485–498. doi:10.1037/a0026609

Unity 3D. (2017). Retrieved June 15, 2017, from https://unity3d.com

Vuforia. (2017). Retrieved June 15, 2017, from https://www.vuforia.com

Williams, K., & Williams, C. (2011). Five key ingredients for improving student motivation. *Research in Higher Education*, 1–23. Retrieved from http://www.jupapadoc.startlogic.com/manuscripts/11834.pdf

Yatim, M. H. M., & Masuch, M. (2007). *Educating children through game making activity*. Retrieved from http://www.learnit.org.gu.se/digitalAssets/862/862887_yatim_masuch.pdf

Yoon, S., Anderson, E., Lin, J., & Elinich, K. (2017). *How Augmented Reality Enables Conceptual Understanding of Challenging Science Content How Augmented Reality Enables Conceptual Understanding of Challenging Science Content*. Academic Press.

Compilation of References

Aberdeen. (2006, September). The Transition from 2D Drafting to 3D Modelling Benchmark Report. *Design Bulletin*.

Adorno, T. W., & Horkheimer, M. (1947). *La industria cultural. Iluminismo como mistificación de las masas*. Retrieved June 26, 2011, from http://www.cholonautas.edu.pe/modulo/upload/Adorno%20Horkheimer.pdf

Adorno, T. W. (1968). Résumé über kulturindustrie. In T. W. Adorno (Ed.), *Ohne leitbild- parva aesthetira* (pp. 60–70). Frankfurt am Main, Germany: Suhrkamp Verlag.

Adorno, T. W. (1970). *Äathetische theorie*. Frankfurt am Main, Germany: Suhrkamp Verlag.

Ağartan, T., Choi, W. Y., & Hyunh, T. (2008). *Kapitalist Dünyanın Dönüşümü:1750-1850.Toplumsal Hareketler 1750-2005 Dipten Gelen Dalgalar* (D. Keskin, Trans.). İstanbul: Versus Yayınları.

Ahmad, F. (1996). *Demokrasi Sürecinde Türkiye.1945-1980*. İstanbul: Hil Yayınları.

Akçayır, M., & Akçayır, G. (2017). Advantages and challenges associated with augmented reality for education: A systematic review of the literature. *Educational Research Review*, *20*, 1–11. doi:10.1016/j.edurev.2016.11.002

Alahuhta, P., Nordbäck, E., Sivunen, A., & Surakka, T. (2014). Fostering team creativity in virtual worlds. *Journal of Virtual Worlds Research*, *7*(3).

Albert, W. (2010). *Measuring the user experience*. Morgan Kaufmann.

Albert, W., & Tullis, T. (2013). *Measuring the user experience: collecting, analyzing, and presenting usability metrics*. Newnes.

Aldrick, P. (2011, June 29). Knowledge economy central to UK growth, says Work Foundation. *The Telegraph*. Retrieved from http://www.telegraph.co.uk/finance/economics/8604057/Knowledge-economycentral- to-UK-growth-says-Work-Foundation.html

Alexander, C., & Ishikawa, S. (1977). *A pattern language: Towns, buildings, construction*. New York: Oxford University Press.

Alexenberg, M. (2008). *Educating Artists for the Future: learning at the intersections of art, science, technology, and culture*. Chicago, IL: The University of Chicago Press.

Alperowitz, L., Weintraud, A. M., Kofler, S. C., & Bruegge, B. (2017). Continuous prototyping: Unified application delivery from early design to code. *CEUR Workshop Proceedings*, *1806*, 36–43.

Anderson, C. (2012). *Makers: The New Industrial Revolution*. New York: Crown Publishing Group.

Anders, P. (1994). The architecture of cyberspace. *Progressive Architecture*, *75*(10), 78.

Andrew, G. (2011). Ceramic Mold Making Techniques (2nd ed.). Retrieved from www.ceramicartsdaily.org

A. Appadurai (Ed.). (1986). *The social life of things: Commodities in cultural perspective*. Cambridge, UK: Cambridge University Press. doi:10.1017/CBO9780511819582

Arendt, H. (1959). *The human condition* (3rd ed.). Chicago, IL: The University of Chicago Press.

Ascott, R. (2008). Cybernetic, Technoetic, Syncretic: The Prospect for Art. *Leonardo*, *41*(3), 204. doi:10.1162/leon.2008.41.3.204

Asymptote Arch. (1999). *Virtual Guggenheim Museum*. Retrieved June, 2017, from http://openbuildings.com/buildings/guggenheim-virtual-museum-profile-2437

Atacan, F. (1990). *Sosyal Değişme ve Tarikat Cerrahiler*. İstanbul: Hil Yayınları.

Audacity. (2017). Retrieved June 15, 2017, from http://www.audacityteam.org

Autodesk. (2013). Digital Prototyping. *Autodesk*. Retrieved from http://usa.autodesk.com/digital-prototyping/

Azlan, N. I. (2013). *Research Methodology in Landscape Architecture*. Singapore: Trafford.

Bachelard, G. (1994). *The Poetics of Space* (M. Jolas, Trans.). Boston: Beacon Press. (Original work published 1958)

Badre, A. (2001). *The effects of cross cultural interface design orientation on world wide web user performance*. Academic Press.

Baratov, K. (2005). *1980'li Yıllarda Türkiye'de Sosyal Sınıflar ve Bölüşüm*. İstanbul: İmge Kitabevi.

Barber, D. B., & Badre, A. (1998). A Year It Was! Culturability: The merging of culture and usability, 1-14.

Bardzell, J. (2009). Interaction Criticism and Aesthetics. In *Proceedings of CHI'09: World Conference on Human factors in Computing Systems*. ACM. doi:10.1145/1518701.1519063

Barkhuus, L., & Rossitto, C. (2016). Acting with technology: rehearsing for mixed-media live performances. In *Proceedings of the 2016 CHI Conference on Human Factors in Computing Systems* (pp. 864-875). ACM. doi:10.1145/2858036.2858344

Barneche-Naya, V., & Hernández-Ibáñez, L. (2014). Evaluating user experience in joint activities between schools and museums in virtual worlds. *Universal Access in the Information Society*, *14*(3), 389. doi:10.1007/s10209-014-0367-y

Barratt, D. (2012). The Geography of Film Viewing: What Are the Implications of Cultural-Cognitive Differences for Cognitive Film Theory? In 2012 Conference for the Society for Cognitive Studies of the Moving Image. SCSMI.

Barthes, R. (1973/1975). *The pleasure of the text* (R. Miller, Trans.). New York: Hill & Wang.

Barthes, R. (1977). Death of the author. In R. Barthes (Ed.), *Image-music-text*. New York, NY: Hill & Wang.

Bartle, R. A. (2004). *Designing Virtual World*. Berkeley, CA: New Riders.

Başer, A. (2000). *Ortadoğu'da Modernleşme ve İslami Hareketler*. İstanbul: Yöneliş Yayınları.

Bayer, J., Bukhari, S. S., Langenhan, C., Liwicki, M., Althoff, K.-D., Petzold, F., & Dengel, A. (2015). Migrating the Classical Pen-and-Paper based Conceptual Sketching of Architecture Plans Towards Computer Tools - Prototype Design and Evaluation. *The Eleventh IAPR International Workshop on Graphics Recognition. IAPR International Workshop on Graphics Recognition (GREC-2015), 11th, Located at 13th International Conference on Document Analysis and Recognition (ICDAR), August 22-23, Nancy, France.* Retrieved from http://www.dfki.de/web/forschung/publikationen/renameFileForDownload?filename=bare_conf.pdf&file_id=uploads_2934

Beichler, J. (2017). *A Physics of Consciousness: The only path to understanding consciousness (since consciousness is its 'self' the path).* Retrieved from https://www.academia.edu/33309104/A_Physics_of_Consciousness_The_only_path_to_understanding_consciousness

Bell, D. (1973). *The coming of post-industrial society: A venture in social forecasting.* New York, NY: Basic Books.

Belting, H. (1994). *Likeness and presence: A history of the image before the era of art.* Chicago, IL: The University of Chicago Press.

Benedickt, M. (1991). *Cyberspace: first steps.* Cambridge, MA: MIT Press.

Bense, M. (1969). *Einführung in die informationstheoretische Ästhetik: Grundlegung und Anwendung in der Textheorie* (Vol. 320). Hamburg, Germany: Rowohlt.

Bertol, B., & Foell, D. (1997). *Designing Digital Space: An Architect's Guide to Virtual Reality.* John Wiley & Sons.

Bier, H. H., & Knight, T. (2010). Digitally-driven architecture. *Footprint: Delft School of Design Journal*, (6).

Billinghurst, M. (2002). Augmented Reality and Education. *New Horizons for Learning, 21*(3), 195-209. <ALIGNMENT.qj></ALIGNMENT>10.4018/jgcms.2011010108

Biloria, N., & Chang, J. R. (2013). Hyper-Morphology: Experimentations with bio-inspired design processes for adaptive spatial re-use. *eCAADe 2013: Computation and Performance Proceedings of the 31st International Conference on Education and Research in Computer Aided Architectural Design in Europe.*

Blender. (2017). Retrieved June 15, 2017, from https://www.blender.org

Bloch, P. H. (1995). Seeking the ideal form: Product design and consumer response. *Journal of Marketing, 59*(3), 16–29. doi:10.2307/1252116

Block, F. W. (2009). *How to Construct the Genre of Digital Poetry. A User Manual.* Retrieved from http://www.netzliteratur.net/block/digital_poetry.pdf

Bloland, H. G. (1995). Postmodernism and Higher Education. *The Journal of Higher Education, 66*(5), 521–559. doi:10.2307/2943935

Boatwright & Cagan. (2010). *Product Emotion: The Way to Captive Customers.* Academic Press.

Boden, M. (1997). *The Creative Mind: Myths and Mechanism.* London: Weidenfeld &Nicholson.

Boellstorff, T. (2008). Coming of Age in Second Life. Princeton University Press.

Bogost, I., Bruns, G. L., Campbell, T., Cixous, H., Conan Doyle, A., Becker, E. J., & Levin, D. (2012). Many Subtle Channels. In *Praise of Potential Literature.* Cambridge, MA: Harvard University Press.

Bök, C. (2006). Aleatory Writing: Notes Toward a Poetics of Chance. Public: Art I Culture I Ideas, 33, 24 - 33.

Bolter, J. D., & Grusin, R. (1999). *Remediation: Understanding new media.* Cambridge, MA: MIT Press.

Bolwerk, A., Mack-Andrick, J., & Frieder, R. L., Arnd Dörler, & Malhöfner. (2013). How Art Changes Your Brain: Differential Effects of Visual Art Production and Cognitive Art Evaluation on Functional Brain Connectivity. *PLoS ONE, 9*(7).

Bosnak, R. (2007). *Embodiment: Creative Imagination in Medicine, Art and Travel.* East Sussex, UK: Routledge.

Bower, M., Lee, M. J., & Dalgarno, B. (2016). Collaborative learning across physical and virtual worlds: Factors supporting and constraining learners in a blended reality environment. *British Journal of Educational Technology.*

Boyle, E., Connolly, T. M., & Hainey, T. (2011). The role of psychology in understanding the impact of computer games. *Entertainment Computing, 2*(2), 69–74. doi:10.1016/j.entcom.2010.12.002

Breuer, N. (2012). *Soziale Netzwerke als Brücke zur Kunst.* Vienna: Thesis.

Bridges, M. M., Para, M. P., & Mashner, M. J. (2011). Control system architecture for the modular prosthetic limb. *Johns Hopkins APL Technical Digest, 30*(3), 217–222.

Brinkman, J. (1998). *Buddhist-Christian Studies* (Vol. 18). Honolulu, HI: University of Hawai'i Press.

Brouchoud, J. (2013). *New Virtual Learning and Workplace Campuses for Avaya with Gronstedt Group*. Retrieved June 2017, from http://archvirtual.com/2013/01/29/new-virtual-learning-and-workplace-campuses-for-avaya-with-gronstedt-group/

Brown, G. (2003) The Moving Camera. Part I. *Zerb,* 58.

Brown, B., & Bell, M. (2006). *Avatars at Work and Play.* In R. Shroeder & A. Axelsson (Eds.), *Play and Sociability in There: Some lessons from online games for collaborative virtual environments* (pp. 227–245). Springer.

Brown, K. W., & Ryan, R. M. (2003). The Benefits of Being Present: Mindfulness and Its Role in Psychological Well-Being. *Journal of Personality and Social Psychology, 84*(4), 822–848. doi:10.1037/0022-3514.84.4.822 PMID:12703651

Brown, K. W., Ryan, R. M., & Creswell, J. D. (2007). Addressing Fundamental Questions about Mindfulness. *Psychological Inquiry, 18*(4), 272–281. doi:10.1080/10478400701703344

Brown, T. (2008). Design thinking. *Harvard Business Review, 86*(6), 84–92, 141. doi:10.5437/08956308X5503003 PMID:18605031

Brunner, J. (1996). The Culture of Education. Cambridge, MA: Harvard College.

Buğra, A. (2010). *Devlet ve İş Adamları* (F. Adaman, Trans.). İstanbul: İletişim Yayınları.

Bullivant, L. (2005). *4dspace: interactive architecture (Architectural Design)*. John Wiley & Sons.

Cahalan, J. M. (1994). The guilty forgiving the innocent: Stanislaus, Shaun, and Shem in Finnegans Wake. *Notes on Modern Irish Literature, 6*, 5–11.

Çaha, Ö. (1999). *Sivil Toplum, Aydınlar ve Demokrasi*. İstanbul: İz Yayınları.

Caillois, R. (1961). *Man, play, and games*. University of Illinois Press.

Çakır, R. (1990). *Ayet ve Slogan*. İstanbul: Metis Yayınları.

Camarinha-Matos, L. M., & Afsarmanesh, H. (1999). The Virtual Enterprise Concept. *Proceedings of the IFIP TC5WG5.3/PRODNET Working Conference on Infrastructures for Virtual Enterprises: Networking Industrial Enterprises*, 3–4. doi:10.1007/978-0-387-35577-1_1

Cañas-Bajo, J., Perälä, P. M., Silvennoinen, J. M., & Saariluoma, P. (2015, September). Affective contents of cross-cultural audiovisual experience. In *Proceedings of the 19th International Academic Mindtrek Conference* (pp. 82-88). ACM. doi:10.1145/2818187.2818294

Cañas-Bajo, J., & Silvennoinen, J. M. (2017). Cross-Cultural Factors in Experiencing Online Video Contents in Product Marketing. *International Journal of Art, Culture and Design Technologies*, *6*(1), 40–56. doi:10.4018/IJACDT.2017010103

Card, S. K., Newell, A., & Moran, T. P. (1983). *The psychology of human computer interaction*. Hillsdale, NJ: Lawrence Erlbaum.

Carnap, R. (1966). *Philosophical foundations of physics: An introduction to the philosophy of science*. New York, NY: Basic Books.

Carroll, N. (2001). *Beyond aesthetics: Philosophical essays*. Cambridge University Press. doi:10.1017/CBO9780511605970

Casey, E. (2000). *Imagining: A Phenomenological Study*. Bloomington, IN: Indiana University Press.

Castronova, E. (2005). *Synthetic Worlds: The Business and Culture of Online Games*. Chicago: University of Chicago Press.

Catenazzi, N., & Sommaruga, L. (2013). Mobile Learning and Augmented Reality: New Learning Opportunities. *Social Media: Challenges and Opportunities*, 9–13.

Cawley, J. D. (Ed.). (1997). *Proc. Int. Gas Turbine and Aeroengine Congress and Exhibition, Orlando, FA, 1-6*. American Society of Mechanical Engineering.

Çayır, K. (1999). *Yeni Sosyal Hareketler*. İstanbul: Kaknüs Yayınları.

Cervieri, M. (2007). *User Generated Content in Second Life with Cory Ondrejka*, . Retrieved from http://www.scribemedia.org/2007/03/20/2nd-life/

Çetinkaya, Y. D. (2008). Tarih ve Kuram Arasında Toplumsal Hareketler. In Y. D. Çetinkaya (Ed.), *Toplumsal Hareketler* (pp. 15–61). İstanbul: İletişim Yayınları.

Chapman, J. (2009). Design for Emotionally Durability. *Design Issues*, *25*(4), 29–35. doi:10.1162/desi.2009.25.4.29

Cheng, K.-H. (2017). Exploring Parents' Conceptions of Augmented Reality Learning and Approaches to Learning by Augmented Reality With Their Children. *Journal of Educational Computing Research*, (1): 1–24. doi:10.1177/0735633116686082

Childers, T. L., Carr, C. L., Peck, J., & Carson, S. (2002). Hedonic and utilitarian motivations for online retail shopping behavior. *Journal of Retailing*, *77*(4), 511–535. doi:10.1016/S0022-4359(01)00056-2

Chua, C. K., Leong, K. F., & Lim, C. C. S. (2010). *Rapid Prototyping: Principles and Applications*. World Scientific Publishing Company. doi:10.1142/6665

Clemmensen, T., & Roese, K. (2010). *An overview of a decade of journal publications about culture and human-computer interaction* (HCI). In D. Katre, R. Orngreen, P. Yammiyavar, & T. Clemmensen (Eds.), *Human Work Interaction Design: Usability in Social, Cultural and Organizational Contexts* (pp. 98–112). Boston, MA: Springer; doi:10.1007/978-3-642-11762-6_9

Cleveland, M., Laroche, M., & Papadopoulos, N. (2009). Cosmopolitanism, consumer ethnocentrism, and materialism: An eight-country study of antecedents and outcomes. *Journal of International Marketing*, *17*(1), 116–146. doi:10.1509/jimk.17.1.116

Cohen, L. J. (1985). Strategy or Identity: New Theoretical Paradigms and Contemporary Social Movements. *Social Research*, *52*(4), 663–716.

Cole, A. (2017, Summer). Class Action: Teaching and Activism. *Artforum*, 304.

Colzato, L. S., van den Wildenberg, W. P., & Hommel, B. (2008). Losing the big picture: How religion may control visual attention. *PLoS One*, *3*(11), e3679. doi:10.1371/journal.pone.0003679 PMID:19002253

Computer Aided Design. (2008). In *The Columbia Encyclopedia*. Columbia.

Cook, N., & Winkler, S. L. (2016). Acceptance, Usability and Health Applications of Virtual Worlds by Older Adults: A Feasibility Study. *JMIR Research Protocols*, *5*(2), e81. doi:10.2196/resprot.5423 PMID:27256457

Coyette, A., Kieffer, S., & Vanderdonckt, J. (2007). *Multi-fidelity Prototyping of User Interfaces*. Academic Press.

Creswell, J. W. (2007). *Qualitative Inquiry and Research Design*. Thousand Oaks, CA: SAGE Publications.

Creutzfeldt, J., Hedman, L., Medin, C., Heinrichs, W. L., & Felländer-Tsai, L. (2010). Exploring virtual worlds for scenario-based repeated team training of cardiopulmonary resuscitation in medical students. *Journal of Medical Internet Research*, *12*(3), e38. doi:10.2196/jmir.1426 PMID:20813717

Cyr, D., Head, M., & Larios, H. (2010). Colour appeal in website design within and across cultures: A multi-method evaluation. *International Journal of Human-Computer Studies*, *68*(1), 1–21. doi:10.1016/j.ijhcs.2009.08.005

Cyr, D., & Trevor-Smith, H. (2004). Localization of Web design: An empirical comparison of German, Japanese, and United States Web site characteristics. *Journal of the American Society for Information Science and Technology*, *55*(13), 1199–1208. doi:10.1002/asi.20075

D'Angour, A. (2013). Plato and Play: Taking Education Seriously in Ancient Greece. *American Journal of Play*, *5*(3), 293–307.

Dalgarno, B., & Lee, M. (2010). What are the learning affordances of 3-D virtual environments? *British Journal of Educational Technology*, *41*(1), 10–32. doi:10.1111/j.1467-8535.2009.01038.x

Darwin, C. (1872). The Descent of Man and Selection in Relation to Sex. D. Appleton.

Davison, A. (2002). *Türkiye'de Sekülerizm ve Modernlik* (T. Birkan, Trans.). İstanbul: İletişim Yayınları.

Dawkins, R. (1976). *The selfish gene*. New York, NY: Oxford University Press.

de Bie, M. H., & Lipman, L. J. A. (2012). The use of digital games and simulators in veterinary education: An overview with examples. *Journal of Veterinary Medical Education*, *39*(1), 13–20. doi:10.3138/jvme.0511.055R PMID:22433739

de Castro Salgado, L., de Souza, C., & Leitão, C. (2011). Using metaphors to explore cultural perspectives in cross-cultural design. In P. Rau (Ed.), *Internationalization, design and global development* (pp. 94–103). Berlin: Springer; doi:10.1007/978-3-642-21660-2_11

De Monchaux, N. (2016). Local code: 3,659 proposals about data, design & the nature of cities. New York: Princeton Architectural Press.

DeLappe, J. (2009). Kritical Works in SL II. Wolverhampton, UK: CADRE (Centre for Art, Design, Research and Experimentation).

DeLappe, J. (2010). *Art and the Emergent Imagination in Avatar-Mediated Online Space* (PhD Thesis). University of East London.

Dempster, A. (2006). Risky business: The art of managing creative ventures. In *Cultural Industries: The British Experience in International Perspective*. Humboldt-Universität zu Berlin.

Desmet & Hekkert. (2002). *The Basis of Product Emotions, no. 1988*. Academic Press.

Desmet, P. (2012). *Faces of Product Pleasure: 25 Positive Emotions in Human-Product Interactions*. Academic Press.

Deterding, S., Dixon, D., Khaled, R., & Nacke, L. (2011). From game design elements to gamefulness: defining gamification. *Proceedings of the 15th International Academic MindTrek Conference on Envisioning Future Media Environments - MindTrek '11*, 9–15. doi:10.1145/2181037.2181040

Dicheva, D., Dichev, C., Agre, G., & Angelova, G. (2015). Gamification in Education : A Systematic Mapping Study Gamification in Education : A Systematic Mapping Study. *Journal of Educational Technology & Society, 18*, 75–88.

Dichev, C., Dicheva, D., Angelova, G., & Agre, G. (2014). From Gamification to Gameful Design and Gameful Experience in Learning. *Cybernetics and Information Technologies, 14*(4), 80–100. doi:10.1515/cait-2014-0007

Dixon, S. (2017). Cybernetic-existentialism in interactive performance: Strangers, being-for-others and autopoiesis. *International Journal of Performance Arts and Digital Media*, 1–22. doi:10.1080/14794713.2017.1301173

Dourish, P. (2001). *Where the action is: The foundations of embodied interaction.* Cambridge, MA: MIT Press.

Dourish, P. (2017). Commentary: Interactivity as cultural. *Human-Computer Interaction, 32*(3), 153–154. doi:10.1080/07370024.2016.1257388

Doyle, D. (2014). Phenomenologies of practice: The artist in virtual Worlds. *International Journal of Art, Culture and Design Technologies, 4*(1), 1–12. doi:10.4018/ijacdt.2014010101

Dreyfus, H. L. (1992). *What computers still can't do: A critique of artificial reason.* Cambridge, MA: MIT Press.

Drucker, J. (1994). The Visible Word: Experimental Typography and Modern Art, 1909–1923. University of Chicago Press.

Duncan, I., Miller, A., & Jiang, S. (2012). A taxonomy of virtual worlds usage in education. *British Journal of Educational Technology, 1*, 1–16.

Durkheim, E. (1965). *The Elementary Forms of the Religious Life* (J. W. Swain, Trans.). New York: Free Press.

Eckler, P., & Rodgers, S. (2014). Viral Advertising. The handbook of international advertising research, 184-202.

Eckler, P., & Bolls, P. (2011). Spreading the virus: Emotional tone of viral advertising and its effect on forwarding intentions and attitudes. *Journal of Interactive Advertising, 11*(2), 1–11. doi:10.1080/15252019.2011.10722180

Ehsani, E., & Chase, S. (2009). Using virtual worlds as collaborative environments for innovation and design: lessons learned and observations from case studies in architectural projects. In *Computation: The New Realm of Architectural Design* (pp. 523-531). Estambul: eCAADe.

Ekman, P., & Friesen, W. V. (1971). Constants across cultures in the face and emotion. *Journal of Personality and Social Psychology, 17*(2), 124–129. doi:10.1037/h0030377 PMID:5542557

J. Elkins, K. McGuire, M. Burns, & A. Chester (Eds.). (2013). *Theorizing Visual Studies: writing through the discipline.* New York, NY: Routledge.

Engelbart, D. C. (2001). Augmenting human intellect: A conceptual framework. In R. Packer & K. Jordan (Eds.), *Multimedia: From Wagner to virtual reality* (pp. 64–90). New York, NY: WW Norton & Company. (Original work published 1962)

Enzensberger, H. (1974). *The consciousness industry: On literature, politics and the media.* New York, NY: Seabury Press.

Erbas, A. K., & Yenmez, A. A. (2011). The effect of inquiry-based explorations in a dynamic geometry environment on sixth grade students' achievements in polygons. *Computers & Education, 57*(4), 2462–2475. doi:10.1016/j.compedu.2011.07.002

Fei, C. (2008). *RMB City. Second Life* Sim. Available at: http://rmbcity.com/

Fischnaller, F., & Hill, A. (2005). CITYCLUSTER. From the Renaissance to the Megabyte Networking Age: A Virtual Reality and High-Speed Networking Project. *Presence (Cambridge, Mass.), 14*(1), 1–19. doi:10.1162/1054746053890251

Foo, H., & Mårtensson, O. (2016). *Designing for engagement using gamification in mobile applications.* Academic Press.

Foucault, M. (1970). *The order of things: An archaeology of the human sciences.* London, UK: Tavistock publications.

Fredriksson, B. (1994, November). Holistic Systems Engineering in Product Development. *The Saab-Scania Griffin.*

Freud, S. (1964). Mourning and melancholia. In The standard edition of the complete psychological works of Sigmund Freud: Vol. 14. On the history of the psycho-analytic movement, papers on meta-psychology and other works (pp. 237-258). London, UK: Hogarth Press. (Original work published 1917)

Gable, R. K., & Wolf, M. E. (1993). *Instrument Development in the Affective Domain* (2nd ed.). Boston: Kluwer Academic. doi:10.1007/978-94-011-1400-4

Gagnebin, J.M. (1982). *Walter Benjamin os Cacos da História.* Sao Paulo, Brazil: Editora Brasiliense s.a.

Gamberini, L., Barresi, G., Majer, A., & Scarpetta, F. (2008, Summer). A game a day keeps the doctor away: A short review of computer games in mental healthcare. *Journal of Cyber Therapy and Rehabilitation*, *1*(2), 127–145.

Gardner, H. (2006). *Five minds for the future*. Boston, MA: Harvard Business School.

Garrett, J. J. (2010). *Elements of user experience, the: user-centered design for the web and beyond*. Pearson Education.

Gary, H. (2012). *The Denby Pottery Company Denby Pottery Extracts Unanticipated Benefits from 3D Printing*. Retrieved from www.printin3d.com

Gary, H., & Thomas, A. (2012). 3D Printing at Denby Pottery. In *Towards a New Ceramic Future: CFPR Symposium*. Victoria and Albert Museum, Hochhauser Auditorium.

Geertz, C. (1973). *The interpretation of cultures* (Vol. 5019). Basic Books.

Ghanbarzadeh, R., Ghapanchi, A. H., & Blumenstein, M. (2015). Characteristics of Research on the Application of Three-Dimensional Immersive Virtual Worlds in Health. In *International Conference on Health Information Science* (pp. 213-224). Springer International Publishing. doi:10.1007/978-3-319-19156-0_22

Gibson, W. (1984). *Neuromancer*. New York: Ace Edition, Berkley Publishing Group.

Gladwell, M. (2002). *The tipping point*. London, UK: Little, Brown, and Company.

Göle, N. (1998). *Modern Mahrem*. İstanbul: Metis Yayınları.

Göle, N. (2000). *Melez Desenler*. İstanbul: Metis Yayınları.

Gomringer, E. (1968). *The book of hours, and Constellations: Being poems of Eugen Gomringer*. Something Else Press.

Goodman, N. (1976). *The Language of Art*. Indianapolis, IN: Hackett.

Goodman, N. (1978). *Ways of Worldmaking*. Indianapolis, IN: Hackett Publishing.

Gorlenko, K. (2010). *Art to Collect Art: Acquisition Policies of Museums of Modern Art from 1980s to the Present*. Vienna: Thesis.

Greenaway, P. (1992). *Hundert Objekte zeigen die welt* [Hundred objects to represent the world]. Verlag erd Hatje.

Gregory, S., Scutter, S., Jacka, L., McDonald, M., Farley, H., & Newman, C. (2015). Barriers and enablers to the use of virtual worlds in higher education: An exploration of educator perceptions, attitudes and experiences. *Journal of Educational Technology & Society*, *18*(1), 3–12.

Gross, J. J., & Levenson, R. W. (1995). Emotion elicitation using films. *Cognition and Emotion*, *9*(1), 87–108. doi:10.1080/02699939508408966

Grosz, E. (2001). *Architecture From the Outside: Essays on Virtual and Real Space*. Cambridge, MA: MIT Press.

Guynup, S. L. (2003). *Bespace, GaTech*. Retrieved June, 2017, from http://bespace.lmc.gatech.edu/single/

GXSC. (2011). *uPrintSE 3d Print Pack*. Retrieved from http://store.gxsc.com/s.nl/it.A/id.1040/.f

Hack, C. J. (2016). The benefits and barriers of using virtual worlds to engage healthcare professionals on distance learning programmes. *Interactive Learning Environments*, *24*(8), 1836–1849. doi:10.1080/10494820.2015.1057743

Hahn, C. (2010). What do reliquaries do for relics? *Numen*, *57*(3-4), 284–316. doi:10.1163/156852710X501324

Hair, J. F., Anderson, R. E., Tatham, R. L., & Black, W. C. (1998). *Multivariate Data Analysis* (5th ed.). Prentice-Hall International.

Hall Edward, T. (1966). *The hidden dimension*. Garden City, NY: Anchor.

Hamari, J., Koivisto, J., & Sarsa, H. (2014). Does gamification work? - A literature review of empirical studies on gamification. *Proceedings of the Annual Hawaii International Conference on System Sciences*, 3025–3034. doi:10.1109/HICSS.2014.377

Hannerz, U. (1990). Cosmopolitans and locals in world culture. *Theory, Culture & Society*, *7*(2), 237–251. doi:10.1177/026327690007002014

Hannes, K. (2004). *Geometry Education with Augmented Reality*. Retrieved from http://gretchen.ims.tuwien.ac.at/media/documents/publications/kaufmann_diss.pdf

Harel, D. (1987). Statecharts: A visual formalism for complex systems. *Science of Computer Programming*, *8*(3), 231–274. doi:10.1016/0167-6423(87)90035-9

Harms, J. B. (1981, October). Reason and Social Change in Durkheim's Thought: The Changing Relationship between Individuals and Society. *Pacific Sociological Review*, *24*(4), 393–410. doi:10.2307/1388775

Hartley, J. (2008). From the consciousness industry to creative industries: Consumer-created content, social network markets and the growth of knowledge. In J. Holt & A. Perren (Eds.), *Media industries: History, theory and methods.* Oxford, UK: Blackwell. Retrieved June 29, 2011, from http://cultural-science. org/ FeastPapers2008/JohnHartleyBp1.pdf

Hartley, J. (2009). From the consciousness industry to creative industries: Consumer-created content, social network markets and the growth of knowledge. In J. Hartley (Ed.), *The uses of digital literacy* (pp. 39–70). St. Lucia, Australia: University of Queensland.

Hayles, K. (2012). *How We Think: Digital Media and Contemporary Technogenesis.* Academic Press.

Hayles, N. (1999). *How we became posthuman: Virtual bodies in cybernetics, literature, and informatics*. Chicago, IL: University of Chicago. doi:10.7208/ chicago/9780226321394.001.0001

Heidegger, M. (1976). Brief über den Humanismus. *Gesamtausgabe*, *9*, 313–364.

Heidegger, M. (1977). *The question concerning technology and other essays*. New York: Harper & Row.

Heise, D. R. (1970). The Semantic Differential and Attitude Research. In G. F. Summers (Ed.), Attitude Measurement (pp. 235–253). Chicago: Rand McNally. Retrieved from http://www.indiana.edu/~socpsy/papers/AttMeasure/attitude.htm

Helfenstein, S. (2012). Increasingly emotional design for growingly pragmatic users? A report from Finland. *Behaviour & Information Technology*, *31*(2), 185–204. doi:10.1080/01449291003793777

Hernández-Ibáñez, L., & Barneche-Naya, V. (2013). The Building as the Interface: Architectural Design for Education in Virtual Worlds. In *International Conference on Virtual, Augmented and Mixed Reality* (pp. 152-161). Springer Berlin Heidelberg. doi:10.1007/978-3-642-39420-1_17

Hernández-Ibáñez, L., & Barneche-Naya, V. (2013). Vitruvius in Cyberspace. *International Journal of Art, Culture and Design Technologies*, *3*(1), 36–47. doi:10.4018/ijacdt.2013010104

Hershman, L. (2007). *Life to the Second Power (L2)*. Available at: https://exhibits. stanford.edu/women-art-revolution/about/life-squared

Herzig, I. (2010). *Sammlungspraxis im Wandel der Zeit*. Vienna: Thesis.

Hesius, G. (2002). *Emblemata sacra: De fide, spe, charitate*. Hidesheim, Germany: Georg Olms Verlag.

Hew, K. F., & Cheung, W. S. (2010). Use of three-dimensional (3-D) immersive virtual worlds in K-12 and higher education settings: A review of the research. *British Journal of Educational Technology*, *41*(1), 33–55. doi:10.1111/j.1467-8535.2008.00900.x

Hill, J. (2001). *Architecture: the subject is matter*. Psychology Press.

Hintikka, J., & Remes, U. (1974). *The method of analysis: Its geometrical origin and its general significance*. Dordrecht: D. Reidel Publishing Company; doi:10.1007/978-94-010-2296-5

Hoberg, C. (2006). *Komplexitets Max*. Stockholm, Sweden: Dialoger Förlag.

Hofstede, G. (2010). *National cultural dimensions*. Academic Press.

Hough, S. (2009, December 16). I want you in my room [Web log post]. Retrieved June 29, 2011, from http://stephaniehough.wordpress.com/

Hsiu-Feng, W. (2014). Picture perfect: Girls' and boys' preferences towards visual complexity in children's websites. *Computers in Human Behavior*, *31*, 551–557. doi:10.1016/j.chb.2013.05.033

Hsu, Y.-S., Lin, Y.-H., & Yang, B. (2017). Impact of augmented reality lessons on students' STEM interest. *Research and Practice in Technology Enhanced Learning*, *12*(1), 2. doi:10.1186/s41039-016-0039-z

Huang, H.-M., Rauch, U., & Liaw, S.-S. (2010). Investigating learners' attitudes toward virtual reality learning environments: Based on a constructivist approach. *Computers & Education*, *55*(3), 1171–1182. doi:10.1016/j.compedu.2010.05.014

Hul, C.H. (1984). *Apparatus for Production of 3D Objects by Stereolithography*. Academic Press.

Huson, D. (2012). 3D Printing of Ceramics for Design Concept Modelling CFPR 3D Lab. In *CFPR Symposium*. London: Arts and Humanities Research Council (AHRC).

Husserl, E. (1970). *The crisis of European sciences and transcendental phenomenology: An introduction to phenomenological philosophy* (4th ed.). Evanston, IL: Northwestern University Press.

Hutchins, E. L., Hollan, J. D., & Norman, D. A. (1985). Direct manipulation interfaces. *Human-Computer Interaction*, *1*(4), 311–338. doi:10.1207/s15327051hci0104_2

Huvila, I., Ek, S., & Widen, G. (2014). Information sharing and the dimensions of social life in Second Life. *Journal of Information Science*, *40*, 237-248. DOI: 0165551513516711

Ihde, D. (1991). *Instrumental realism: The interface between philosophy of science and philosophy of technology*. Bloomington, IN: Indiana University Press.

Ihde, D. (1993). *Philosophy of Technology*. Paragon House.

Ihde, D. (2002). *Bodies in Technology*. Minneapolis, MN: University of Minnesota Press.

Ito, M., & Nakakoji, K. (1996, August). Impact of culture on user interface design. In *International users interface* (pp. 105–126). John Wiley & Sons, Inc.

James, A., & Brad, A. (2001). *Sketching Interfaces: Toward More Human Interface Design*. Academic Press.

Jan, W., & Konrad, W. (2006). Rapid Manufacturing of Ceramic Components for Medical and Technical Applications via Selective Laser Melting. Steinbachstr. Germany: Frauhofer Institute for Laser Technology (ILT).

Janlert, L. E., & Stolterman, E. (2017). The meaning of interactivity: Some proposals for definitions and measures. *Human-Computer Interaction*, *32*(3), 103–138. doi: 10.1080/07370024.2016.1226139

Jenkins, H. (2008). *Convergence culture: Where old and new media collide*. New York, NY: NYU Press.

Jennifer, H. (2010). 3D Model-Based Design: Setting the Definitions Straight. *MCADCafe*. Retrieved from http://www10.mcadcafe.com/nbc/articles/2/867959/3D-model-based-design-setting-definitions-straight

Jiang, Z., & Benbasat, I. (2004). Virtual product experience: Effects of visual and functional control of products on perceived diagnosticity and flow in electronic shopping. *Journal of Management Information Systems*, *21*(3), 111–147. doi:10.1080/07421222.2004.11045817

Johansson, M., & Linde, P. (2004). *Fieldasy*. Sheffield, UK: Fieldasy. Pixel Raiders.

Jokinen, J. P., Silvennoinen, J. M., Perälä, P. M., & Saariluoma, P. (2015, April). Quick affective judgments: Validation of a method for primed product comparisons. In *Proceedings of the 33rd Annual ACM Conference on Human Factors in Computing Systems* (pp. 2221-2230). ACM. doi:10.1145/2702123.2702422

Jordan, P. (2002). *Designing Pleasurable Products: An Introduction to the New Human Factors*. London: Taylor & Francis.

Jorge, F., Rui, J., Ricardo, P., & Ana, M. (2004). Rapid Prototyping and Rapid Tooling Applied in Product Development of Ceramic Components. *Materials Science Forum II, 455–456*, 835–38.

Jung, C. G. (1951). *Foreword to Richard Wilhelm's translation of the I Ching*. Routledge & Kegan Paul.

Y. Kafai, K. Peppler, & R. Chapman (Eds.). (2009). *The Computer Clubhouse: constructionism and Creativity in Youth Communities*. New York, NY: Teachers College Press.

Kai, Y., & El-Haik Basem, S. (2009). Design for Six Sigma: A Roadmap for Product Development (2nd ed.). McGraw-Hill.

Kaikati, A. M., & Kaikati, J. G. (2004). Stealth marketing: How to reach consumers surreptitiously. *California Management Review*, *46*(4), 6–22. doi:10.2307/41166272

Kandalaft, M. R., Didehbani, N., Krawczyk, D. C., Allen, T. T., & Chapman, S. B. (2013). Virtual reality social cognition training for young adults with high-functioning autism. *Journal of Autism and Developmental Disorders*, *43*(1), 34–44. doi:10.1007/s10803-012-1544-6 PMID:22570145

Kaplan, S., & Kaplan, R. (1982). *Cognition and Environment: Functioning in an Uncertain World*. Westport, CT: Praeger.

Kasemsap, K. (2016). Cultural Perspectives and Cultural Dynamics: Advanced Issues and Approaches. *International Journal of Art, Culture and Design Technologies*, *5*(1), 35–47. doi:10.4018/IJACDT.2016010103

Kay, A., & Goldberg, A. (1977). Personal dynamic media. *Computer*, *10*(3), 31–41. doi:10.1109/C-M.1977.217672

Keagan, R. (1982). The Evolving Self: problem and process in human development. Cambridge, MA: Harvard College.

Keeler, M. A. (1990). Investigating transparency in the conditions of mediation from a semeiotic view. *Semiotica*, *82*(1-2), 15–42. doi:10.1515/semi.1990.82.1-2.15

Keeler, M. A., & Denning, S. M. (1991). The challenge of interface design for communication theory: From interaction metaphor to contexts of discovery. *Interacting with Computers*, *3*(3), 283–301. doi:10.1016/0953-5438(91)90018-W

Kember, S., & Zylinska, J. (2009). Performance, invention, critique. In M. Chatzichristodoulou, J. Jefferies, & R. Zerihan (Eds.), *Interfaces of performance* (pp. 27–43). Farnham, UK: Ashgate.

Keyder, Ç. (1999). *Türkiye'de Devlet ve Sınıflar*. İstanbul: İletişim Yayınları.

Kırık, H. (2005). *Kamusal Alan ve Demokrasi*. İstanbul: Salyangoz Yayınları.

Kirriemuir, J., & McFarlane, A. (2004). Literature Review in Games and learning. *Futurelab Series*, (8), 39. Retrieved from https://telearn.archives-ouvertes.fr/hal-00190453%0Ahttp://telearn.archives-ouvertes.fr/hal-00190453/

Köroğlu, M. A. (2012). *Özerk Kamusal Alanların Oluşumunda Dinin Rolü* (Unpublished doctoral dissertation). Necmettin Erbakan University, Konya, Turkey.

Kozel, S. (2007). *Closer: Performance, Technologies: Performance, Technologies, Phenomenology*. Cambridge, MA: MIT Press.

Krauss, R. (1979). Sculpture in the expanded field. *October, 8*, 30-44.

Krauss, R. (2000). *A voyage on the North Sea: Art in the age of the post-medium condition*. London, UK: Thames and Hudson.

Krentz, A. (1998). Play and Education in Plato's Republic. *Twentieth World Congress of Philosophy*, (2). Retrieved from https://www.bu.edu/wcp/Papers/Educ/EducKren.htm

Lacey, E. (2009). *Contemporary Ceramic Design for Meaningful Interaction*. Academic Press.

Lanham, R. A. (1993). *The Electronic Word: Democracy*. University of Chicago Press. doi:10.7208/chicago/9780226469126.001.0001

Latour, B. (2005). *Reassembling the social: An introduction to actor-network-theory*. Oxford, UK: Oxford University Press.

Laurel, B. (1993). *Computers as theatre*. Reading, MA: Addison-Wesley.

Lazar, A., Edasis, C., & Piper, A. M. (2017). Supporting people with dementia in digital social sharing. In *Proceedings of the 2017 CHI Conference on Human Factors in Computing Systems* (pp. 2149-2162). ACM. doi:10.1145/3025453.3025586

Le Corbusier. (1923). *Vers une architecture*. G. Crès et Cie.

Leikas, J. (2009). *Life-based design. A holistic approach to designing human-technology interaction*. Espoo, Finland: VTT Publications.

Leikas, J., Saariluoma, P., Heinilä, J., & Ylikauppila, M. (2013). A methodological model for life-based design. *International Review of Social Sciences and Humanities*, *4*(2), 118–136.

Lei, S. A. (2010). Intrinsic and Extrinsic Motivation : Evaluating Benefits and Drawbacks from College Instructors '. *Perspectives*, *37*(2), 153–160. Retrieved from http://web.a.ebscohost.com/ehost/pdfviewer/pdfviewer?sid=1b4f8e6a-7ae0-444d-9d6c-b0e343f3df1f%40sessionmgr4003&vid=0&hid=4106

Leitão, R., Rodrigues, J. M. F., & Marcos, A. F. (2014). Game-Based Learning: Augmented Reality in the Teaching of Geometric Solids. *International Journal of Art, Culture and Design Technologies*, *4*(1), 63–75. doi:10.4018/ijacdt.2014010105

Lessig, L. (2007). *How creativity is being strangled by the law* [Video clip]. Retrieved June 29, 2011, from http://www.youtube.com/watch?v=7Q25-S7jzgs

Lessig, L. (2008). *Remix: Making art and commerce thrive in the hybrid economy*. New York, NY: Penguin Group. doi:10.5040/9781849662505

Levy, P. (1998). *Becoming Virtual: Reality in the Digital Age*. New York: Plenum Trade.

Liao, C., Motter, J. L., & Patton, R. M. (2016, July). Tech Savy Girls: Learning 21st-Century Skills Through Digital Artmaking. *Art Education*.

Lichty, P. (2008). *Why Art in Virtual Worlds? E-Happenings, Relational Milieux & "Second Sculpture." CIAC Electronic Magazine*.

Liedtka, J. (2013). Design Thinking: What it is and why it works. *Design at Darden Working Paper Series*. Retrieved from http://batten.squarespace.com/storage/books-journals-articles/DesignThinking_WhatitIs_Liedtka.pdf

Lindgaard, G., Fernandes, G., Dudek, C., & Brown, J. (2006). Attention web designers: You have 50 milliseconds to make a good first impression! *Behaviour & Information Technology*, *25*(2), 115–126. doi:10.1080/01449290500330448

Lin, H.-C. K., Chen, M.-C., & Chang, C.-K. (2015). Assessing the effectiveness of learning solid geometry by using an augmented reality-assisted learning system. *Interactive Learning Environments*, *23*(6), 799–810. doi:10.1080/10494820.2013.817435

Lin, R. T. (2007). Transforming Taiwan Aboriginal Cultural Features into Modern Product Design: A Case Study of a Cross- Cultural Product Design Model. *International Journal of Design*, *1*(2), 45–53.

Lippe, C. (2002). Real-time interaction among composers, performers, and computer systems. *Information Processing Society of Japan SIG Notes, 2002*(123), 1-6.

Lovelace, M., & Brickman, P. (2013). Best Practices for Measuring Students' Attitudes toward Learning Science. *CBE Life Sciences Education, 12*(4), 606–617. doi:10.1187/cbe.12-11-0197 PMID:24297288

Lubell, S. (Dec. 4, 2008). *Original Sim.* Retrieved June, 2017, from http://www.nytimes.com/2008/12/07/style/tmagazine/07secondlife.html

Luke, W. (2009). *The Structure, Influence and Application of Emotion in Product Design.* Leicester, UK: De Montfort University.

Lupton, E. (1994). A Post-Mortem on Deconstruction? *AIGA Journal of Graphic Design, 12*(2), 45–47.

Lynch, K. (1960). *The Image of the City.* Cambridge, MA: MIT Press.

Lynn, G. (1996). Blobs, or why tectonics is square and topology is groovy. ANY: Architecture New York, (14), 58-61.

Machado, A. (2007). *Arte e mídia.* Rio de Janeiro, Brazil: Jorge Zahar Editor.

Madoff, S. (2009). *Art School (Propositions for the 21st Century).* Cambridge, MA: MIT Press.

Maguire, M. (2001a). Context of Use within usability activities. *International Journal of Human-Computer Studies, 55*(4), 453–483. doi:10.1006/ijhc.2001.0486

Maguire, M. (2001b). Methods to support human-centred design. *International Journal of Human-Computer Studies, 55*(4), 587–634. doi:10.1006/ijhc.2001.0503

Mahçupyan, E. (1998). Osmanlı'dan Günümüze Parçalı Kamusal Alan ve Siyaset. *Doğu- Batı, 5*, 22-47.

Maley, C. J., & Piccinini, G. (2017). The Evolution of Consciousness. In Routledge Handbook of Consciousness. New York: Routledge.

Mandanici, M., Rodà, A., & Canazza, S. (2017). Bodily interactions in motion-based music applications. *Human Technology, 13*(1), 82–108. doi:10.17011/ht/urn.201705272519

Manovich, L. (2001). *The language of new media.* Cambridge, MA: MIT Press.

Marcus, A. (1994). Managing metaphors for advanced user interfaces. In *AVI '94 Proceedings of the Workshop on Advanced Visual Interfaces* (pp. 12-18). New York, NY: ACM. doi:10.1145/192309.192317

Marcus, A., & Gould, E. W. (2000). Crosscurrents: cultural dimensions and global Web user-interface design. *Interactions, 7*(4), 32-46.

Marcus-Roberts, H. M., & Roberts, F. S. (1987). Meaningless Statistics. *Journal of Educational Statistics, 12*(4), 383–394. doi:10.2307/1165056

Mardin, Ş. (2010). *Din ve İdeoloji*. İstanbul: İletişim Yayınları.

Marshall, G. (1999). *Sosyoloji Sözlüğü* (O. Akınbay & D. Kömürcü, Trans.). Ankara: Bilim ve Sanat Yayınları.

Martiniano, C. (2016, Fall). The Scientization of Creativity. *The Journal of the Midwest Modern Language Association, 49*(2), 161–190. doi:10.1353/mml.2016.0033

Massumi, B. (2002). *Parables for the Virtual: Movement, Affect, Sensation*. Durham, NC: Duke University Press. doi:10.1215/9780822383574

Mastandrea, S. (2014). How emotions shape aesthetic experiences. In P. P. L. Tinio & J. K. Smith (Eds.), *The Cambridge Handbook of the Psychology of Aesthetics and the Arts* (pp. 500–518). Cambridge, UK: Cambridge University Press, UK.

Masuda, T., Ellsworth, P. C., Mesquita, B., Leu, J., Tanida, S., & Van de Veerdonk, E. (2008). Placing the face in context: Cultural differences in the perception of facial emotion. *Journal of Personality and Social Psychology, 94*(3), 365–381. doi:10.1037/0022-3514.94.3.365 PMID:18284287

Mattes, E. (2009). *Synthetic Performances*. Available at: http://0100101110101101.org/synthetic-performances/

Maturana, H., & Varela, F. (1987). *The tree of life*. Boston, MA: Shambala.

Mayer, R. E. (2005). Cognitive Theory of Multimedia Learning. In *The Cambridge handbook of multimedia learning* (pp. 31–48). Cambridge University Press. doi:10.1017/CBO9780511816819.004

McCarthy, J., & Wright, P. (2017). Commentary: Making interactivity meaningful for contemporary HCI. *Human-Computer Interaction, 32*(3), 148–152. doi:10.1080/07370024.2016.1254048

McCarty, J., & Zald, M. (1977). Resource Mobilization and Social Movements: A Partial Theory. *American Journal of Sociology, 82*(6), 1212–1241. doi:10.1086/226464

McCullough, M. (1996). *Abstracting Craft: The Practiced Digital Hand*. Boston: MIT Press.

McCullough, M. (2005). *Digital ground: Architecture, pervasive computing, and environmental knowing*. The MIT Press.

McGonigal, J. (2011). *Reality is broken: Why games make us better and how they can change the world*. Penguin Group.

McGonigal, J. (2011). *Reality is broken: Why games make us better and how they can change the world*. Penguin.

McLuhan, M. (1964). *Understanding media: The extensions of man*. Cambridge, MA: MIT Press.

Mehta, S., & Nicholas, P. (2011, June 29). Obama highlights manufacturing revival, jobs in Iowa. *Los Angeles Times*. Retrieved from http://www.latimes.com/news/nationworld/nation/la-na-obamaiowa-20110629,0,5563738.story

Melucci, A. (1985). The Symbolic Challenge of Contemporary Movements. *Social Research*, *52*(4), 789–816.

Merleau-Ponty, M. (1962). *Phenomenology of perception* [Phénoménologie de la perception]. (C. Smith, Trans.). London, UK: Routledge.

Merrick, K., Gu, N., & Wang, X. (2011). Case studies using multiuser virtual worlds as an innovative platform for collaborative design. *Journal of Information Technology in Construction*, *16*, 165–188.

Meuleman, R., & Savage, M. (2013). A field analysis of cosmopolitan taste: Lessons from the Netherlands. *Cultural Sociology*, *7*(2), 230–256. doi:10.1177/1749975512473991

Miller, D., Costa, E., Haynes, N., McDonald, T., Nicolescu, R., Sinanan, J., ... Wang, X. (2016). *How the World Changed Social Media*. UCL Press.

Miller, P. (2004). *Rhythm science*. Cambridge, MA: MIT Press.

Mitchell, W. J. (2003). ME, the cyborg self and the networked city. MIT Press.

Mitchell, W. J. (1996). *City of bits: space, place, and the infobahn*. MIT Press.

Moggridge, B. (2010). *Designing media*. MIT Press.

Mohanty, J. N. (1991). Method of Imaginative Variation in Phenomenology. In T. M. Horowitz (Ed.), Thought Experiments in Science and Philosophy. Rowman & Littlefield Publishers.

Moran, T. P. (1981). The command language grammar: A representation for the user interface of interactive computer systems. *International Journal of Man-Machine Studies, 15*(1), 3–50. doi:10.1016/S0020-7373(81)80022-3

Morie, J. F. (2009). Kritical Works in SL II. Wolverhampton, UK: CADRE (Centre for Art, Design, Research and Experimentation).

Morie, J. F. (2010). *Art and the Emergent Imagination in Avatar-Mediated Online Space* (PhD Thesis). University of East London.

Morley, D. (2006). Unanswered questions in audience research. *Communication Review, 9*(2), 101–121. doi:10.1080/10714420600663286

Morris, J. M. (2013). Collaborating with machines: Hybrid performances allow a different perspective on generative art. In *Proceedings of the 16th Generative Art Conference* (pp. 114-125). Retrieved from http://www.generativeart.com/ga2013xWEB/proceedings1/11.pdf

Morrissey, K., Wood, G., Green, D., Pantidi, N., & McCarthy, J. (2016). "I'm a rambler, I'm a gambler, I'm a long way from home": The place of props, music, and design in dementia care. In *Proceedings of the 2016 ACM Conference on Designing Interactive Systems* (pp. 1008-1020). ACM. doi:10.1145/2901790.2901798

Moss, G. A., & Gunn, R. W. (2009). Gender differences in website production and preference aesthetics: Preliminary implications for ICT in education and beyond. *Behaviour & Information Technology, 28*(5), 447–460. doi:10.1080/01449290802332662

Moustakas, C. (1990). *Heuristic Research: Design, Methodology, and Applications.* Newbury Park, CA: SAGE Publications. doi:10.4135/9781412995641

Moustakas, C. (1994). *Phenomenological Research Methods.* Thousand Oaks, CA: SAGE Publications. doi:10.4135/9781412995658

Muñoz, K., Lunney, T., Mc Kevitt, P., Noguez, J., & Neri, L. (2013). An Emotional Student Model for Game-Based Learning. In *Technologies for Inclusive Education.* Beyond Traditional Integration Approaches. doi:10.4018/978-1-4666-2530-3.ch009

Murugappan, S., Piya, C., Yang, M. C., & Ramani, K. (2017). FEAsy : A Sketch-Based Tool for Finite Element Analysis. *Journal of Computing and Information Science in Engineering, 17*(3), 031009. doi:10.1115/1.4034387

Naismith, L., Lonsdale, P., Vavoula, G., & Sharples, M. (2004). *Literature Review in Mobile Technologies and Learning Literature Review in Mobile Technologies and Learning*. Slough: FutureLab. Retrieved from http://www.futurelab.org.uk/resources/documents/lit_reviews/Mobile_Review.pdf

Navon, D. (1977). Forest before trees: The precedence of global features in visual perception. *Cognitive Psychology*, *9*(3), 353–383. doi:10.1016/0010-0285(77)90012-3

Nelson, H. G., & Stolterman, E. (2003). The design way: Intentional change in an unpredictable world: foundations and fundamentals of design competence. Englewood Cliffs, NJ: Educational Technology Publications.

Newell, A., & Simon, H. A. (1972). Human problem solving: Vol. 104. *No. 9*. Englewood Cliffs, NJ: Prentice-Hall.

Nielsen, J. I. (2005). Bordwell on Bordwell. *Part IV-Levels of Engagement*, *16*, 9.

Norman, D. A. (2014). Commentary on: Svanaes, Dag (2014): Philosophy of interaction. In M. Soegaard & R. F. Dam (Eds.), The Encyclopedia of Human-Computer Interaction (2nd ed.). Aarhus, Denmark: The Interaction Design Foundation. Retrieved from https://www.interaction-design.org/encyclopedia/philosophy_of_interaction.html

Norman, D. (2004). *Emotion Design: Why We Love (or Hate) Everyday Things*. New York, NY: Basic Books.

Norman, D. A. (2002). *The design of everyday things*. New York, NY: Basic Books.

Norman, D. A. (2005). *Emotional design: Why we love (or hate) everyday things*. Basic books.

Norman, D. A. (2010). *Living with complexity*. Cambridge, MA: MIT Press.

Novak, M. (1996, December 9). *TransArchitecture. Building the edge of thought*. Retrieved June, 2017 from https://www.heise.de/tp/features/transArchitecture-3445869.html

Novak, M. (1991). Liquid architectures in cyberspace. In M. Benedikt (Ed.), *Cyberspace: First Steps* (pp. 225–254). Cambridge, MA: MIT Press.

O'Connell, R. M. (2014). *Visualizing Culture: Analyzing the Cultural Aesthetics of the Web*. Peter Lang Publishing.

Offe, C. (1985). New Social Movements: Challenging the Boundaries of Institutional Politics. *Social Research*, *52*(4), 817–868.

Olalere, F. E., & Ab, S. A. (2014). Computer-Aided Ceramic Design: Its Viability for Building User-Centered Design. In Contemporary Advancements in Information Technology Development in Dynamic Environments. Hershey, PA: IGI Global. doi:10.4018/978-1-4666-6252-0

Olalere, F. E., & Ab, S. B. A. (2012). The Use of Computer Aided Tools for Prototyping in Traditional Ceramics Production in Malaysia. *Proceedings of the 1st International Conference on Design and Innovation, ICDI2012*.

Oogjes, D., & Wakkary, R. (2017). Videos of things: Speculating on, anticipating and synthesizing technological mediations. In *Proceedings of the 2017 CHI Conference on Human Factors in Computing Systems* (pp. 4489-4500). ACM. doi:10.1145/3025453.3025748

Oosterhuis, K. (1998). *Spacestation. Hyperbody*. Retrieved June, 2017, from http://www.onl.eu/projects/space-station

Oosterhuis, K., & Jaskiewicz, T. (2007). 798 Multiplayer Design Game. A New Tool for Parametric Design. In Space Time Play. Computer Games, Architecture and Urbanism: The Next Level (pp. 358-361). Basel: BirkhäuserPublishing.

Oppegaard, B., & Still, B. (2013). Bodystorming with Hawkins's block: Toward a new methodology for mobile media design. *Mobile Media & Communication, 1*(3), 356–372. doi:10.1177/2050157913496421

Oppenheimer, T. (2004). *The flickering mind: Saving education from the false promise of technology*. New York, NY: Random House.

Osborn, A. F. (1963). Applied imagination: Principles and procedures of creative problem-solving (3rd ed.). New York : Scribner's.

Osgood, C. E., Suci, G. J., & Tannenbaum, P. H. (1957). *The Measurement of Meaning*. University of Illinois Press.

Ostler, T. (1994). Architecture in cyberspace. *Architects'. Journal, 199*(21), 33–35.

Oulasvirta, A., Kurvinen, E., & Kankainen, T. (2003). Understanding contexts by being there: Case studies in bodystorming. *Personal and Ubiquitous Computing, 7*(2), 125–134. doi:10.1007/s00779-003-0238-7

Overbeeke, C. J., & Hekkert, P. (1999). Editorial. In *Proceedings of the First International Conference on Design and Emotion*. Delft: Department of Industrial Design.

Parish, N. (2008). Between Text and Image, East and West: Henri Michaux's Signs and Christian Dotremont's 'Logogrammes'. *RiLUnE, 8*, 67–80.

Partnership for 21st Century Skills. (2011). *U.S. government framework for 21st-century literacies.* Retrieved June 29, 2011, from http://www.p21.org/index.php?option=com_content&task=view&id=254&Itemid=119

Paterson, B., Winschiers-Theophilus, H., Dunne, T. T., Schinzel, B., & Underhill, L. G. (2011). Interpretation of a cross-cultural usability evaluation: A case study based on a hypermedia system for rare species management in Namibia. *Interacting with Computers, 23*(3), 239–246. doi:10.1016/j.intcom.2011.03.002

Pavlou, P. A., & Chai, L. (2002). What Drives Electronic Commerce across Cultures? Across-Cultural Empirical Investigation of the Theory of Planned Behavior. *Journal of Electronic Commerce Research, 3*(4), 240–253.

Pineda, R. G. (2014). *Technology in culture: A theoretical discourse on convergence in human-technology interaction* (Unpublished doctoral dissertation). University of Jyväskylä, Jyväskylä, Finland.

Pink, D. (2005). *A whole new mind: Why rightbrainers will rule the future.* New York, NY: Penguin Books.

Poggenpohl, S., & Sato, K. (2009). *Design Integration.* Bristol, UK: Intellect.

Polanyi, M. (1958). *Personal knowledge: Towards a post-critical philosophy.* London, UK: Routledge.

Postman, N. (1985). *Amusing ourselves to death: Public discourse in the age of show business.* New York, NY: Penguin Group.

Prakash, A. (2012). The Rules for Eating with Your Hands in India, Africa and the Middle East. *Food Republic.* Retrieved from http://www.foodrepublic.com/2012/11/19/the-rules-for-eating-with-your-hands-in-india-africa-and-the-middle-east/

Prensky, M. (2005, October). Engage Me or Enrage Me: What Today's Learners Demand. *EDUCAUSE Review*, 60–64. Retrieved from http://cff.wiki.elanco.net/file/view/Engage+Me+or+Enrage+Me.pdf/84842561/Engage+Me+or+Enrage+Me.pdf

Przybylski, A. K., Rigby, C. S., & Ryan, R. M. (2010). A motivational model of video game engagement. *Review of General Psychology, 14*(2), 154–166. doi:10.1037/a0019440

Purcell, K. (2010). *The state of online video.* Washington, DC: Pew Internet & American Life Project.

Quine, W. V. O. (1953). *From a logical point of view: Nine logico-philosophical essays*. Cambridge, MA: Harvard University Press.

Razak, A. A., Connolly, T., Baxter, G., Hainey, T., & Wilson, A. (2012). The use of Games-Based Learning at Primary Education Level Within the Curriculum for Excellence: A Combined Result of two Regional Teacher Surveys. In *Proceedings of the 6th European Conference on Games Based Learning* (p. 52). Academic Press.

Reeves, B., & Nass, C. (1996). *The media equation: How people treat computers, television, and new media like real people and places*. Stanford, CA: CSLI Publications.

Reinecke, K., & Bernstein, A. (2011). Improving performance, perceived usability, and aesthetics with culturally adaptive user interfaces. *ACM Transactions on Computer-Human Interaction, 18*(2), 8. doi:10.1145/1970378.1970382

Reinecke, K., & Gajos, K. Z. (2014, April). Quantifying visual preferences around the world. In *Proceedings of the SIGCHI Conference on Human Factors in Computing Systems* (pp. 11-20). ACM.

Rheingold, H. (2005). *On collaboration* [Video clip]. Retrieved June 29, 2011, from http://www.ted.com/talks/howard_rheingold_on_collaboration.html

Risset, J. C. (2003). The perception of musical sound. Computer music: Why? *France-University of Texas Institute for Interdisciplinary Studies*. University of Texas at Austin. Retrieved from http://liberalarts.utexas.edu/france-ut/_files/pdf/resources/risset.pdf

Risset, J. C. (1996). Real-world sounds and simulacra in my computer music. *Contemporary Music Review, 15*(1-2), 29–47. doi:10.1080/07494469600640341

Robinson, A. (2009). Kritical Works in SL II. Wolverhampton, UK: CADRE (Centre for Art, Design, Research and Experimentation).

Robinson, A. (2010). *Art and the Emergent Imagination in Avatar-Mediated Online Space* (PhD Thesis). University of East London.

Roseman, I. J., & Smith, C. A. (2001). Appraisal theory. In K. R. Scherer, A. Schorr, & T. Johnstone (Eds.), *Appraisal processes in emotion: theory, methods, research* (pp. 3–19). Oxford, UK: Oxford University Press.

Rosler, M. (2010). Take the money and run? Can political and socio-critical art 'survive'? *e-Flux Journal, 12*. Retrieved June 26, 2011, from http://www.e-flux.com/journal/view/107

Rössel, J., & Schroedter, J. H. (2015). Cosmopolitan cultural consumption: Preferences and practices in a heterogenous, urban population in Switzerland. *Poetics*, *50*, 80–95. doi:10.1016/j.poetic.2015.02.009

Rostami, A., McMillan, D., Segura, E. M., Rossitto, C., & Barkhuus, L. (2017). Bio-sensed and embodied participation in interactive performance. *Proceedings of the TEI*, 17. doi:10.1145/3024969.3024998

Rotman, B. (2000). *Mathematics as Sign: Writing, Imagining, Counting*. Stanford, CA: Stanford University Press.

Rowe, R. (1993). *Interactive music systems: Machine listening and composing*. Cambridge, MA: MIT Press.

Ryan, M. L. (2003). *Narrative as Virtual Reality. Immersion and Interactivity in Literature and Electronic Media*. Baltimore, MD: Johns Hopkins University Press.

Saariluoma, P. (1997). *Foundational analysis: Presuppositions in experimental psychology*. London, UK: Routledge.

Saariluoma, P., Cañas, J. J., & Leikas, J. (2016). *Designing for Life: A Human Perspective on Technology Development*. Springer. doi:10.1057/978-1-137-53047-9

Saariluoma, P., & Jokinen, J. P. (2014). Emotional dimensions of user experience: A user psychological analysis. *International Journal of Human-Computer Interaction*, *30*(4), 303–320. doi:10.1080/10447318.2013.858460

Salter, C. (2009). Environments, interactions and beings: The ecology of perfomativity and technics. In M. Chatzichristodoulou, J. Jefferies, & R. Zerihan (Eds.), *Interfaces of performance* (pp. 44–62). Farnham, UK: Ashgate.

Sanchez Ramon, O., Garcia Molina, J., Sachez Cuadrado, J., & Vanderdonckt, J. (2013). GUI Generation from Wireframes. In *Actas del XIV Congreso Internacional de Interacción Persona-Ordenador*. Madrid: Academic Press. Retrieved from http://hdl.handle.net/2078/153911

Sanlı, L. (2003). Türkiye'de Toplumsal Hareketler. In L. Sanlı (Ed.), *Toplumsal Hareketler Konuşuyor* (pp. 9–32). İstanbul: Alan Yayınları.

Scherer, K. R. (2009). The dynamic architecture of emotion: Evidence for the component process model. *Cognition and Emotion*, *23*(7), 1307–1351. doi:10.1080/02699930902928969

Schlagenhaufer, C., & Amberg, M. (2015). A Descriptive Literature Review and Classification Framework for Gamification in Information Systems. *ECIS 2015 Completed Research*, 0–15. <ALIGNMENT.qj></ALIGNMENT>10.18151/7217466

Schmidt, A. (2017). Understanding and researching through making-A Plea for Functional Prototypes. *Interaction*, *24*(3), 78–81. doi:10.1145/3058498

Schmidt, M., Laffey, J. M., Schmidt, C. T., Wang, X., & Stichter, J. (2012). Developing methods for understanding social behavior in a 3D virtual learning environment. *Computers in Human Behavior*, *28*(2), 405–413. doi:10.1016/j.chb.2011.10.011

Schultz, R. (2007). *Designing the Architecture of Architectural Design Collaboration: Studio Wikitecture 2.0*. Retrieved June, 2017, from http://studiowikitecture.wordpress.com/2007/04/15/wikitecture-20-designing-the-architecture-of-architectural-design-collaboration/

Schwartz, S., & Ishofsky, R. (2011, August 29). How Solar Stores are helping to meet rural Uganda's energy needs. *PV Magazine*.

Seevinck, J. (2017). Interaction in art and computing. In *Emergence in Interactive Art* (pp. 31–46). Springer International Publishing. doi:10.1007/978-3-319-45201-2_3

Serafijn, Q.S., & Spuybroek, L. (2004). Emotional landscapes in the City. D-Tower in Doetinchem. *(In) Security*, (6), 142-150.

Sermon, P. (2007). *Liberate your Avatar. Public Performance Installation*. Available at: http://www.paulsermon.org/liberate/

Sermon, P., & Gould, C. (2012). *Mirror on the Screen*. Interactive *Second Life* Installation. Available at: http://www.paulsermon.org/mirror/

Sharp, R., McKinney, P., & Ross, S. (2007). *Visual text: Concrete poetry, hyperfiction and the future of the narrative form*. Glasgow, UK: Humanities Advanced Technology and Information Institute. Retrieved from http://www.ed.ac.uk/iash/vkpublication/Sharp_McKinney_Ross.pdf

Sharples, M. (1996). An Account of Writing as Creative Design. In A Writer's Reference (6th ed.). Academic Press.

Shaw, C. D., Gromala, D., & Seay, A. F. (2007). The meditation chamber: Enacting autonomic senses. *Proceedings of ENACTIVE/07 4th International Conference on Enactive Interfaces.*

Shirky, C. (2009). *Here Comes Everybody*. New York, NY: Penguin Press.

Shneiderman, B., & Plaisant, C. (2005). Designing the user interface: Strategies for effective human-computer interaction (4th int. ed.). Boston, MA: Addison-Wesley.

Silva, D., & Silveira, I. (2012). Jogos Educacionais com Realidade Aumentada: Desafios de Integração e Possibilidades na Construção de Objetos de Aprendizagem. In Conferencias LACLO 3.1. São Paulo: Academic Press.

Silvennoinen, J. M., Rousi, R., Jokinen, J. P., & Perälä, P. M. (2015, September). Apperception as a multisensory process in material experience. In *Proceedings of the 19th International Academic Mindtrek Conference* (pp. 144-151). ACM. doi:10.1145/2818187.2818285

Silvia, P. J. (2005). Cognitive appraisals and interest in visual art: Exploring an appraisal theory of aesthetic emotions. *Empirical Studies of the Arts*, *23*(2), 119–133. doi:10.2190/12AV-AH2P-MCEH-289E

Silvia, P. J. (2005). Emotional responses to art: From collation and arousal to cognition and emotion. *Review of General Psychology*, *9*(4), 342–357. doi:10.1037/1089-2680.9.4.342

Silvia, P. J. (2005). What is interesting? Exploring the appraisal structure of interest. *Emotion (Washington, D.C.)*, *5*(1), 89–102. doi:10.1037/1528-3542.5.1.89 PMID:15755222

Silvia, P. J. (2006). *Exploring the psychology of interest*. Oxford University Press. doi:10.1093/acprof:oso/9780195158557.001.0001

Silvia, P. J., & Berg, C. (2011). Finding movies interesting: How appraisals and expertise influence the aesthetic experience of film. *Empirical Studies of the Arts*, *29*(1), 73–88. doi:10.2190/EM.29.1.e

Sivaji, A., Nielsen, S. F., & Clemmensen, T. (2017). A textual feedback tool for empowering participants in usability and UX evaluations. *International Journal of Human-Computer Interaction*, *33*(5), 357–370. doi:10.1080/10447318.2016.1243928

Skrbis, Z., & Woodward, I. (2007). The ambivalence of ordinary cosmopolitanism: Investigating the limits of cosmopolitan openness. *The Sociological Review*, *55*(4), 730–747. doi:10.1111/j.1467-954X.2007.00750.x

Sliwinski, J., Katsikitis, M., & Jones, C. M. (2017). A Review of Interactive Technologies as Support Tools for the Cultivation of Mindfulness. *Mindfulness*, 1–10. doi:10.1007/s12671-017-0698-x

Sloterdijk, P. (2011). *Spheres* (Vol. 1). Bubbles Microspherology.

Small, D. (1996). Navigating large bodies of text. *IBM Systems Journal*, *35*(3/4), 514–525. doi:10.1147/sj.353.0514

Smith, C. A., & Kirby, L. D. (2001). *Toward delivering on the promise of appraisal theory*. Academic Press.

Smith, S. (2013). *Determining Sample Size: How to Ensure You Get the Correct Sample Size*. Retrieved from https://www.qualtrics.com/blog/determining-sample-size/

Soegaard, M. (2010). *Prototyping*. Retrieved from http://www.interaction-design.org/encyclopedia/prototyping.html

Spiller, N. (1998). *Architecture in Cyberspace II*. Academy Editions.

Spiller, N. (2009). Plectic architecture: Towards a theory of the post-digital in architecture. *Technoetic Arts: a Journal of Speculative Research*, *7*(2), 95–104. doi:10.1386/tear.7.2.95/1

Squire, K. D., & Jan, M. (2007). Mad city mystery: Developing scientific argumentation skills with a place-based augmented reality game on handheld computers. *Journal of Science Education and Technology*, *16*(1), 5–29. doi:10.1007/s10956-006-9037-z

Stevens, W. (1965). *The Necessary Angel: Essays on Reality and the Imagination*. Vintage Books.

Stichter, J. P., Laffey, J., Galyen, K., & Herzog, M. (2014). iSocial: Delivering the social competence intervention for adolescents (SCI-A) in a 3D virtual learning environment for youth with high functioning autism. *Journal of Autism and Developmental Disorders*, *44*(2), 417–430. doi:10.1007/s10803-013-1881-0 PMID:23812663

Stratasys. (2007). ABS. *Stratasys*. Retrieved from www.nrri.umn.edu/NLTC/ABS07.pdf

Sullivan, G. (2005). *Art Practice As Research: inquiry in the visual arts*. Thousand Oaks, CA: Sage Publications.

Svanaes, D. (2014). Philosophy of Interaction. In M. Soegaard & R. F. Dam (Eds.), The Encyclopedia of Human-Computer Interaction (2nd ed.). Aarhus, Denmark: The Interaction Design Foundation. Retrieved from https://www.interaction-design.org/encyclopedia/philosophy_of_interaction.html

Sweeney, R. (2010). *Inter/Actions/Inter/Sections: art education in a digital visual culture*. Reston, VA: National Art Education Association.

Sweeny, R. (2005). Three funerals and a wedding: Art education, digital images, and an aesthetics of cloning. *Visual Arts Research, 31*(61), 26–37.

Sweller, J. (2010). Element interactivity and intrinsic, extraneous, and germane cognitive load. *Educational Psychology Review, 22*(2), 123–138. doi:10.1007/s10648-010-9128-5

Szerszynski, B., & Urry, J. (2002). Cultures of cosmopolitanism. *The Sociological Review, 50*(4), 461–481. doi:10.1111/1467-954X.00394

Tan, E. S. (1996). *Emotion ond the Strtucture of Narrative film: Film as an Emotion Machine, prijev*. B. Fasting.

Tarvainen, J., Sjöberg, M., Westman, S., Laaksonen, J., & Oittinen, P. (2014). Content-based prediction of movie style, aesthetics, and affect: Data set and baseline experiments. *IEEE Transactions on Multimedia, 16*(8), 2085–2098. doi:10.1109/TMM.2014.2357688

Tea Cup Communication. (2008). *Tea Cup Communication Enters Second Life Business Partnership with Atelier Ten*. Retrieved June, 2017, from https://www.gmo.jp/en/news/article/?id=261

Tilly, C. (2008). *Toplumsal Hareketler* (O. Düz, Trans.). İstanbul: Babil Yayınevi.

Tobias, S., & Fletcher, J. D. (2011). *Computer games and instruction*. Charlotte, NC: Information Age Publishing.

Tomlinson, J. (1999). *Globalization and culture*. University of Chicago Press.

Touraine, A. (1999). *Toplumdan Toplumsal Harekete. Toplumsal Hareketler* (K. Çayır, Trans.). İstanbul: Kaknüs Yayınları.

Trompenaars, F., & Hampden-Turner, C. (2010). *Riding the waves of innovation: Harness the power of global culture to drive creativity and growth*. McGraw Hill Professional.

Tschumi, B. (1994). *Architecture, and disjunction*. Cambridge, MA: MIT Press.

Turing, A. M. (1950). Computing machinery and intelligence. *Mind, 59*(236), 433-460. doi:/LIX.236.43310.1093/mind

Turkle, S. (1995). *Life on the screen: Identity in the age of the Internet*. New York, NY: Simon & Schuster.

Turkle, S. (2009). *Simulation and Its Discontents*. Cambridge, MA: MIT Press.

Turkle, S. (2011). *Alone together: Why we expect more from technology and less from each other*. New York, NY: Basic Books.

Um, E., Plass, J. L., Hayward, E. O., & Homer, B. D. (2012). Emotional design in multimedia learning. *Journal of Educational Psychology, 104*(2), 485–498. doi:10.1037/a0026609

Underwood, J. A. (2008). *'The Work of Art in the Age of Mechanical Reproduction' first published in 1936*. England: Penguin Books.

Unity 3D. (2017). Retrieved June 15, 2017, from https://unity3d.com

uPrintSE. (2012). *uPrint 3d Printer*. Author.

Van Manen, M. (2007). Phenomenology of Practice. *Phenomenology & Practice, 1*, 10.

Van Slyke, C., Lou, H., Belanger, F., & Sridhar, V. (2010). The influence of culture on consumer-oriented electronic commerce adoption. *Journal of Electronic Commerce Research, 11*(1), 30.

Venuvinod, P. K., & Ma, W. (2004). *Rapid Prototyping: Laser-Based & Other Technologies*. Boston, MA: Kluwer Academic. doi:10.1007/978-1-4757-6361-4

Vidyarthi, J., Riecke, B. E., & Gromala, D. (2012). Sonic Cradle: designing for an immersive experience of meditation by connecting respiration to music. In *Proceedings of the designing interactive systems conference* (pp. 408-417). ACM. doi:10.1145/2317956.2318017

Virilio, P. (2005). *The original accident.* Cambridge, UK: Polity Press.

Visi, F., Coorevits, E., Schramm, R., & Miranda, E. R. (2017). Musical instruments, body movement, space, and motion data: Music as an emergent multimodal choreography. *Human Technology*, *13*(1), 58–81. doi:10.17011/ht/urn.201705272518

Vitruvius Pollio, M. (23-27 adC). *De Architecturae, Liber primus*. Retrieved June, 2017, from http://www.thelatinlibrary.com/vitruvius1.html

Vuforia. (2017). Retrieved June 15, 2017, from https://www.vuforia.com

Wallerstein, I., Hopkins, T. K., & Arrighi, G. (2004). *Sistem Karşıtı Hareketler* (C. Kanat, B. Somay, & S. Sökmen, Trans.). İstanbul: Metis Yayınları.

Walters, P. (2012). Ceramic 3D Printing : A Design Case Study. In *CFPR Symposium*. London: Arts and Humanities Research Council (AHRC).

Wang, H. L., & Cheong, L. F. (2006). Affective understanding in film. *IEEE Transactions on Circuits and Systems for Video Technology*, *16*(6), 689–704. doi:10.1109/TCSVT.2006.873781

Wardrip-Fruin, N. (2009). *Expressive processing: digital fictions, computer games, and software studies*. Cambridge, MA: The MIT Press.

Weber, R. P. (1990). *Basic content analysis (No. 49)*. Sage. doi:10.4135/9781412983488

Welsh, T. N., Chandrasekharan, S., Ray, M., Neyedli, H., Chua, R., & Weeks, D. J. (2012). Perceptual-motor interaction: Some implications for human-computer interaction. In J. A. Jacko (Ed.), *The human-computer interaction handbook: Fundamentals, evolving technologies, and emerging applications* (3rd ed.; pp. 3–20). Boca Raton, FL: CRC Press; doi:10.1201/b11963-3

Wilhelmsson, U. (2008). Game Ego presence in video and computer games. In Extending Experiences. Structure, analysis and design of computer game player experience (pp. 58-72). Lapland University Press.

William, P. (1998). *Rapid Prototyping Primer*. Retrieved from http://www.me.psu.edu/lamancusa/rapidpro/primer/chapter2.htm

Williams, D. (2010). The mapping principle, and a research framework for virtual worlds. *Communication Theory*, *20*(4), 451–470. doi:10.1111/j.1468-2885.2010.01371.x

Williams, K., & Williams, C. (2011). Five key ingredients for improving student motivation. *Research in Higher Education*, 1–23. Retrieved from http://www.jupapadoc.startlogic.com/manuscripts/11834.pdf

Winograd, T., & Flores, F. (1988). *Understanding computers and cognition: A new foundation for design*. Reading, MA: Addison-Wesley.

Wittgenstein, L. (1958). *Philosophical investigations* [Philosophische Untersuchungen] (G. E. M. Anscombe, Trans.). Oxford, UK: Blackwell.

Wittgenstein, L. (1958). *Preliminary Studies for the "Philosophical Investigations"*. Generally Known as The Blue and the Brown Books.

Wodehouse, A., & Marks, J. (2013). Gestural product interaction. *International Journal of Art, Culture and Design Technologies*, *3*(2), 1–13. doi:10.4018/ijacdt.2013070101

Woodward, C. (1996). *Computer Aided Industrial Design for Ceramics and Glass Industries*. Dest Artes Oy.

Woodward, I., Skrbis, Z., & Bean, C. (2008). Attitudes towards globalization and cosmopolitanism: Cultural diversity, personal consumption and the national economy. *The British Journal of Sociology*, *59*(2), 207–226. doi:10.1111/j.1468-4446.2008.00190.x PMID:18498592

Yatim, M. H. M., & Masuch, M. (2007). *Educating children through game making activity*. Retrieved from http://www.learnit.org.gu.se/digitalAssets/862/862887_yatim_masuch.pdf

Yavuz, M. H. (2003). Türkiye'de Çoğulculuk ve Kamusal Alanın Kullanımı. In EDAM (Ed.), Sivil Bir Kamusal Alan (pp. 163–192). İstanbul: Kaknüs Yayınları.

Yılmaz, H. (2006). Kamu, Kamu Otoritesi ve Devlet; Habermas'ın Işığında Türkiye'yi Düşünmek. In A. Karadağ (Ed.), *Kamusal Alan ve Türkiye* (pp. 192–209). İstanbul: Asil Yayınları.

Yoon, S., Anderson, E., Lin, J., & Elinich, K. (2017). *How Augmented Reality Enables Conceptual Understanding of Challenging Science Content How Augmented Reality Enables Conceptual Understanding of Challenging Science Content*. Academic Press.

Yue, W. Y., & Zhi, F. Z. (2012). Rapid Development of New Sanitary Ware Products of Ceramics and Application Study with Computer. *Advanced Materials Research*, *466–467*, 201–206.

About the Contributors

Mehdi Khosrow-Pour, D.B.A., received his Doctorate in Business Administration from the Nova Southeastern University (Florida, USA). Dr. Khosrow-Pour taught undergraduate and graduate information system courses at the Pennsylvania State University – Harrisburg for almost 20 years. He is currently Executive Editor at IGI Global (www.igi-global.com). He also serves as Executive Director of the Information Resources Management Association (IRMA) (www.irma-international.org), and Executive Director of the World Forgotten Children's Foundation (www.world-forgotten-children.org). He is the author/editor of more than 100 books in information technology management. He is also the Editor-in-Chief of the *International Journal of Open Source Software and Processes*, *International Journal of Green Computing*, *International Journal of Digital Library Systems*, *International Journal of E-Entrepreneurship and Innovation*, *International Journal of Art, Culture, and Design Technologies*, *International Journal of Signs and Semiotic Systems*, and *International Journal of Disease Control and Containment for Sustainability*, and has authored more than 50 articles published in various conference proceedings and scholarly journals.

* * *

Elif Ayiter is a designer, educator and researcher whose creative interests are based in three dimensional online virtual worlds and their avatars, as well as in developing and implementing hybrid educational methodologies between art & design and computer science. She teaches full time at Sabanci University in Istanbul. Her texts have been published at academic journals such as the Leonardo Electronic Almanac, the Journal of Consciousness Studies, the Journal of Gaming & Virtual Worlds, International Journal of Art, Culture and Design Technologies, and Technoetic Arts, and she has authored many book chapters in edited academic books. She has presented creative as well as research output at conferences such as Consciousness Reframed, Siggraph, Creativity and Cognition, SPIE, Computational Aesthetics and Cyberworlds. http://www.elifayiter.com/.

Viviana Barneche-Naya received her architectural degree from the Universidad de la República (Udelar -Uruguay) and her PhD in Architecture from the Universidade da Coruña (Spain) where she now works as a professor and researcher at the Advanced Visualization Group (Videalab). She writes and presents widely on issues regarding Ciberarchitecture, Virtual Archaeology, and applications of video games and virtual worlds technology to architectural design. She was awarded the Architectural 3D Award in 2011 for her work on the visualization of the Cathedral of Santiago de Compostela.

Jose Cañas-Bajo graduated in Audiovisual Communication at the University of Malaga and specialised in Script Writing at ESCAC (Escola Superior de Cinema i Audiovisuals de Catalunya) and the University of Barcelona. He recently obtained his doctoral degree at the University of Jyväskylä. His research focuses on viewers' emotional experiences of audio-visual products in different cultures. Currently, he is participating with Menestys research group (Alto's University) as a post doctoral researcher in a project regarding cross-cultural factors of full scripted films experience from an appraisal theory of interest. His aim is to identify factors and properties that makes a featured film successful in different cultural contexts.

Denise Doyle is a Senior Lecturer in Digital Media at the University of Wolverhampton, UK, Adjunct Professor of Virtual Worlds and Digital Practice at Ontario College of Art and Design University, Toronto and Editor in Chief of the Journal of Virtual Creativity (formally Metaverse Creativity), Intellect, UK. With a background in Fine Art Painting and Digital Media, Denise has contributed research in the fields of art and technology, phenomenology, performance, video games, art and consciousness, virtual worlds, and digital arts practice through numerous book chapters and articles and recently edited the artist led book New Opportunities for Artistic Practice in Virtual Worlds (2015). She sits on the editorial boards for the International Journal of Performance Art and Digital Media (Routledge) and the Journal of Gaming and Virtual Worlds, (Intellect).

Roger Gacula Pineda received his PhD in cognitive science from the Faculty of Information Technology at the University of Jyväskylä, Finland in 2014. Dr. Pineda is an independent researcher whose work focuses on how cultural diversity in cognitive and affective processes can be embedded into the design of technology to better support living environments. His industry experience includes Nokia, Oracle and Sonera where he held management roles in design, development and marketing of telecommunication products.

Luis A. Hernández-Ibáñez received his architectural degree from the Universidad Politécnica de Madrid (Spain) and his Ph.D. in Architecture from the Universidade da Coruña (Spain) where he now teaches and leads the Advanced Visualization Group (Videalab). His line of research include Digital Architecture, Virtual Archaeology, and all kinds of applications of computer graphics to Historical Heritage and Museology.

Cemile Zehra Köroğlu has been an Assistant Professor in Social Work Department at Uşak University, Turkey. She obtained her Masters Degree from Selçuk University, Philosophy and Religious Sciences Department in 2002. She completed her Ph.D. in the same department in 2009. She focused on relations of consumption culture and religion in her Ph.D. researches. Her research interests are Sociology of religious groups, disadvantaged groups, social change, and gender studies.

Muhammet Ali Köroğlu has been an Assistant Professor at Uşak University, Social Work Department. He obtained his Masters Degree from Selçuk University Philosophy and Religious Sciences Department in 2004. And he obtained his Ph.D. from Necmettin Erbakan University in 2012. He focused on relations of public sphere and religion in his Ph.D. researches. His research interests are social and religious movements, social work with elderly and relations of autonomous public spheres and religion.

Rui Leitão is a communication designer and his work spans from print to interactive displays. He is concerned with how media communicates with people in order to inform, educate, and engage them. Rui's research is in the field of education technologies, with particular interest in gamification in science teaching and learning. More information: http://www.lboro.ac.uk/departments/design-school/staff/rui-leitão/.

Adérito Fernandes Marcos is graduated in Computer Science Engineering from the Nova University of Lisbon; holds a Ph.D. Suma cum laude in Engineering (Computer Graphics) from the Technical University of Darmstadt, Germany; and Habilitation in Technology and Information Systems from the University of Minho. Dr. Marcos is Full Professor at Aberta University (Portuguese Open University), Department of Sciences and Technology, Lisbon, Portugal. He is the founder and current Director of the Doctoral Program in Digital Media-Art, a joint offer of Universidade Aberta and University of Algarve. Previously, he was an Ass. Professor at University of Minho, where he was responsible for the creation of the Masters Course in Computer Graphics and Virtual Environments; and Masters Course in Technology and Digital Art. He has been the (co)founder of the several series of scientific events, namely, the CoopMedia series (Workshop of Cooperative and Distributed Multimedia Systems); the SIACG series (Ibero-American Symposium on Computer Graphics); and the Artech series (International Conference in Digital Arts) in 8th edition. He is author/co-author of more than one hundred articles in refereed Journals, Conference Proceedings, Book chapters and Science Promotion Booklets. His current research interests embrace the areas of computer/digital art, applied computer graphics and creative media for e-learning, digital/computer artifacts in general. He integrated the Executive Board of the Eurographics from Jan. 2002 until Ago. 2005; and again, since October 2008. From Oct. 2008 until Feb.2013 he was President of the Executive Board of the Eurographics Portuguese Chapter. Since Feb. 2013, he is the President of the Artech-Int - the International Association for Computer Arts (www.artech-international.org) with activities in all over the world.

Folasayo Enoch Olalere holds M.Arts and PhD in product design from the Universiti Malaysia Kelantan (UMK). Currently, he is a Postdoctoral Research Fellow at the Department of Visual Arts and Design, Vaal University of Technology South Africa, where he is exploring on the commodification of Indigenous arts and crafts. He is a recipient of several academic and innovative awards, which includes Malaysian International Scholarship 2014, Excellent Postgraduate Researcher's Award 2016 and Gold Medalist at ITEX'15. His primary research interests are cultural design, computer-aided design & additive manufacturing.

João Rodrigues graduated in Electrical Engineering in 1993, he got his M.Sc. in Computer Systems Engineering in 1998 and Ph.D. Electronics and Computer Engineering in 2008 from University of the Algarve, Portugal. He is Adjunct Professor at Instituto Superior de Engenharia, also in the University of the Algarve, where he lectures Computer Science and Computer Vision since 1994. He is member of associative laboratory LARSyS (ISR-Lisbon), CIAC and the Associations APRP, IAPR and ARTECH. He participated in 15 financed scientific projects, and he is co-author more than 140 scientific publications. His major research interests lie in computer vision, human-computer interaction, augmented reality and assistive technologies.

Claudia Sandoval Romero is a Colombian artist and theoretician currently based in Vienna, Austria. She graduated in Social Communication and Journalism from Valle University. She received her Master in Multimedia Arts from the Sao Paulo University in Brazil with scholarships Capes and Fapesp and Master in Critical Studies from the Academy of Fine Arts in Vienna where she is also a candidate for the title of Doctor in Philosophy with a dissertation about latino women artists in Vienna from the perspective of art sociology and history, under supervision of Univ.-Prof. Mag. Dr. Sabeth Buchmann. Claudia Sandoval Romero has published her inquiries about art in the current time, as well as about photography, in both academic and cultural magazines in Colombia, Spain, Mexico, England, Germany, the USA and Brazil. More of her work can be seen at www.claudiasandovalromero.com.

Index

F

G

H

I

J

L

M

N

O

P

R

S

T

U

V

www.ingramcontent.com/pod-product-compliance
Lightning Source LLC
LaVergne TN
LVHW062116200926
848484LV00007B/1688